Nematodes as Environmental Indicators

Edited by

Michael J. Wilson

*Institute of Biological and Environmental Sciences,
The University of Aberdeen, Aberdeen, Scotland, UK*

Thomais Kakouli-Duarte

*EnviroCORE Department of Science and Health,
Institute of Technology,
Carlow, Ireland*

www.cabi.org

CABI is a trading name of CAB International

CABI Head Office
Nosworthy Way
Wallingford
Oxfordshire OX10 8DE
UK

Tel: +44 (0)1491 832111
Fax: +44 (0)1491 833508
E-mail: cabi@cabi.org
Website: www.cabi.org

CABI North American Office
875 Massachusetts Avenue
7th Floor
Cambridge, MA 02139
USA

Tel: +1 617 395 4056
Fax: +1 617 354 6875
E-mail: cabi-nao@cabi.org

A catalogue record for this book is available from the British Library, London, UK.

Library of Congress Cataloging-in-Publication Data

Nematodes as environmental indicators / edited by Michael J.
Wilson, Thomais Kakouli-Duarte.
p. cm.
Includes bibliographical references and index.
ISBN 978-1-84593-385-2 (alk. paper)
1. Nematodes–Ecology. 2. Indicators (Biology) I. Wilson,
Michael J. (Michael John), 1964- II. Kakouli-Duarte, Thomais. III. Title.

QL391.N4N382 2009
592'.5717--dc22

2008049111

ISBN-13: 978 1 84593 385 2

Typeset by SPi, Pondicherry, India.
Printed and bound in the UK by the MPG Books Group.

The paper used for the text pages in this book is FSC certified. The FSC (Forest Stewardship Council) is an international network to promote responsible management of the world's forests.

Contents

Contributors vii

Preface xi

Part I Nematodes and Bioindication – General Considerations

1 The Role of Nematodes in Ecosystems 1
G.W. Yeates, H. Ferris, T. Moens and W.H. Van der Putten

2 Nematode Diversity in Terrestrial, Freshwater Aquatic and Marine Systems 45
M. Hodda, L. Peters and W. Traunspurger

3 Molecular Markers, Indicator Taxa, and Community Indices: the Issue of Bioindication Accuracy 94
K. Ekschmitt and G. Korthals

Part II Analysis of Nematode Assemblages in Environmental Samples

4 General Community Indices that can be used for Analysis of Nematode Assemblages 107
D.A. Neher and B.J. Darby

5 Indices Developed Specifically for Analysis of Nematode Assemblages 124
H. Ferris and T. Bongers

6 Case Studies Using Nematode Assemblage Analysis in Aquatic Habitats 146
R. Danovaro, C. Gambi, S. Höss, S. Mirto, W. Traunspurger and A. Zullini

7 Case Studies Using Nematode Assemblage Analysis in Terrestrial Habitats 172
P. Nagy

8 Molecular Tools for Analysing Nematode Assemblages 188
R. Neilson, S. Donn, B. Griffiths, T. Daniell, K.D. Rybarczyk, S. van den Elsen, P. Mooyman and J. Helder

Part III Nematode Ecotoxicity Tests and Gene Expression Analysis

9 Ecotoxicity Testing with Nematodes 208
S. Höss and P.L. Williams

10 Transgenic *Caenorhabditis elegans* as Biosensors 225
C. Lagido

11 *Caenorhabditis elegans* and Chemical Environmental Stressors: from DNA Microarrays and Data Analysis to Genomic Data of True Ecotoxicological Relevance 253
R. Menzel, C.E.W. Steinberg and S.R. Stürzenbaum

Part IV Commercial Use of Nematodes as Bioindicators

12 Commercial Aspects of the Use of Nematodes as Bioindicators 275
M.W. Trett, B.C. Urbano, S.J. Forster and S.P. Trett

Index 315

Contributors

Tom Bongers, Laboratory of Nematology, Department of Plant Sciences, Wageningen University, Binnenhaven 5, 6709 PD Wageningen, The Netherlands (Email: Tom.Bongers@wur.nl)

Brian J. Darby, Department of Plant and Soil Science, University of Vermont, 105 Carrigan Drive, Burlington, Vermont, 05405, USA (Email: brian.darby@uvm.edu)

Tim Daniell, Environment Plant Interactions Programme, SCRI, Invergowrie, Dundee, DD2 5DA, UK (Email: tim.daniell@scri.ac.uk)

Roberto Danovaro, Dipartimento di Scienze del Mare, Università Politecnica delle Marche, Via Brecce Bianche, 60131, Ancona, Italy (Email: r.danovaro@univpm.it)

Suzanne Donn, Environment Plant Interactions Programme, SCRI, Invergowrie, Dundee, DD2 5DA, UK (Email: suzanne.donn@scri.ac.uk)

Klemens Ekschmitt, Justus Liebig University, IFZ – Department of Animal Ecology, H.-Buff-Ring 26-32, D-35392 Giessen, Germany (Email: Klemens.Ekschmitt@uni-giessen.de)

Howard Ferris, Department of Nematology, University of California Davis, Davis, CA 95616, USA (Email: hferris@ucdavis.edu)

Simon J. Forster, Physalia Limited, Consultant & Forensic Ecologists, Sedgefen House, Meadow Walk, Harpenden, Hertfordshire AL5 5TF, UK (Email: forensic@physalia.com)

Cristina Gambi, Dipartimento di Scienze del Mare, Università Politecnica delle Marche, Via Brecce Bianche, 60131, Ancona, Italy (Email: c.gambi@univpm.it)

Bryan Griffiths, Teagasc, Environmental Research Centre, Johnstown Castle, Wexford, Ireland (Email: bryan.griffiths@teagasc.ie)

Johannes Helder, Laboratory of Nematology, Department of Plant Sciences, Wageningen University, Binnenhaven 5, 6709 PD Wageningen, The Netherlands (Email: Hans.Helder@wur.nl)

Mike Hodda, Nematode Biosystematics & Ecology, Australian National Insect Collection, CSIRO Entomology, GPO Box 1700, Canberra ACT 2601, Australia (Email: Mike.Hodda@csiro.au)

Sebastian Höss, Ecossa, Giselastr. 6, 82319 Starnberg, Germany (Email: hoess @ecossa.de)

Thomais Kakouli-Duarte, EnviroCORE Department of Science and Health, Institute of Technology, Carlow, Kilkenny Road, Carlow, Ireland (Email: thomais.kakouli@itcarlow.ie)

Gerard Korthals, Wageningen UR, Applied Plant Research, P.O. Box 430, NL-8200 AK Lelystad, The Netherlands (Email: Gerard.Korthals@wur.nl)

Cristina Lagido, School of Medical Sciences, University of Aberdeen, Aberdeen AB25 2ZD, UK (Email c.lagido@abdn.ac.uk)

Ralph Menzel, Department of Biology, Freshwater and Stress Ecology, Humboldt University at Berlin, Unter den Linden 6, 10099 Berlin, Germany (Email: ralph.menzel@biologie.hu-berlin.de)

Simone Mirto, Istituto per l'Ambiente Marino Costiero (IAMC), CNR, Spianata San Raineri, 86, 98122, Messina, Italy (Email: simone.mirto@ iamc.cnr.it)

Tom Moens, Marine Biology Section, Department of Biology, Ghent University, Krijgslaan 281/S8, B-9000 Ghent, Belgium (Email: Tom.Moens@ UGent.be)

Paul Mooyman, Laboratory of Nematology, Department of Plant Sciences, Wageningen University, Binnenhaven 5, 6709 PD Wageningen, The Netherlands (Email: Paul.Mooyman@wur.nl)

Péter Nagy, Department of Zoology and Animal Ecology, Szent István University, Gödöllő, Hungary (Email: nagy.peter@mkk.szie.hu)

Deborah A. Neher, Department of Plant and Soil Science, University of Vermont, 105 Carrigan Drive, Burlington, VT 05405, USA (Email: deborah. neher@uvm.edu)

Roy Neilson, Environment Plant Interactions Programme, SCRI, Invergowrie, Dundee, DD2 5DA, UK (Email: roy.neilson@scri.ac.uk)

Lars Peters, Department of Animal Ecology, University of Bielefeld, Morgenbreede 45, 33615 Bielefeld, Germany (Email: lars.peters@ uni-bielefeld.de)

Katarzyna D. Rybarczyk, Laboratory of Nematology, Department of Plant Sciences, Wageningen University, Binnenhaven 5, 6709 PD Wageningen, The Netherlands (Email: Kasia.Rybarczyk@wur.nl)

Christian E.W. Steinberg, Department of Biology, Freshwater and Stress Ecology, Humboldt University at Berlin, Unter den Linden 6, 10099 Berlin, Germany (Email: christian_ew_steinberg@web.de)

Stephen R. Stürzenbaum, School of Biomedical and Health Sciences, King's College London, 150 Stamford Street, London SE1 9NH, UK (Email: stephen.sturzenbaum@kcl.ac.uk)

Walter Traunspurger, Department of Animal Ecology, University of Bielefeld, Morgenbreede 45, 33615 Bielefeld, Germany (Email: traunspurger@ uni-bielefeld.de)

Marcus W. Trett, Physalia Limited, Consultant & Forensic Ecologists, Sedgefen House, Meadow Walk, Harpenden, Hertfordshire AL5 5TF, UK (Email: physalia@terra.es)

Simon P. Trett, Physalia Limited, Consultant & Forensic Ecologists, Sedgefen House, Meadow Walk, Harpenden, Hertfordshire AL5 5TF, UK (Email: forensic@physalia.com)

Beatriz Calvo Urbano, Nebalia S.L. Consultoría Medioambiental, Calle de Amor de Dios, 4, bajo A, 28014 Madrid, Spain (Email: nebalia@terra.es)

Sven van den Elsen, Laboratory of Nematology, Department of Plant Sciences, Wageningen University, Binnenhaven 5, 6709 PD Wageningen, The Netherlands (Email: Sven.vandenElsen@wur.nl)

Wim H. Van der Putten, NIOO-KNAW, Centre for Terrestrial Ecology, PO Box 40, 6666 ZG Heteren, The Netherlands and Laboratory of Nematology, Department of Plant Sciences, Wageningen University, Binnenhaven 5, 6709 PD Wageningen, The Netherlands (Email: w.vanderputten@nioo.knaw.nl)

Phillip L. Williams, College of Public Health, University of Georgia, Paul D. Coverdell Center, Athens, GA 30602-7396, USA (Email: pwilliam@uga.edu)

Michael J. Wilson, Institute of Biological and Environmental Sciences, University of Aberdeen, Aberdeen AB24 3UU, UK (Email: m.j.wilson@abdn.ac.uk)

Gregor W. Yeates, P.O. Box 1758, Palmeston North 4440, New Zealand (Email: gregor.yeates@gmail.com)

Aldo Zullini, Dipartimento di Biotecnologie & Bioscienze, Università di Milano-Bicocca, Piazza della Scienza 2, 20126 Milan, Italy (Email: aldo.zullini@unimib.it)

Preface

Both of us are relatively new to the field of nematodes as biological indicators. We have backgrounds in crop protection and particularly the use of parasitic nematodes as biological control agents. In the late 1990s, we both obtained tenure track positions in institutions where there was a strong research emphasis on environmental diagnostics and bioremediation of contaminated land. Attendance at seminars given by students and visitors combined with conversations with colleagues made us both acutely aware of the potential nematodes offer as biological indicators. This potential exists from the sub-cellular to landscape scale.

Perusing the literature made us realize that much high quality work had been done to exploit nematodes as biological indicators, but that information sources were diffuse with two problems (common to many fields of biology) being apparent: marine scientists rarely communicate with terrestrial scientists, and molecular biologists (particularly the *Caenorhabditis elegans* community) have little contact with ecologists. If nematodes are to realize their true potential, these groups must communicate and collaborate. As a means of opening avenues of communication we organized the First International Symposium on Nematodes as Environmental Bioindicators, held in Edinburgh in June 2007. The current volume, written by many of the key participants from the symposium, represents the next stage.

Several reviews have considered the ideal characteristics that organisms need to be effective bioindicators. Amongst other features, ideal organisms should be:

- Abundant and easily manipulated.
- Easily sampled and sorted.
- Cheap and easy to husband in the laboratory.
- Representative of their habitat, be it terrestrial or aquatic.
- Known to exhibit well-defined responses to environmental challenges.

In light of the above criteria, we believe that no other group of organisms offers as much potential as nematodes. Nematodes can be used to study gene expression in relation to environmental challenges, they can be used in laboratory ecotoxicity tests, or their assemblages can be studied *in situ* to monitor the environmental impact of a wide range of challenges.

The book comprises four parts. The first part (Chapters 1–3) deals with background issues relating to nematodes themselves and to bioindicators in general. We were delighted that such an eminent group of authors agreed to write Chapter 1, which we believe for the first time tries to define the importance of nematodes in ecosystems. We hope this chapter will convince any doubters that as well as being good indicators of environmental health, nematodes are also key drivers of many ecosystem functions.

Chapter 2 deals with nematode diversity. Earth is currently losing biodiversity due to habitat loss, spread of invasive species, climate change etc. This loss is likely to be reflected in nematode biodiversity, but without at least some knowledge of the current state of global nematode biodiversity, such losses will go unnoticed. Furthermore, changes in nematode diversity at local scales can be used to indicate changes in environmental quality. Chapter 2 represents a comprehensive survey of current knowledge of nematode biodiversity and also tackles key issues regarding the meaning of biodiversity and challenges in its measurement. Chapter 3 is a timely warning on the mathematical limitations of using bioindicators. Much work has been done over the last two decades on the potential of bioindicators to measure environmental stresses. In laboratory studies, it is fairly easy to predict what effects, for example the addition of copper to soils, will have on indicator organisms. However, if we look at the same organisms in the field, estimating levels of copper from their response is much more challenging.

Part II (Chapters 4-8) describes practical aspects of using nematode assemblages as bioindicators in real world settings. Chapter 4 describes the calculation and use of many general biodiversity indices that can be applied to nematodes and also introduces the concepts that underlie multivariate analysis. This approach to analysing nematode assemblages maximizes information use and thus overcomes some of the limitations of univariate analysis described in Chapter 3. Chapter 5 deals with the calculation, use and scope of the numerous indices developed specifically for nematode assemblages, including the widely-used maturity index. Indeed, it is the availability and broad scope of nematode-specific indices that make nematodes so attractive as biological indicators.

Chapters 6 and 7 review the current state of knowledge regarding use of nematode assemblages as bioindicators in aquatic and terrestrial ecosystems respectively and Chapter 8 deals with the potential of molecular methods as tools to analyse nematode assemblages. Nobody doubts that nematode assemblages can give much valuable information about soils and sediments, but there are few people with the skills necessary to carry out such analyses. As taxonomic training becomes increasingly rare, and many students prefer to avoid hours of routine microscope work, this situation is likely to worsen and limit the widespread uptake of nematode assemblage analysis. To counter this, many workers are turning to the broad array of techniques developed by environmental microbiologists to analyse assemblages of non-cultivable microorganisms.

Part III (Chapters 9-11) describes the potential of nematodes, particularly the model organisms *Caenorhabditis elegans*, to be used in laboratory-based toxicity assays. It is precisely the same features (short life span, ease of culture, transparent cuticle) that made *C. elegans* the first model animal that make it so suitable for laboratory-based bioassays. Furthermore, because of the vast amount of knowledge and technical know-how generated by the *C. elegans* community, we can generate and interpret data from *C. elegans* assays better than from any other animal-based biosensor. Chapters 9, 10 and 11 respectively deal with whole animal ecotoxicity testing, use of transgenic *C. elegans* biosensors and use of microarrays to study gene expression in response to environmental stressors. These three types of test allow rapid accumulation of data on nematode response to environmental stresses. A future challenge for the field of nematode-bioindicators will be determining whether such laboratory assays can be used to predict the effects stressors will have on nematode assemblages *in situ*.

The final part (Chapter 12) of the book deals with commercial aspects of using nematodes as environmental bioindicators. Physalia Ltd have been established as consulting forensic ecologists since 1990 and along with their sister company Nebalia S.L. in Spain, specialize in studies of marine benthos and particularly nematodes. The chapter outlines the clear practical and scientific benefits perceived by industry (both consultants and their customers) of using nematodes. The chapter describes several case studies that convincingly demonstrate the utility of nematodes in a wide range of contrasting applications, including both terrestrial and aquatic habitats.

Physalia and Nebalia do not rely solely on nematodes and any study of environmental health will benefit from analysis of a wide range of contrasting organisms. However, we hope that the contents of this book will convince readers of the many benefits of using nematodes as environmental indicators, and hope its publication will lead to more widespread uptake of these technologies.

Michael Wilson and Thomais Kakouli-Duarte

1 The Role of Nematodes in Ecosystems

Gregor W. Yeates,[1] Howard Ferris,[2] Tom Moens[3] and Wim H. Van der Putten[4,5]

[1]*Landcare Research, Palmerston North, New Zealand;* [2]*Department of Nematology, University of California Davis, California, USA;* [3]*Marine Biology Section, Department of Biology, Ghent University, Belgium;* [4]*NIOO-KNAW, Centre for Terrestrial Ecology, Heteren, The Netherlands;* [5]*Laboratory of Nematology, Wageningen University, The Netherlands*

Introduction

Human society is entirely dependent on a variety of ecosystem services (Wall, 2004; Millennium Ecosystem Assessment, 2005). Nematodes play a major role in component processes of most ecosystem services, such as the provision of food, fibre, clean water and air, pest and disease regulation. As nematodes are heterotrophic, differences in, for example, C:N:P ratios between them and their food lead inevitably to their excreting minerals. In soil food webs, nematodes are involved in the transformation of organic matter into mineral and organic nutrients which can be taken up by plants, as well as in influencing plant growth and crop productivity (Ingham *et al.*, 1985; Ferris *et al.*, 1998, 2004a). Nematode feeding activity contributes to soil food web stability. When soils are degraded there may be adverse effects on ground- and surface water quality. Nematodes cause animal and human diseases, and nematodes can influence pest damage to plants by regulating insect abundance (Viglierchio, 1991). In natural ecosystems, nematodes contribute to spatial and temporal diversity in plant communities and, therefore, to the diversity of plant-associated communities both above- and below-ground (De Deyn and Van der Putten, 2005). However, plant community development is not necessarily indicative of below-ground nematode assemblage development (Kardol *et al.*, 2005; Sánchez-Moreno *et al.*, 2008), so that restoration and conservation of nematode assemblages in the soil cannot be inferred from plant community composition. On the other hand, because nematodes are so abundant and omnipresent in ecosystems, they serve as elegant indicators of environmental disturbance (Bongers 1990; Ferris *et al.*, 2001; Yeates, 2003; Höss *et al.*, 2004; Schratzberger *et al.*, 2006; Heininger *et al.*, 2007). Finally, one nematode, *Caenorhabditis elegans* (Rhabditina), has become a central model for genomic studies that aim to relate gene expression to the development and functioning of organisms. Indeed *C. elegans* was the first multicelleular organism whose genome was sequenced (CESC, 1998). Many of the advances in molecular biology are underpinned by studies on *C. elegans*. The apparent determinate development and cell constancy (eutely) of nematode species were among the reasons that *C. elegans* was selected for study and the phylum continues to be a focus of studies in basic biology, with cell differentiation, moulting, and ageing being of particular interest.

There are estimates of between 40,000 and 10,000,000 species in the phylum Nematoda (Blaxter, 1998; Yeates and Boag, 2006). One provocative estimate speculates that there might be as many as 100,000,000 species, even before considering the cryptic diversity among morphologically indistinguishable taxa (Lambshead, 1993; Hodda *et al.*, Chapter 2, this volume). The key roles of nematodes in agricultural and natural ecosystems, as well as their usefulness for indicator and molecular studies, make them an important focus for taxonomic, ecological, physiological and molecular research.

Nematodes are built on a simple plan, with the gut and body wall being concentric tubes. The typical nematode has a mouth (= oral aperture) leading to a stoma (= buccal cavity), connected to the lumen of a muscular and/or glandular pharynx (= oesophagus) that leads to a tubular intestine, a rectum and posterior anus. The body is cylindrical or spindle-shaped and the tail beyond the anus varies enormously in length. Adult females range in length from 0.3 mm to 8 m.

The great diversity of nematodes is paralleled by a vast range of head structures which facilitate food ingestion and can be useful indicators of feeding habits (Fig. 1.1). The tubular gonads lie in the cavity between the digestive tract and body wall, and they, together with the associated genital structures, are further sources of species differentiation. The typical life history of nematodes includes an egg and typically four juvenile stages (= 'larvae') preceding the adults. The morphology of the juvenile stages converges with that of the adult; there is no metamorphosis.

Throughout the 20th century, the nature of the nematode body cavity was debated, the supposed lack of complete mesodermal lining suggesting the terminology 'pseudocoelom' (Hyman, 1951). Cilia were not detected in nematodes until the advent of the electron microscope (Roggen *et al.*, 1966). While cilia are now known to occur widely and there have been studies of their sensory role in the behaviour of taxa as diverse as *Caenorhabditis* spp., *Haemonchus contortus* (Rhabditina) and *Onchocerca volvulus* (Spirurina), ciliated epithelia remain unknown in the phylum.

Molecular techniques have improved understanding of the phylogenetic relationships of nematodes with other animal groups. However, there is still vigorous debate about their most likely position and the significance of the ecdysozoan theory based on the occurrence of the *Hox* gene in various animal groups (de Rosa *et al.*, 1999; Gutierrez and Sommer, 2004) although this is disputed by studies that included a larger number of genes in the analysis (Blair *et al.*, 2002). Nematodes are now generally classified as the Phylum Nematoda in the Acoelomata of the Metazoa (See outline classification in Appendix 1).

Nematodes were recorded as human parasites by the Chinese around 2700 BC and as plant parasites by Shakespeare in 1594 (Viglierchio, 1991) and

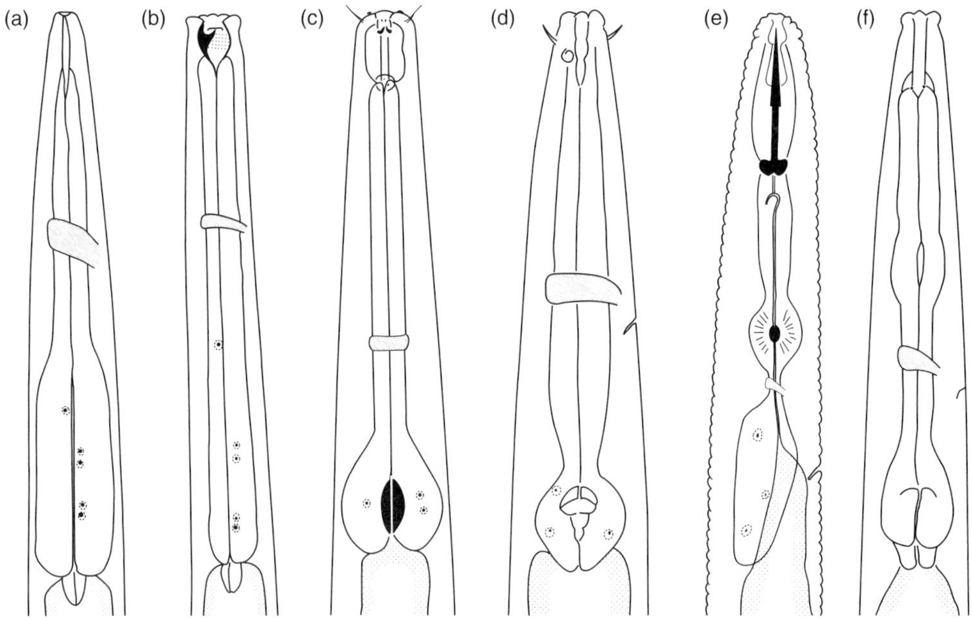

Fig. 1.1. Head and pharyngeal regions of six nematodes found in soil and water environments. (a) Dorylaimida: Dorylaimidae, *Dorylaimus*, (b) Mononchida, Mylonchulidae, *Mylonchulus*, (c) Chromadorida, Ethmolaimidae, *Ethmolaimus*, (d) Plectida, Plectidae, *Plectus*, (e) Tylenchina, Hoplolaimidae, *Rotylenchus*, (f) Rhabditina, Rhabditidae, *Rhabditis*. In most, the location of the circum-pharyngeal nerve ring, pharyngeal gland nuclei and amphid are shown. The diameter of the stylet lumen is ~5 μm in (a) and ~0.1 μm in (e). Diagrammatic and not to scale. (Compiled from various sources and observations.)

more formally by Needham (1744). At a time when there is concern about potential extinction of as yet undescribed species – of both nematodes and other phyla – our ignorance of nematode diversity, habitat range and ecological amplitude continues to be challenged. While European workers have contemplated the extinction in the wild of the vinegar eelworm, *Turbatrix aceti* – the first 'free-living' nematode reported (1656) – it has recently been recorded from spoiled vinegar in Brazil (De Moura *et al.*, 2006). On the other hand, there is a strong desire to eliminate *Onchocerca volvulus*, microfilarial juveniles of which cause river blindness in humans.

In the latter part of the 20th century the roles of nematodes in biological turnover in soil came to be appreciated (Yeates, 1984; Ingham *et al.*, 1985). Until that time, so-called free-living nematodes were little more than scientific curiosities. Awareness of the impact of nematode activity on other populations, and thus on ecosystem processes, developed concomitant with awareness of 'biodiversity', a term which first appeared in print in 1988 (Wilson, 1988), and a range of ways in which the abundant, diverse nematode assemblages found in various habitats could be summarized in, and interpreted through, information-rich 'indices', were proposed (Bongers, 1990; Ferris *et al.*, 2001; Yeates, 2003).

Nematode Biology – Individuals and Populations – an Outline

Feeding by nematodes

Most nematodes utilize the energy fixed by plant photosynthesis. They may feed directly on primary producers, such as higher plants (e.g. *Aphelenchoides* on foliar parts, *Ditylenchus* on stems, *Pratylenchus* and *Meloidogyne* on roots) and unicellular algae (e.g. *Chromadorita, Pareudiplogaster* and *Daptonema* on diatoms), or on microbes associated with decomposing plant material (e.g. *Aphelenchus, Filenchus* on fungal hyphae; *Rhabditis, Plectus, Leptolaimus* and many Monhysteridae on bacteria). Both animal faeces and cadavers are important resources for microbes and thus for microbial-feeding nematodes. Higher trophic levels feed as predators of nematodes and other microinvertebrates (e.g. *Mononchus, Nygolaimus, Enoploides, Sphaerolaimus*) or as parasites of invertebrate and vertebrate animals (e.g. *Thelastoma, Ascaris*) themselves dependent on plants. Nematodes feeding at more than one trophic level are termed omnivores. There is increasing awareness of atypical nematodes found in anoxic environments that utilize the energy bound in chemicals as their resource, either directly or via symbiotic bacteria (Polz *et al.*, 2000; Van Gaever *et al.*, 2006).

Substrates occupied by nematodes

Nematodes are influenced by the nature of their physical environment (e.g. soil or sediment texture, water chemistry) and by gradients within it (e g. redox potential, plant root distribution). As far as we know there have not been any formal, quantitative indices proposed to measure nematode response to such differences in environmental conditions.

There is a clear distinction between the substrate occupied by a nematode and the food resource it uses, and each may vary with the nematode life stage. In typical bacterial-feeding, hyphal-feeding, and predacious nematodes, all post-hatching stages feed and typically live entirely in non-living substrates such as soils, sediments and decaying vegetation while feeding. Plant-feeding nematodes such as *Pratylenchus* and *Radopholus* feed solely on plant tissue but migrate between living plant and non-living soil substrates; they typically hatch as second-stage juveniles. *Heterodera* and *Meloidogyne* also emerge as second-stage juveniles, and migrate, without feeding, into roots, where they develop into saccate females feeding on plant resources via highly specialized transfer cells.

In terrestrial habitats, eggs of ascarids (e.g. *Ascaris*) do not hatch outside the invertebrate or vertebrate host and the complete life history occurs within the living host; adults inhabit the stomach and intestine of the definitive host and consume food ingested by the host. In contrast, while adult hookworms (e.g. *Ancylostoma*), strongylids (e.g. *Strongylus*) and trichostrongyles (e.g. *Trichostrongylus, Haemonchus*) also feed and develop within their vertebrate hosts, commonly evidenced by blood in host intestines, their eggs hatch in faeces and grow as bacterial-feeding stages, in non-living substrate, before reinfecting the living substrate of new vertebrate hosts.

Nematode response to stimuli

Behavioural responses to stimuli are essential for the success of nematodes in locating food resources, hosts and mates. While thermotaxis, phototaxis and geotropism are responses to basic environmental cues, they are important in host location in taxa as diverse as *Ancylostoma*, *Stephanurus*, *Anguina*, Mermithidae and Trichostrongylidae. Chemotaxis is commonly involved in food-finding and food-selection by free-living nematodes. Responses to different kinds of chemicals and/or bacterial food tend to be highly species-specific and potentially contribute to the small-scale patchiness of nematode populations (Venette and Ferris, 1998; Moens *et al.*, 1999; Höckelmann *et al.*, 2004). Some of the attractants that nematodes use as chemical guides in locating roots have been identified and characterized. For example CO_2 attracts many species of plant-parasitic nematodes (McCallum and Dusenbery, 1992; Robinson, 1995), ethylene attracts *Heterodera* juveniles (Wubben *et al.*, 2001), and onion root exudates attract *Ditylenchus* juveniles (Spiegel *et al.*, 2003). Rühm *et al.* (2003) found that an unknown kairomone in *Sinapsis alba* exudates served as an attractant to *H. schachtii* juveniles. Some of these interactions seem to be species-specific; tobacco root exudates increase egg hatching of tobacco cyst nematodes (Wang *et al.*, 1997) and exudates of cotton roots increase egg hatching in *Rotylenchulus reniformis* (Sankaralingam and McGawley, 1994). However, although plant-parasitic nematodes are highly efficient in locating plant roots, not very many substances have been positively identified as nematode attractants. Amongst the early work on sex attractants was that of Green and Plumb (1970) with *Heterodera* spp. and *Globodera* spp., with other work including studies on *Panagrolaimus*, *Rhabditis*, *Panagrellus*, *Ditylenchus*, *Belonolaimus* and Trichostrongylidae. The sensitivity of nematode response to chemical stimuli in their environment is demonstrated by the ability of *Caenorhabditis elegans* to detect and distinguish between olfactory and taste odour compounds at picomolar (10^{-12}) concentrations (Bargmann and Mori, 1997).

Abundance and diversity of nematodes

In ecological terms 'communities' are made up of interacting species. In the various habitats they occupy, nematodes interact with the species providing their food (e.g. bacteria, fungi, plants etc.), other organisms using the same food resource (e.g. bacterial-feeding protozoa, rotifers), and there are predator:prey interactions (e.g. protozoa, fungi and mites as well as nematodes prey on soil nematodes), and the myriad of other organisms inhabiting their environment. The nematode species occurring in a sample do not themselves comprise an ecological community and we prefer to term the collection of populations as an assemblage. We reserve the term 'fauna' for the list of taxa present in a sample, field or landscape.

Each square metre of soil, forest litter or aquatic sediment may contain millions of individual nematodes belonging to over 400 species. Nematodes may also be abundant and diverse within living substrates. N.A. Cobb recorded 40,000 individual nematodes from the stomach of a wallaby, and radiation within the Macropodinae (kangaroos and wallabies) has led to a swarm of 112 species of *Cloacina* (Strongyloidea) (Cobb, 1915; Beveridge *et al.*, 2002). In New Zealand, domestic cattle and sheep are infected by 27 and 29 nematode species respectively (McKenna, 1997).

Identifying the taxa in each nematode assemblage, and assessing the absolute and relative contribution of each taxon to the nematode assemblage would be a prodigious task. Unless a study is restricted to a particular group (e.g. comparing pre-plant and post-harvest populations of plant-feeding nematodes; assessing infective juveniles of Trichostrongylidae on pasture herbage) it is common to present results in information-rich indices. Such indices may be either those widely used in ecological studies such as the Shannon–Wiener Index of diversity (H'), Margalef Index of richness (D or SR) and Pielou's evenness (J') (Pielou 1975; Yeates, 1984; Magurran, 1988; Neher and Darby, Chapter 4, this volume) or indices specifically developed for nematodes (Ferris and Bongers, Chapter 5, this volume). The former have been widely applied across organisms and ecosystems and can be regarded as robust, while the latter need to be carefully examined for circularity. The validity of all interpretations need to be tested against current understanding of population dynamics and ecological processes and interactions (Wardle, 2002; Bardgett, 2005; Schratzberger *et al.*, 2007). Nematode grazing on microbes can significantly increase nutrient cycling and plant response in localized patches (Ingham *et al.*, 1985). Although there have been various studies, there is a need for understanding of the factors governing the distribution of patches, when scaling up from a roughly homogeneous patch (e.g. a pedon or uniform agricultural field) to the landscape scale (Coleman *et al.*, 1992; Müller and Lenz, 2006; Sánchez-Moreno *et al.*, 2008). This is also important if one tries integrating across levels by applying frameworks such as the holistic Eco-Energy concept of Jørgensen and Mejer (1979).

Effects of nematodes on their resources

Populations of nematodes may affect the growth rates, health, and yield of plants or animals (e.g. infection of soybeans or potatoes with cyst nematodes or grazing mammals with trichostrongylids). In contrast, grazing and bioturbation by microbial feeding nematodes may stimulate nutrient cycling and plant yield (Ingham *et al.*, 1985; Alkemade *et al.*, 1992; Aller and Aller, 1992; Fu *et al.*, 2005). Such effects are dependent on the ecological setting and are usually greater when the system is subject to other stresses. Conversely, adequate levels of nutrition can mask the consequences of infections.

By its very nature, parasitism involves mutual adjustment between host and parasite to permit coexistence without serious harm to either component of the interaction. Typically, there is loss of production, or thriftiness, when the parasite burden increases, or the host is additionally stressed. Indeed, low burdens of plant-feeding nematodes have been found to stimulate plant growth under favourable conditions. Examples of the range of interactions were found in a study of *Heterodera glycines*-resistant and susceptible soybean (*Glycines max*) cultivars across ten states in the United States. Yield loss due to *H. glycines* was greatly confounded by other stress factors, including temperature and moisture extremes (Donald *et al.*, 2006). In terms of beneficial effects of nematode activity, when plant nutrients are non-limiting in the soil nutrient mineralization by bacterial-feeding nematodes will not be reflected in plant yield (Ferris *et al.*, 2004a). Interpretation of such relationships involves a functional, rather than taxonomic, analysis of the populations.

Nematode activities cannot be considered in isolation from abiotic environmental conditions. Physical conditions are important and, for both soil-inhabiting and aquatic nematodes, several studies have shown that soil type is more important than time of year or management practices in determining the overall makeup of the nematode assemblage (Yeates, 1984; Heip *et al.*, 1985; Griffiths *et al.*, 2003). A classic experiment by Griffiths *et al.* (2000) demonstrated that, because of their effects on soil biodiversity, combinations of stressful conditions may heavily impact the stability of ecosystem services. Similarly, damaging effects of plant-feeding nematodes may depend on external conditions, such as drought (Haverkort *et al.*, 1992) or the presence of other pathogens or symbionts, which may lead to synergistic (De Rooij-Van der Goes, 1995) or antagonistic (Brinkman *et al.*, 2005; Hol *et al.*, 2007) interactions. Nematodes may differentially stress one plant species, thereby indirectly benefiting other plant species. For example, clover cyst nematodes may selectively influence clover roots resulting in the leakage of nitrogen compounds, which benefit grasses that are not affected by the nematodes (Bardgett *et al.*, 1999). Feeding rates of the predacious marine nematode *Enoploides longispiculosus*, which may exert a strong top-down control over nematode and ciliate prey communities, are strongly reduced or even completely impeded by subtle shifts in silt content, mean grain size and water content of intertidal flats (Gallucci *et al.*, 2005).

Traditionally, agricultural nematologists have considered single nematode species as causes of crop losses. More recently, however, ecologists have become concerned with spatial and temporal patterns in communities of organisms in soils and sediments (Blanchard, 1990; Traunspurger, 2000; Ettema and Wardle, 2002; Ettema and Yeates, 2003; Fisher, 2003; Van Gaever *et al.*, 2004; Bardgett *et al.*, 2005; Michiels and Traunspurger, 2005). Even in agricultural fields, which are considered to have relatively homogeneous soils, there is considerable variability in the spatial and temporal plant parasitic nematode species composition and abundance; a variety of approaches has been used in efforts to understand this heterogeneity (Goodell and Ferris,

1980; Ferris *et al.*, 1990; Robertson and Freckman, 1995). These spatial and temporal patterns also are important when studying the role of nematodes in natural ecosystems, which are far less homogeneous than agricultural soils, and when using nematodes as environmental indicators. In these conditions, it is crucial to understand the drivers and consequences of nematode diversity within samples (α-diversity), between samples within fields (β-diversity) and between fields within landscapes (γ-diversity).

The contribution of nematodes to foodweb interactions and ecosystem processes, such as the cycling of nitrogen and carbon, may not require species level identification, if the complementary activities of species with similar ecological adaptations and feeding at the same trophic level are considered (i.e. uses the functional guild approach). In terms of ecosystem structure and function, Laakso and Setälä (1999) regarded diversity among functional groups as more important than diversity within them. However, when the parasitic nature of nematodes is considered, and when parasitism is host or even race species-specific, such as that of potato cyst nematodes (*Globodera* spp.), the functional approach requires identifications down to the species, or even pathotype level (Folkertsma *et al.*, 2001). Even taxonomically very closely related bacterial-feeding nematode species may have differential effects on phytodetritus decomposition and on the activity and community composition of the associated bacteria (De Mesel *et al.*, 2003, 2004). Therefore, functional interpretations depend on the specific nature of the functions and the nematodes involved in performing them.

Nematode Feeding and its Ecosystem Consequences

Food resources of nematodes

Many of the ecosystem functions and services provided by nematodes are direct consequences of their feeding activity and of the physiological processes of digestion and metabolism. Across the diversity of the phylum are representatives that ingest a vast array of resources to drive their metabolic processes. We might divide them into general categories of *grazers* or browsers, which feed on food resources that continue producing, and *predators* whose feeding results in the death of their prey. Nematodes that appear to be feeding on dead organic matter such as detritus and cadavers are typically grazing or preying on organisms associated with that dead matter.

Within these two general categories we may separate nematodes according to the nature of the food resource: (i) *herbivores* feeding on living tissues of higher plants; (ii) *carnivores* feeding on animal tissues, vertebrate (as parasites) or invertebrate (as predators or parasites); (iii) *fungivores* feeding on fungi; (iv) *bacterivores* feeding on prokaryotic organisms; and (v) *unicellular eukaryote feeders*, feeding on ciliates, other protozoans or diatoms and unicellular algae. Some nematodes (e.g. ascarids, thelastomatids) that inhabit the gastrointestinal tract of vertebrates or invertebrates, graze on microbes rather than being carnivores.

In each case, and as detailed below, there are nematodes that are *specialists* in their feeding habits with stoma structures, behavioural attributes or specific biochemical requirements (e.g. transfer cells of *Heterodera* and *Meloidogyne*) that are adaptations to feeding on a narrow range of food sources; others are *generalists* and capable of obtaining resources from a wider range of sources. Some of the latter are omnivores, crossing feeding type boundaries and feeding at more than one trophic level. For example, some may be carnivorous as well as bacterial feeding (e.g. Mononchida), or bacterial feeding as well as feeding on unicellular eukaryotes (e.g. *Thalassomonhystera, Daptonema*). Omnivory appears to be quite common in some nematodes, and food sources from different trophic levels may be utilized simultaneously (e.g. many Dorylaimida), in different life stages (e.g. bacterial-feeding juveniles of hookworms whose adults occur in the gastrointestinal tract of mammals), or follow temporal or environmental fluctuations in the availability of different resources (e.g. the marine nematode *Enoplus brevis* feeding on cyanobacteria, diatoms, oligochaetes, nematodes and rotifers (Hellwig-Armonies *et al.*, 1991)).

As might be expected in a group with enormous diversity of habitat, there are exceptions to these simplistic categorizations of feeding types. Some nematodes appear to be able to transport, or allow diffusion, of dissolved organic molecules across the cuticle, as an alternative to, or perhaps in addition to, stomal ingestion. Cuticular microvilli are used in the insect-inhabiting *Bradynema*; *C. elegans* has been cultured on chemically defined media, without other living organisms (Vanfleteren, 1980); dissolved organic matter may be ingested via the stoma (Chia and Warwick, 1969). Several marine nematodes, including some recently discovered species of the anoxic oceanic abyss, lack any orifice for the digestive system and derive their resources either directly or via symbiotic bacteria (e.g. *Astomonema* and several Stilbonematinae) (Hentschel *et al.*, 1999; Dover, 2000). The characterization of such unusual feeding strategies, as well as some of the difficulties in modelling the roles and functions of nematodes in a food web context, may benefit from stable isotope approaches, using both natural and experimentally enriched abundances of, mainly, ^{13}C and ^{15}N (Moens *et al.*, 2002, 2005a). Chemosynthetically produced carbon, for instance, is typically very depleted in ^{13}C, while ^{15}N fractionates significantly with trophic level and is thus a useful tracer of trophic position. The technology has been used successfully in tracking the fate of carbon from plant residues into the microbial biomass and into nematodes (Minoshima *et al.*, 2007), or from tidal flat microalgae and bacteria into nematodes (Moens *et al.*, 2002; van Oevelen *et al.*, 2006). The relative ^{15}N enrichment of *Graphidium strigosum* and *Passalurus ambiguus* in rabbits, compared with depletion in cestodes, has been used to suggest differing trophic relationships (Boag *et al.*, 1998).

Stoma morphology in the phylum Nematoda ranges from simple apertures of fixed diameter, which apparently limit the size of the ingested material, to permanently cavernous features or structures that can be opened to enormous size for ingestion of large prey. Even simple tubular stomata adapted for ingestion of bacteria (e.g. in the Rhabditidae) or diatoms (e.g. in

Praeacanthonchus, Gonionchus) often have small teeth or denticles, presumably to abrade, crush or filter ingested particles as was demonstrated by Bird and Ryder (1993) in *Acrobeloides nanus* (Cephalobidae). Additionally, there may be other cuticular structures in the stoma and/or pharynx that serve to abrade or rupture food before it enters the intestine. Intestinal parasites of vertebrates are often equipped with teeth that allow tearing of mucosa to provide access to tissue and blood (e.g. *Ancylostoma*). Many nematodes are equipped with fixed or moveable teeth (Mononchida, Diplogastridae, Chromadoridae, Enoplidae) or with hollow spears or stylets (Dorylaimida, Tylenchida) for piercing a resource and withdrawing contents. In most Tylenchida and the plant-feeding Longidoridae, the spear lumen is very narrow (~0.1 and 0.5 μm in diameter respectively), significantly limiting what is ingested. Many omnivorous and predatory Dorylaimida have a wide stylet lumen (≥ 6 μm) or even just a single mural tooth (e.g. *Nygolaimus*). The stoma is, however, not a simple indicator of food resource, with some *Seinura* spp. (Aphelenchidae) being predacious and ingesting their requirements through a <0.5 μm aperture, while other predators (e.g. *Anatonchus tridentatus* (Anatonchidae)) are frequently observed to have intact nematode prey in their intestine (Small, 1987).

Microbial feeding in soils and sediments

The ecosystem functions associated with location and ingestion of food by nematodes include the redistribution of resources so that they are more available to other consumers, and the transport of prey organisms to other locations where they gain access to new resources and stimulation of grazed populations (Fu *et al.*, 2005). Their ability to locate suitable feeding patches from a distance, e.g. by sensing cues of decomposition-associated end-products such as CO_2 (Klingler, 1965; Pline and Dusenbery, 1987; Riemann and Schrage, 1988), renders nematodes efficient vectors in transferring microorganisms between suitable resource patches (e.g. Jatala *et al.*, 1974). Generalist obligate bacterial feeders include those 'enrichment opportunists' (*sensu* Ettema and Bongers, 1993) which appear to simply draw in aqueous suspensions of their food, with larger particles restricted only by oral diameters of 2 μm or less, but with apparently little other restriction on the types of bacteria ingested (Venette and Ferris, 1998; Salinas *et al.*, 2007). Rather than the dimensions of the adult mouth or stoma, it is the oral dimensions of the first post-hatching (i.e. feeding) stage that limits food size for species population maintenance. Other bacterial feeders are more specialist, feeding actively with, for example, simple to elaborate head probolae, muscular contractions to open a closed (or 'collapsed') stoma during ingestion and sweeping motions of the head which allow them access to food by stirring sediments or disrupting adherence of the bacteria to surfaces (Paul de Ley in Moens *et al.*, 2004).

Several important ecosystem functions and services are well-documented for bacterial-feeding nematodes in soils and aquatic systems. Nematodes grazing on microbes may result in greater metabolic activity in their prey populations – essentially nematode grazing on a bacterial population may keep it young and reproductively active. In addition, around 30% of the bacteria ingested by bacterial-feeding nematodes are not digested and assimilated; in fact, they may still be alive when defaecated (Yeates, 1969a, b; Ingham *et al.*, 1985; Bird and Ryder, 1993; Fu *et al.*, 2005; Ghafouri and McGhee, 2007). The proportion of ingested bacteria that survive passage through the nematode intestine may vary with food availability and bacterial density (Moens *et al.*, 2006). Some bacteria adhere to the cuticle of bacterial-feeding nematodes. As nematodes move away from the bacterial colony into other areas of the substrate, bacteria are transferred to new areas and to new substrate both as surface 'passengers' and through nematode defaecation. Such 'resource farming' results in the bacterial food resource for the nematodes increasing until predator:prey dynamics are such that overgrazing occurs (Fu *et al.*, 2005). Another example of such nematode-aided bacterial farming is the migration of entomopathogenic nematodes into arthropods where they release bacteria toxic to the arthropod which multiply in the arthropod cadaver and provide food for perhaps two generations of the nematode (Boemare, 2002).

It has also been suggested that nematodes can discriminate among soil microbes. Rodger *et al.* (2004) cultured bacterial-feeding nematodes (*C. elegans*, *Coarctadera cystilarva*; Rhabditidae) on four species of bacteria. They then measured migration towards, that is their attraction to, those bacteria. On agar plates there seemed to be a 'substrate legacy' affecting the subsequent ability of these bacterial-feeding nematodes to locate the same food. They also carried out experiments in sand and, as verified by Fu *et al.* (2005), commented that the migration that did occur presumably had a role in dispersing bacteria through the soil.

Excretion by bacterial-feeding soil nematodes

Another important ecosystem service emerges from the nature of the substrate ingested and subsequently assimilated across the intestinal wall. Bacterial-feeding soil nematodes may excrete materials assimilated, but in excess of their needs, in forms that are available to other organisms. A familiar example of the mineralization of digested (i.e. simplified) organic molecules is the participation of most organisms in the carbon cycle. In liberating energy from ingested materials, nematodes have been calculated to release, across the cuticle, about 40% of the ingested carbon in the form of CO_2 (Klekowski *et al.*, 1972; Ferris *et al.*, 1995). The CO_2 returns to the atmosphere and is available to plants to be once again fixed into complex molecules through the process of photosynthesis. But, the ingested molecules from which the respired carbon is derived may also contain other elements in excess of the needs of the nematode for maintenance, growth and reproduction. Such excess minerals are presumably excreted in mineral form, rather than defaecated.

The best studied example is the excretion of excess nitrogen in the form of ammonium, which is then available for uptake by plants or for bacterial transformation to either nitrates or to atmospheric nitrogen. The mineralization service resulting from nematode digestion of organic molecules may be enhanced by differences in the molecular ratios of the food and the consumer. The C:N ratio of the biomass of many bacteria is in the range of 4:1 while that of their nematode predators is around 5.9:1 (Ferris *et al.*, 1997). The C:N ratio of some fungi averaged 8.5:1 while that of their nematode grazers was 9.1:1 (Chen and Ferris, 1999). Although some of these C:N ratios may be revised after further work, they suggest that for every molecule of C ingested, the bacterial-feeding nematode ingests about 8% more than its body requirement for N. After respiratory mineralization of around 40% of the ingested C, the nematode has, in total, about 18% more N than required for body structure and reproductive output. The excretions of bacterial-feeding nematodes alone may enhance available mineral N in the soil by 20% or more (Ferris *et al.*, 1998). Even though the C:N ratios of fungal-feeding nematodes are apparently similar to those of their food substrate, the N associated with respired C (around 5% of N intake) is excreted as excess, varying with fungal substrate (Chen and Ferris, 1999; Okada and Ferris, 2001). The combined excretions of all consumer organisms in the soil food web may account for 80% of total mineralized N (Sánchez-Moreno *et al.*, 2008).

Differences in elemental ratios among organisms and their differential excretion can lead to uncoupling of conventional C:N:P ratios and element cycling – an emerging area which merges with the stoichiometry of biogeochemical cycling (Sterner and Elser, 2002).

Phoresy or vectoring by and of nematodes

Feeding activities of stylet-bearing nematodes, in addition to facilitating nitrogen mineralization, contribute to other ecosystem services. First, they may provide avenues of ingress into host or prey tissues, for example, access to bacteria and fungi through the migration and feeding activities of endoparasitic herbivores in roots and plant storage organs. This may be associated with weakened physical barriers and defence mechanisms allowing fungal and bacterial infections of plant tissue (e.g. *Fusarium*, *Phytophthora parasitica* and *Pseudomonas solanacearum* in tobacco (Powell, 1971)). Second, they may provide phoretic transport of bacteria or viruses into hosts (Hewitt *et al.*, 1958; Brown *et al.*, 1993). The reported instances seem to provide no obvious advantage to the nematode, in fact it may be to the detriment of the nematode and represent exploitation of nematode behaviour and biology by other organisms. For example, the transport of *Clavibacter* bacteria by *Anguina* spp. to their developing seed galls in Gramineae, where the nematode may be outcompeted for the resource by the bacteria (Bird and Stynes, 1977). In vectoring of plant viruses, *Longidorus* spp. and *Xiphinema* spp, may weaken the host and render it a poorer resource for the nematode (Brown *et al.*, 1995). A mutually beneficial example of such phoresy is provided by the transport of *Bursaphelenchus xylophilus* by long-horned beetles (Coleoptera: Cerambycidae: *Monochamus*) to healthy pine trees where they enter through feeding wounds on young twigs. Invasion of the nematodes into the pine tree, and the damage they cause in resin canals, renders the tree less vigorous and more favorable for oviposition by the beetles, followed by larval development and pupation. The emerging adults become contaminated by nematodes nictating in the pupal chamber and transfer them to healthy trees (Togashi and Shigesada, 2006).

Nematodes as Regulators of Populations, Succession and Production

The feeding of nematodes on their food sources, besides providing the required energy and nutrient resources, has the potential to regulate or even suppress the magnitude of those resources and consequently impact on ecosystem structure and function. Such effects are in addition to excretion arising from the metabolic costs of nematodes building and maintaining somatic tissue and reproduction. Some general examples are shown in Fig. 1.2 and the following sections give more specific findings.

Plant-feeding nematodes may affect plant community composition

In addition to the much-studied effects of plant-feeding nematodes on plant performance and crop yield, there are many examples in which the reduction in the rate of growth and fitness of higher plants potentially decreases the exclusionary competitiveness of those plants and confers relatively greater fitness on their competitors. Consequently, ecosystem succession increases and plant diversity is increased (Van der Putten, 2003). The result in natural systems is that the susceptible and less fit are reduced or even eliminated from the plant community. More constrained examples are provided in agricultural systems, where aggressive strains of plant-feeding nematodes have often been introduced with their susceptible, but agronomically desirable, hosts. A seeding rate designed for maximizing production in the absence of nematodes will provide a less vigorous stand of the crop in the presence of nematode herbivores. This opens up the canopy and reduces competitiveness with weeds which may out-compete the crop so that the resultant losses of yield can be enormously magnified (Alston *et al.*, 1991; Schroeder *et al.*, 2005). Interestingly, applied ecologists have seldom exploited the effects of nematode herbivory on plant competition by designing cropping systems that enhance herbivores that render the weeds less competitive. Some examples suggest that seed and bud-feeding nematodes may render certain weeds less effective and this clearly affects the plant community. For example, the effects of *Anguina amsinckiae* (Anguinidae) on the growth of coast fiddleneck (*Amsinckia intermedia*: Boraginaceae) in wheat was modelled by Pantone *et al.* (1989a, b), and in a field trial in Texas, *Ditylenchus phyllobia* (Anguinidae) reduced the density of the target silverleaf nightshade (*Solanum elaeagnifolium*: Solanaceae) plants by 66% (Northam and Orr, 1982).

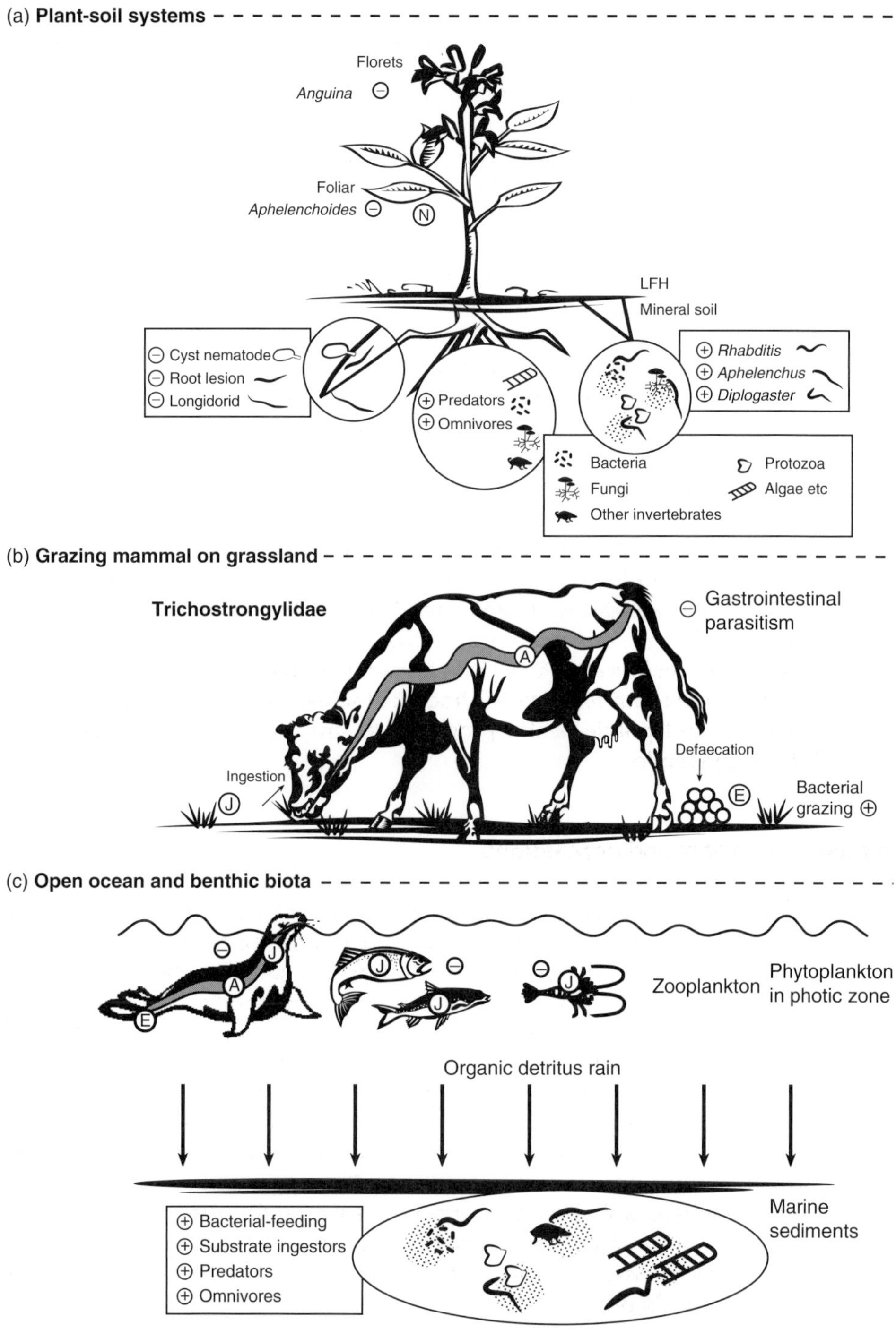

Fig. 1.2. Potential consequences of nematode feeding on food resources [code + being positive effects of grazing on microbes and nutrient availability; – being potential loss of plant or animal production] in (a) plant/soil systems, (b) grassland with grazing mammals, and (c) open ocean pelagic and benthic biota. Nematode metabolic activities and excretion are not specifically shown. For simplicity only selected nematode groups are included. Sites of nematode activity are labelled E, J, A or N to represent occurrence of egg, juveniles, adults, or the species as a whole; in soils and sediments developmental stages are not separated (inspiration from Wardle *et al.*, 2004).

Nematodes in vertebrates

Similar examples of reduced competitiveness certainly occur among vertebrates, including humans. Individuals with genetic characteristics or behaviour patterns that render them more susceptible to nematode infections may be selected against, resulting in greater average fitness or tolerance. Multigenerational selection by farmers of those sheep in a flock performing better under uniform parasite challenge can lead to selection of sheep less susceptible to nematode infection (Bisset and Morris, 1996; Morris *et al.*, 2000). In this work, nematode faecal egg counts are useful as an indicator.

The functional role of nematode parasitism in human populations stimulates the societal response of sanitation and other public health measures and the search for appropriate parasite management tools. River blindness in humans, caused by *Onchocerca volvulus* (Filarioidea), can be controlled by annual treatment with the orally administered microfilaricidal drug, ivermectin (Mectizan®) (Cupp and Cupp, 2005). Elephantiasis or lymphatic filariasis, caused by *Wuchereria bancrofti* (Filarioidea), can be locally eliminated using two drugs: ivermectin (Mectizan®) and albendazole in sub-Saharan Africa or diethylcarbamazine and albendazole elsewhere (Richard-Lenoble *et al.*, 2003). However, for dracunculiasis, caused by Guinea Worm (*Dracunculus medinensis*: Dracunculoidea), no vaccine or medication is available, and public health programmes are used to reduce exposure to infective juveniles.

Regulation of arthropod populations

Several families of nematodes include associates of arthropods. Among those most studied because of their potential as biocontrol agents have been families Mermithidae, Tetradonematidae, Allantonematidae, Phaenopsitylenchidae, Sphaerulariidae, Steinernematidae, and Heterorhabditidae (Kaya and Stock, 1997; Lacey *et al.*, 2001). The regulatory potential of nematodes for arthropods has been most evidently demonstrated by the exploitation of heterorhabditid and steinernematid nematodes that carry toxic bacteria, *Photorhabdus* and *Xenorhabdus*, respectively, into their hosts. Mass production of these nematodes on artificial diets has fostered an industry dedicated to biological approaches in management of those insects that have a life stage in the soil. There have also been attempts to capitalize on the direct parasitism and feeding of *Romanomermis culicivorax* on mosquito larvae (Petersen, 1985) and *Deladenus siricidicola* has been used successfully in biocontrol of the woodwasp *Sirex* in New Zealand and Australia (Bedding, 1993).

Influence on aquatic bacterial community diversity

Less spectacular, but with potentially important consequences for instance in decomposition processes, are observations that mucus secretions by aquatic nematodes may selectively favour settlement of specific strains of bacteria (Moens *et al.*, 2005b), and that even low grazing rates by nematodes may significantly affect bacterial community composition, while high grazing rates may depress bacterial community diversity (De Mesel *et al.*, 2004).

Suppression of mycelial growth

The rate of spread of saprophytic and mycorrhizal fungi in Petri dishes may be reduced by the feeding of aphelenchid nematodes (Riffle, 1967; Sutherland and Fortin, 1968; Ruess and Dighton, 1996) (Table 1.1). Under standard conditions, different fungi have been found to differ in their effects on both the populations and morphometrics of particular Aphelenchidae, Tylencholaimidae and Tylenchidae feeding on them (Faulkner and Darling, 1961; Townshend and Blackith, 1975; Ruess and Dighton, 1996; Okada and Kadota, 2003). It should be noted that Okada and Kadota (2003), Okada *et al.* (2005) found that while *Pleurotus ostreatus* (Basidiomycetes) supported population increase of *Filenchus misellus* and *Tylencholaimus parvus*, in contrast, *Aphelenchus avenae* populations were reduced as a result of predation by *P. ostreatus*. Recent success in maintaining certain Tylenchidae and Tylencholaimidae on fungi provides the opportunity for more specific studies on interactions between mycelial-feeding nematodes and their food supply, including the complexities of decomposer fungi in mineral and organic soil horizons and mineralization of nutrients as a result of nematodes grazing in those horizons as initially explored by Okada and Ferris (2001).

Nematode grazing on mycelia may play other roles in complex interactions in soils. Laboratory studies suggest that feeding by *Aphelenchoides* spp. on the biocontrol fungus *Trichodema harzianum* may constrain its efficacy as a biocontrol agent (Bae and Knudsen, 2001) and Table 1.1 suggests that nematode feeding on mycorrhizal fungi could also reduce exploration of the soil body by their mycelia. *Aphelenchoides* spp. have been found to suppress colonization of *Pinus ponderosa* seedlings by the ectomycorrhiza *Suillus granulatus*

Table 1.1. Colony size of two ectotrophic mycorrhizal fungi grown for 3 weeks at 25 °C after addition of five initial densities of *Aphelenchus avenae* (Aphelenchidae) (after Sutherland and Fortin, 1968).

	Amanita rubescens		*Suillus granulatus*	
Initial nematode population	Colony diameter (mm)	Final nematode population	Colony diameter (mm)	Final nematode population
0	49a[a]	0a	40a	0a
5	47a	18 166b	34b	5 322b
10	43a	57 055c	26c	18 011bc
25	34b	57 555c	21d	34 811cd
50	25c	81 360d	16e	37 011d

[a]In each column numbers not followed by the same letter differ significantly at $P < 0.01$.

(Riffle, 1975). In a pot study, Hussey and Roncadori (1981) demonstrated that grazing by *A. avenae* on the vesicular-arbuscular endomycorrhizae *Glomus margarita* and *G. etunicatus* could retard shoot and root growth of cotton. However, they concluded that the large populations of *A. avenae* required probably precluded any significant interactions under field conditions. *Pochonia chlamydosporia* is being assessed as a potential biocontrol agent for economically important plant-feeding nematodes such as *Globodera* spp., *Heterodera* spp. and *Meloidogyne* spp. However, its intraspecific variants differ in their host preference and its persistence in soil requires saprophytic activity (Mauchline *et al.*, 2004), raising further questions about its relationships to the soil biota.

Predacious and omnivorous nematodes

The nematode faunal structure of undisturbed soils often has an abundance of specialist predators of nematodes, for example Mononchida and Diplogastrida, as well as an abundance of generalist predators, mainly Dorylaimida. Observational evidence of such systems, at least partially, supports the hypothesis that the predators have some regulatory, or even suppressive, effect on the relatively low abundance and temporal stability of nematodes occupying lower trophic levels in the food web, including herbivores, fungivores and bacterivores (e.g. Wardle *et al.*, 1995). Specialist and generalist predators are, however, quite sensitive to soil disturbance and chemical amendments (Korthals *et al.*, 1996; Berkelmans *et al.*, 2003; Tenuta and Ferris, 2004). They are at relatively low abundance in agricultural systems, where their potential prey increase to high levels of abundance and biomass and exhibit unregulated population increase when provided with suitable resources and environmental conditions. Interestingly, conversion of such disturbed systems to reduced tillage and organic production does not immediately result in a more structured soil food web with greater connectance to higher trophic levels. The predators may be slow colonizers, have longer life cycles and lower productivity. Colonization and regulatory balance in the soil food web may require considerable time (Korthals *et al.*, 1996; Yeates *et al.*, 1999a; Sánchez-Moreno *et al.*, 2006).

In some marine sediments, predacious nematodes may reach high abundances and dominate biomass and even densities of nematode assemblages. The genus *Enoploides*, abundant in many fine to medium sandy sediments along coasts and estuaries in NW Europe, is a voracious and selective predator, including oligochaetes, nematodes and ciliates among its prey. Both laboratory and field evidence indicate that it exerts substantial top-down control over prey density and community composition (Moens *et al.*, 2000; Hamels *et al.*, 2001; Gallucci *et al.*, 2005). Many other aquatic nematodes are presumed (based on stoma morphology) to be predators, but lack of empirical evidence on their actual food sources and feeding rates hampers proper assessment of the importance of top-down regulation within the meiobenthic compartment of aquatic food webs.

Nematodes in a Community / Ecosystem Context

Economic crop loss due to plant-feeding nematodes

Galled wheat florets ('cockles') and patches of stunted crop plants in sugar-beet and potato fields were among the first noted signs of plant nematode activity. Over the past 150 years there have been many investigations of such plant-feeding nematodes and how their potential pathogenic effects can be managed. Diverse examples can be found in books such as Lee (2002), Luc *et al.* (2005) and Perry and Moens (2006).

Management of microbial-feeding nematode function in an agricultural context

In the Mediterranean climate of the central valley of Northern California, annual crops are grown during the spring and summer months with the aid of irrigation. Following harvest in August or September, fields are left fallow, irrigation ceases, and the soil becomes very dry prior to the start of winter rainfall in November. During the September to November fallow period, soil temperatures would be conducive to biological activity if water was available. Cover crops, when used, are usually planted with the first rain in late November. In a study of the transition from conventional to organic farming practices, crops planted in the spring following incorporation of a winter-grown legume cover crop but without application of mineral fertilizers exhibited symptoms of nitrogen deficiency. Microbial biomass was at high levels following cover crop incorporation, but population levels of bacterial-feeding nematodes were very low (Fig. 1.3). About six weeks after planting the summer crop, bacterial-feeding and fungal-feeding nematodes had increased on the newly available resources and the nitrogen deficiency symptoms in the crop disappeared, with crop yield and mineral N concentrations being strongly correlated (Fig. 1.4).

In microcosm experiments, soil mineral nitrogen levels could be increased by 20% or more when bacteria were grazed upon by bacterial-feeding nematodes (Ferris *et al.*, 1998) and fungal-feeding nematodes may also contribute to N mineralization (Chen and Ferris, 1999). The abundance of bacterial-feeding nematodes was considered an indicator of concomitant abundance of other bacterial grazers, especially protozoa. The window of opportunity for increasing the abundance of bacterial- and fungal-feeding nematodes and other organisms in field soil in the spring, at the time of cover crop incorporation, was during the warm soil temperature period of September to November since winter soil temperatures were too cool for nematode reproduction and biological activity.

In field plots where the soil was irrigated during the September to November period, there were higher abundances of bacterial-feeding nematodes in the spring and greater nitrogen availability at the time of establishment of the new crop. Nitrogen deficiency symptoms were not seen in those

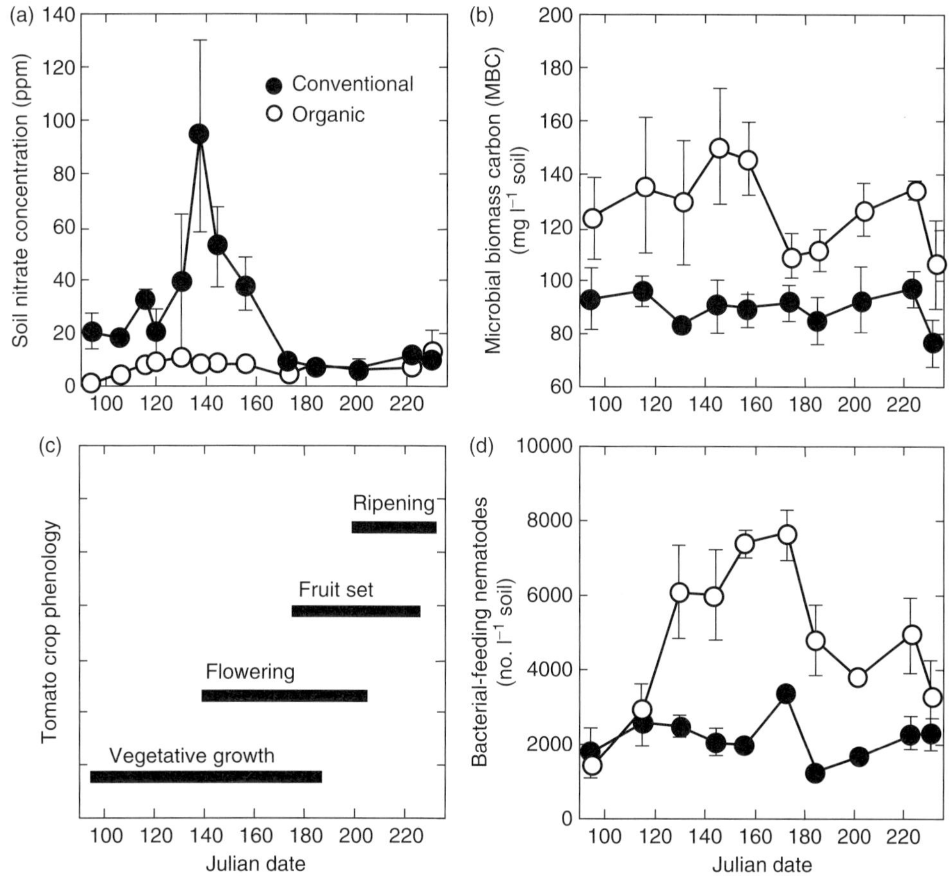

Fig. 1.3. Relationship of soil nitrate availability, microbial biomass and bacterial-feeding nematode abundance in soils under organic and conventional farming systems to tomato crop phenology. Data are means and standard errors across four replicates; Julian days run from 1 January. (a) Extractable soil nitrate concentration (from Temple, 1993). (b) Microbial biomass expressed as microbial biomass-carbon (data from Gunapala and Scow, 1998). (c) Tomato crop phenology (data from Flint, 1985). (d) Abundance of bacterial-feeding nematodes (redrawn from Ferris *et al.*, 1996).

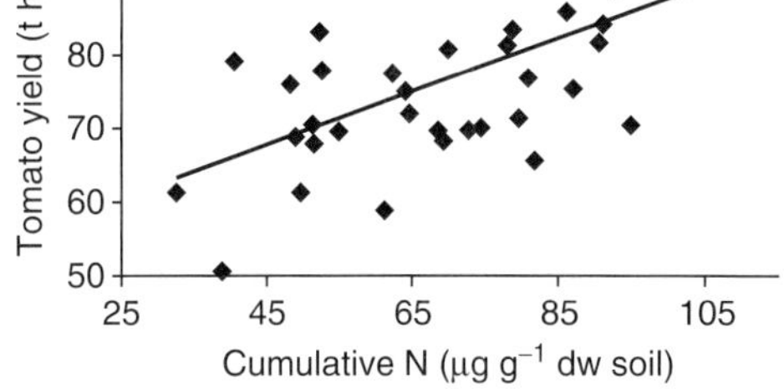

Fig. 1.4. Tomato crop yield in August 1996 in relation to cumulative soil N measurements over five sampling dates in April and May, prior to planting (redrawn from Ferris *et al.*, 1996).

plots. As a caveat, it should be noted that there is a downside to the late summer irrigation. The costs of water and pumping are additional financial overheads in the operation and weeds that would otherwise be dormant at that time may grow actively and require management (Ferris *et al.*, 2004a).

Nematode control of game bird population cycles

Cyclic fluctuations in vertebrate numbers have puzzled observers for generations. In Britain red grouse (*Lagopus lagopus scoticus*) is a game bird on which a sizeable industry is built, but it suffers irregular population crashes. The large caeca of red grouse are infected with *Trichostrongylus tenuis* (Rhabditina, Trichostrongylidae) and eggs are passed in the faeces, their development giving rise to infective third-stage juveniles on heather and in the soil; this development is affected by temperature and desiccation (Shaw *et al.*, 1989). Grouse are most likely infected when they feed on heather, their main food plant. The growth rate of the grouse population is negatively related to nematode infection ($r = -0.676$) while grouse brood mortality is positively related to such infection ($r = +0.641$) (Figure 1.5. a, b). Treating adult birds with levamisole hydrochloride (a standard anthelmintic) before breeding increased the number of game birds shot later that year, and reduced the cyclic fluctuation in grouse population compared with untreated populations (Fig. 1.5. c, d). Treatment with anthelmintic in two years gave the most stable game bird populations (Fig. 1.5 e) (Hudson *et al.*, 1998).

Coastal sand dune nematodes are controlled by a multitude of factors

Vegetation succession in coastal foredunes is driven by plant species-specific negative feedbacks with the soil community (Van der Putten *et al.*, 1993). Initially, it was assumed that nematodes play a major role in the soil pathogen complexes, because of the specific occurrence of specialized root-feeders in association with the different dominant plant species (Van der Putten and Van der Stoel, 1998). In other continents, however, the pioneer species marram grass (*Ammophila arenaria*) has been introduced without the specialized nematodes which correlated with plant invasiveness in Europe (Van der Putten *et al.*, 2005). However, all nematodes appeared to be controlled in a species-specific way by competition (Brinkman *et al.*, 2005), arbuscular mycorrhizal fungi (de la Peña *et al.*, 2006), endophytes (Hol *et al.*, 2007), soil microorganisms (Piśkiewicz *et al.*, 2007), or by the plants themselves (Fig. 1.6). Therefore, the current view is that in the root zone of marram grass, plant-feeding nematodes, albeit some that are potentially harmful to marram grass, are controlled in a multi-factorial manner. This interesting phenomenon may stimulate new thinking about sustainable nematode control in agriculture based on principles derived from natural systems (Van der Putten *et al.*, 2006).

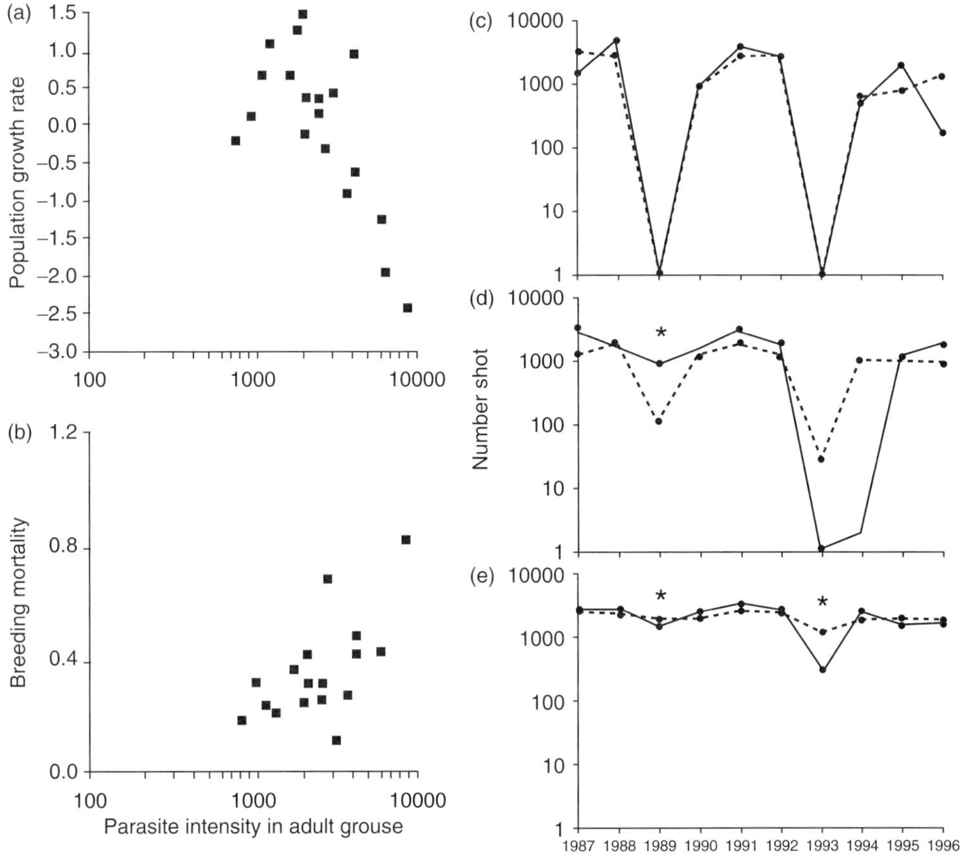

Fig. 1.5. Relationship between populations of red grouse (*Lagopus lagopus scoticus*) in northern England and their infection with the nematode *Trichostrongylus tenuis*. (a) Annual population growth rate [rt = (ln Nt+1 – ln Nt)] against mean log worm intensity in breeding adult grouse. (b) Breeding mortality [log maximum clutch size (12) – log mean brood size at 6 weeks] against mean log worm intensity in breeding adult grouse. Population changes of red grouse as represented by bag records at (c) two control sites, (d) two populations receiving a treatment with levamisole in 1989, and (e) two populations each receiving treatment with levamisole in 1989 and 1993. * indicate the years when treatment reduced nematode burdens in adult grouse (after Hudson *et al.*, 1998).

Intertidal nematode populations governed by environmental modulation of top-down control

Nematodes inhabiting intertidal habitats often display vertical migrations in relation to the tidal cycle. The genus *Enoploides*, abundant in many fine to medium sandy sediments along coasts and estuaries in NW Europe, is a voracious and selective predator of, for example, oligochaete, nematode and ciliate prey, and potentially exerts significant top-down control over its prey

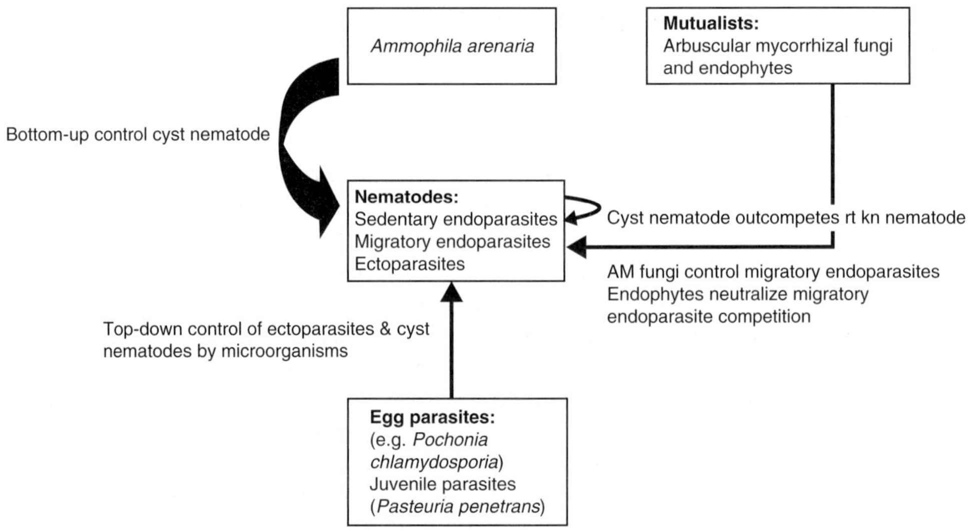

Fig. 1.6. Schematic of multiple controls on root-feeding nematodes on marram grass (*Ammophila arenaria*) in coastal sand dunes in north-western Europe (modified after Van der Putten, 2003).

communities (Moens *et al.*, 2000; Hamels *et al.*, 2001). *Enoploides* is generally restricted to the upper 3cm of the sediment, with clear density peaks in the upper 2cm. Total abundance of all other nematode species in a fine to medium sandy sediment on the Molenplaat, Schelde Estuary (SW Netherlands), peak at a depth of 4–5cm, which is unusual for any type of marine sediment (Steyaert, personal communication). These sediments are porous, and groundwater rapidly drains at low tide, leaving the upper sediment relatively dry after a few hours of exposure. Laboratory experiments have demonstrated that the predation efficiency of *Enoploides* on nematode prey is strongly impaired by even mild sediment desiccation (Gallucci *et al.*, 2005). Since this and similar intertidal sites are exposed 8–18 hours per day, optimal foraging by *Enoploides* must be restricted to relatively short periods during, and shortly after, inundation. Probably as an avoidance of sediment drying, *Enoploides* migrate a short distance into the sediment during low tide. Remarkably, some of its most abundant prey species at the Molenplaat exhibit the opposite vertical migration behaviour, peaking at or near the sediment surface only upon low tide exposure (Steyaert *et al.*, 2001). Many of these prey species rely on diatoms as their principal food. Since diatom production on intertidal flats is largely restricted to the very surface of the sediment (i.e. the photic zone), the upward migration of prey nematodes during low tide at the Molenplaat is likely a strategy to optimize feeding. In contrast, their downward migration upon inundation may be interpreted as an avoidance of predation by *Enoploides*. *Enoploides* distribution, in turn, appears to be largely controlled by sediment effects on the activity of the predator (Gallucci *et al.*, 2005; Steyaert *et al.*, unpublished data).

Pollution experiments suggest multifactorial control over nematode abundances in a salt marsh benthic food web

Experimentally imposed diesel contamination of a *Spartina alterniflora* salt marsh caused high mortality of meiobenthic harpacticoid copepods and concomitant (transient) increases in both microphytobenthos (MPB) biomass and nematode abundance and in meiofaunal grazing rates on MPB. Total meiobenthic (copepods + nematodes) grazing was lower in diesel-contaminated than in unaffected plots, suggesting: (i) that MPB biomass is top-down controlled by meiobenthic (here mainly harpacticoid) grazing; and (ii) that harpacticoid copepods and nematodes compete for this limiting resource (Carman *et al.*, 1997; Fleeger *et al.*, 2006). The system is, however, yet more complex in that inclusion of the naked goby, *Gobiosoma bosc*, a burrowing fish predacious on meiofauna, affects both MPB and meiofauna in a multifactorial manner. Specifically, presence of the naked goby reversed the diesel-induced increase in nematode and MPB abundance, while at the same time enhancing abundance of cyanobacteria (Fleeger *et al.*, 2006). Bioturbation of the sediment by *G. bosc* physically disrupts the sediment surface and MPB patches, while at the same time limiting light availability to the benthos through an increased turbidity. Hence, the diesel-induced release of the MPB from grazing control by harpacticoid copepods is counteracted by a fish-induced light limitation (which in turn improves the competitive ability of cyanobacteria over diatoms), and nematodes are again bottom-up controlled by a limited availability of MPB.

Nematodes of lakes and seas

In intertidal and shallow sediments, local primary production by microphytobenthos often appears to be the predominant carbon source fuelling nematode assemblages (Riera *et al.*, 1996; Moens *et al.*, 2002, 2005a). In the absence of such local primary production, nematodes in deeper (or light limited) environments are primarily dependent on the rain of organic debris from the productive photic zone above as well as input from rivers and streams. Both in shallow and deep environments, the relative importance of direct grazing on primary production or on associated bacteria and/or heterotrophic remains (especially those of protozoa) is still largely unclear. *Rhabditis marina* is almost uniquely associated with seaweed wrack on beaches (Sudhaus, 1976), and decomposition of dead fish (Gerlach, 1977) or marine mammals (Debenham *et al.*, 2004) has quantifiable impacts on nematode assemblages. Littoral macrophytes and their associated periphyton provide limited *in situ* resources for nematodes (Peters and Traunspurger, 2005).

As in soils, meiobenthic nematodes contribute to ecological processes (Alkemade *et al.*, 1992; Aller and Aller, 1992; Montagna 1995; Traunspurger *et al.*, 1997). Energy availability is generally negatively correlated with the depth through which phytodetrital food sinks (Suess, 1980). Tietjen *et al.* (1989)

found strong correlations between deep-sea meiofaunal abundance (dominated by nematodes) and fluxes of both organic carbon ($r^2 = 0.982$) and nitrogen ($r^2 = 0.971$). Nematode species richness in the deep sea of the North Atlantic Ocean increases with latitude, in contrast to patterns of molluscan and isopod diversity. This is likely to be related to the increase in primary productivity with latitude in this area (Lambshead *et al.*, 2000), a correlation which was also confirmed in the central equatorial Pacific Ocean (Lambshead *et al.*, 2002).

In the deep-sea hydrothermal vents and seeps, as well as in some anoxic coastal sediments (such as subsurface sediments in mangroves) with an abundance of sulfur-reducing bacteria, chemoautotrophic production may replace sedimentation of particulate food from the euphotic zone as the primary carbon and energy source for particular nematode genera or assemblages (Dover, 2000). Chemoautotrophic bacteria may be directly grazed upon, as in the case of *Halomonhystera disjuncta* feeding in mats of the sulphide-oxidizing *Beggiatoa* at a 1280 m deep arctic mud volcano (Van Gaever *et al.*, 2006). However, more specialized relationships between chemoautotrophic bacteria and nematodes have evolved: the adaptation of *Stilbonema* and *Laxus* (Desmodoridae: Stilbonematinae) to sulphide-rich sediments involves bacterial ectosymbionts capable of respiratory reduction of nitrate to nitrite (Hentschel *et al.*, 1999); the nematodes act as vectors for the bacteria, but probably also graze upon them (Polz *et al.*, 2000). The unrelated mouthless nematode *Astomonema* (Siphonolaimidae) lacks a functional stoma, and probably derives at least part of its nutrition from bacterial endosymbionts (Giere *et al.*, 1995). These endosymbionts are gammaproteobacteria which do appear closely related to the bacterial ectosymbionts on stilbonematinid nematodes and in gutless oligochaetes (Musat *et al.*, 2007). The marine nematode *Oncholaimus campylocercoides* (Oncholaimidae) is adapted to sulphidic sediments by the development of polysulfur chains and S-8 rings in the epidermis; these disappear on return to normal oxygen levels (Thiermann *et al.*, 2000). While populations of deep-sea mussels have been used as bioindicators (Jones *et al.*, 2006), it will be some time before a comparable data set is available to utilize nematodes in a similar way.

Diversity within vertebrate hosts

There has been radiation/speciation of nematodes in the stomach of kangaroos and wallabies with 55 species of host containing 112 species of the genus *Cloacina* alone. Up to nine nematode species commonly occurred in a host (*Macropus dorsalis*), while particular nematodes occurred in up to 11 hosts (Beveridge *et al.*, 2002). A cladistic analysis of nematode species suggested that the aggregations within a given host are polyphyletic and probably evolved by host switching or colonization rather than by co-speciation. This radiation differs from that in ruminants where overlapping distribution of hosts and human management has confounded host distributions. There are, however, species flocks of strongylid nematodes in equids (Bucknell *et al.*, 1996).

Nematodes as Environmental Modifiers

Sediment agitation by nematodes

We have already described how microbial grazing can enhance nutrient cycling. In addition, several aquatic studies have demonstrated bioturbation by nematodes, that is physical disturbance of the mineral and organic particles comprising the substrate, to be of major importance through direct physical enhancement of fluxes of oxygen and nutrients, and indirectly through the stimulatory effect of such enhanced fluxes on microbial activity (Alkemade *et al.*, 1992; Aller and Aller, 1992). Alkemade *et al.* (1992) demonstrated that a 30% enhancement of saltmarsh grass (*Spartina anglica*) decomposition in sediments with an abundant *Diplolaimella dievengatensis* population was almost entirely due to bacterial stimulation by an increased oxygen availability as a result of bioturbation by nematodes. Riemann and Schrage (1978) observed that aquatic nematodes rapidly aggregate both sediment and organic matter particles when dispersed in a little water. They hypothesized an underlying trophic relationship, where mucus secreted by the nematodes stimulates growth of bacteria, which could in turn be grazed upon. This hypothesis was later modified into the 'enzyme sharing' concept: nematodes may produce exo-enzymes that start the decomposition of complex molecules and promote establishment and growth of bacteria, which then take over the organic matter decomposition. Both nematodes and bacteria then feed on the nutritious dissolved organic matter 'soup' released from this shared use of enzymes (Riemann and Helmke, 2002). Several chromadorid nematodes from intertidal mudflats (e.g. the widespread and often abundant *Ptycholaimellus ponticus*), build tubes by agglutinating sediment and small organic particles through mucus produced by the ventral gland. This affects sediment stability, enlarges the surface available to decomposers (the total inner surface area of the tube may be up to five times greater than the sediment surface), and probably offers the nematodes shelter against the effects of water current as well as predation (Nehring *et al.*, 1990; Nehring, 1993).

Nematodes influence ecological succession in grasslands

Effects of nematodes on ecological succession have been studied mainly in terrestrial grassland systems. Very few studies have related the occurrence of plant parasitic nematodes to the performance of aquatic macrophytes (plants) (Fritz *et al.*, 2004) and some studies have investigated the role of nematodes in the decline of wild nitrogen-fixing shrubs (Oremus and Otten, 1981; Maas *et al.*, 1983; Zoon *et al.*, 1993). As explained before, plant-feeding nematodes in the root zone of the foredune marram grass (*Ammophila arenaria*) all appear to be controlled. However, in secondary grassland succession following removal of intensive farming practices, nematodes have been shown to influence plant performance (Verschoor *et al.*, 2001) and plant succession

(De Deyn *et al.*, 2003). Usually, abundance of plant-feeding nematode species is relatively low, but their high abundance in local hot spots may allow substantial effects on plant productivity (Verschoor, 2002). Selectivity of the nematodes, or their focusing on dominant, fast-growing plant species, may explain their contribution to plant community composition (van Ruijven *et al.*, 2005), or succession (De Deyn *et al.*, 2003). Plant species can influence the composition of nematode assemblages in soil; however, the actual species present are more important than the diversity of the plant community per se (De Deyn *et al.*, 2004; Viketoft *et al.*, 2005). Some plant species that enhance specific root-feeding nematodes may indirectly enhance biotic resistance of grassland communities against nematode-sensitive plant species when invading the existing grassland communities (van Ruijven *et al.*, 2005). Most studies mentioned above have involved inoculation experiments. Studies in prairie ecosystems in the 1970s concluded that nematodes use substantial amounts of the net primary productivity (Stanton, 1988). However, these conclusions were mostly based on selective biocides and the results have not been verifiable by inoculation studies, which casts doubt on the assertion that grassland nematodes account for losses up to one quarter of the net primary production.

Nematodes may influence succession of aquatic microbial communities

The transport of, and grazing on, bacteria by nematodes may affect settlement (Moens *et al.*, 2005b), community composition (De Mesel *et al.*, 2004), densities and activity (Traunspurger *et al.*, 1997) of bacteria, and hence their likely regulation of important ecosystem processes such as organic matter mineralization. Microcosm studies using an estuarine example of the duality of enrichment opportunists (e.g. *Rhabditis marina*, *Panagrolaimus paetzoldi*) and general opportunists (e.g. several species of Monhysteridae) indicate that the succession often observed from the former group to the latter does not passively follow the organic matter and bacterial dynamics. Instead, overgrazing of bacteria by rhabditid nematodes may facilitate monhysterids by suppressing bacterial densities to levels which provide them with optimal feeding conditions (T. Moens, Ghent, 1997, personal communication; Santos *et al.*, 2008).

Nematodes may predispose hosts to other organisms

There is widespread belief that root-feeding nematodes predispose their host plants to pathogenic soil fungi (Castillo *et al.*, 2003; Back *et al.*, 2006). Although these effects may occur, other studies have argued the interaction effects to be additive, rather than synergistic. In a critique, Sikora and Carter (1987) argued that most studies that showed synergistic interactions were based on similar model systems and experimental procedures. It is often advocated that nematodes create entry points in the roots for pathogenic soil fungi to colonize but the mechanism of predisposition has not always been identified. However, Van Gundy *et al.* (1977), in an elegant experiment, demonstrated that metabolic leakage from roots infested with root-knot nematodes (*Meloidogyne incognita*) stimulated the transformation of *Rhizoctonia solani* from saprophytic to parasitic growth on uninjured roots.

Nematodes may also predispose hosts to other nematodes, which may lead to competitive or facilitative effects. In general, the more complex the relationship between root-feeding nematodes and their host plants is, the more competitive they are within and among species (Eisenback, 1993). On the other hand, nematodes may avoid direct competition by feeding on different cell layers in the root cortex (Bongers and Bongers, 1998; Siddiqi, 2000). Another issue of competition or facilitation, which has attracted far less attention, is that between root-feeding nematodes and other soil invertebrates. These interactions may lead to contrasting patterns, such as the induction of antagonistic effects by earthworms against cyst nematodes (Blouin *et al.*, 2005), to facilitative effects of wireworms (Elateridae) on root knot nematodes by indirect promotion of *Meloidogyne* abundance in mixed plant communities (De Deyn *et al.*, 2007). Whereas we have already mentioned facilitative effects among aquatic bacterial-feeding nematodes, inhibitory interactions between congeneric species of *Diplolaimelloides* have been demonstrated and would appear not to be primarily due to direct competition for food or space (De Mesel *et al.*, 2006; Santos *et al.*, 2008).

Similarly, inhibition between soil bacterial-feeding nematode species (Bongers *et al.*, 2001; Postma-Blaauw *et al.*, 2005) may not be completely due to competitive interactions but could conceivably be mediated through bacterial defence signals (Phillips *et al.*, 2003). Probably, when explored more intensively, such indirect effects among nematodes and between nematodes and other organisms will be found to be important determinants of community interactions.

Synoptic Integration to System, Landscape and Biosphere Levels

Nematodes are among the most diverse soil and benthic organisms and are usually the most abundant of the soil or sediment Metazoa. They are the most important secondary consumers within the soil mesofauna (Mulder *et al.*, 2005). There has been less research in aquatic systems but studies suggest at least a small contribution of nematodes to carbon turnover in aquatic sediments (Soetaert *et al.*, 1997; Coull, 1999). Nematodes have been extensively used as indicators of soil diversity and functioning (Neher, 2001; Mulder *et al.*, 2005). While there are some similar studies for aquatic environments there is an underlying problem that the causes for very high local diversity in, for example, deep-sea sediments, are very poorly understood. Thus the use of nematode diversity per se as an indicator in aquatic sediments is not well established. While literature dealing with nematode faunae as soil health indicators in different farming and natural systems is abundant, few studies deal with regional/landscape/ocean floor zone/distribution patterns of nematode faunae. Most studies on nematode assemblages in soil are based on field plots or single-crop farm fields (e.g. Wardle *et al.*, 1995; Berkelmans *et al.*, 2003; Ferris and Matute, 2003). However, feedback loops among aboveground and belowground biota are important ecological drivers in terrestrial ecosystems (Sánchez-Moreno *et al.*, 2008). The spatial patterns of soil biota have important aboveground consequences on both plant community structure and on individual plants (Ettema and Wardle, 2002). The reciprocal is also true.

In general, there is a lack of detailed information on the effects of landscape heterogeneity and soil management practices on nematode spatial patterns. In marine systems, on the other hand, there has been a strong emphasis on comparisons of nematode assemblages at β and γ scales, but there is a lack of process-oriented or mechanistic understanding of drivers of the patterns.

A simple example of the problems of scale and resolution in considering nematode roles and services is in their function as herbivores and entry points for carbon and energy into the soil food web (e.g. Yeates *et al.*, 1999b). Individual nematodes and their activities represent the minimum patch and species population the more general, local patch. In agriculture such patches of plant-feeding nematodes may cause local loss of yield, with field-scale loss being determined by the field-scale nematode population, which reflects the age of the infestation, edaphic conditions, crop history, and management practices. Similarly, strong aggregations of nematodes and other organisms in the root zone sites of uprooted orchard trees can result in death of trees in a subsequent orchard planting.

In another example of functional linkages between aboveground and belowground biota, an abundance of bacterial-feeding nematodes provides important services in well-managed organic farms, whereas in conventional production systems, where nutrients are supplied from external sources, a similar abundance of nematodes may contribute little benefit to crop growth. While these effects occur at the ped/microsite level, scaling up by using GIS and multi-layer mapping techniques will allow a more comprehensive understanding of the functions and services of nematodes at the landscape level. Recent studies indicate that nematodes can enhance the control of aboveground pests through increasing both bottom-up and top-down control activities (Bezemer *et al.*, 2005). This linking of above- and belowground subsystems is a new and promising development indicating that sustainable crop protection, even against aboveground pests, may start with proper soil management.

The shift from mixed cropping and mixed grazing to more intensive livestock farming has often led to significant problems with nematode parasites of grazing animals for which anthelmintic drenches gave relief. However, apart from perceived non-target effects, these drenches have resulted in genetic selection for anthelmintic-resistant nematodes. The use of the nematode-trapping fungus *Duddingtonia flagrans* has reduced parasitism with acceptable economic return (Waller *et al.*, 2004). In contrast, two anthelmintic drenches (ivermectin and albendazole) have proved useful for effective, control of river blindness and elephantiasis, caused by filarioid nematodes.

Humans, Nematodes and Ecosystem Management

Nematodes are intrinsic components of natural ecosystems. The early focus of helminthology and nematology on nematodes as pests and parasites of humans, plants and animals has developed into a realization that parasitism plays a role in natural ecosystems, in controlling species abundance and maintaining species diversity. Moreover, the awareness of the role of nematodes in decomposition processes as well as their potential role in controlling outbreaks of insect pests and their use as environmental indicators has led to a paradigm shift from consideration of nematodes only as undesirable pests to a broader recognition of their contributions in ecosystem services. There is increasing awareness that nematode populations may be optimized by managing nature rather than utilizing broad-spectrum control measures with unknown ecological consequences.

Plant-feeding nematodes may contribute to coexistence and succession in natural plant communities (De Deyn *et al.*, 2003) and the same principles drive the need for crop rotation in agriculture. In the past, when wide crop rotations were applied, nematodes were less of a problem. However, the globalization of world food and commodity markets, and the expensive specialized machinery have led to the need for specialization among growers. Rotations have become so narrow that natural population dynamics cannot be used to reduce parasite population levels. The side effects of chemical nematicides on soil life and neighboring aquatic ecosystems have resulted in awareness that these chemicals threaten the sustainability of crop production and have highlighted the need for biological control. However, biological control, because it represents a series of interlocking biological populations, is much more variable and less predictable, limiting the generalization of findings in one crop at specific site conditions (Van der Putten *et al.*, 2006). This leaves considerable challenges for nematologists, agronomists, entomologists and ecologists to develop farming systems that maximize nematode control while fulfilling the high demands of global trade and economy on cropping systems.

Field studies suggest that when humans modify natural selection, competition and species replacement, through crop breeding for high yield, weed control and other features, the need for understanding impacts throughout the food chain is paramount. When pests are not subjected to natural regulatory pressures of competition, predation and genetic barriers, massive outbreaks may result, similar to those seen after the application of broad-spectrum pesticides. Biological and organic agriculture each aim for conditions that more or less restore biodiversity and habitat complexity, avoiding chemical crop protection and mineral fertilizers. The question of whether maximal yields are achievable with natural equilibria of plant pests and their antagonists, non-chemical fertilizers and sustainable soil tillage, is still open and nematode management is one of the key issues to be solved in order to reach that state. The major question of whether or not to return to cropping practices that were applied in the past or to develop new farming systems has not yet been answered. Almost certainly, the use of cultivars that have not been selected for maximum yield under conditions utilizing pesticides and mineral fertilizers will be necessary. Some studies have shown that nematodes can enhance the control of aboveground pests through enhanced bottom-up and top-down control activities (Bezemer *et al.*, 2005). This linking of above- and belowground subsystems is a new and promising development suggesting that sustainable crop protection, even against aboveground pests, may start with proper soil management.

Suitability of Nematodes as Environmental Indicators

The development of nematodes as bioindicators in soil and aquatic systems required determination of appropriate ways to assess and quantify their contributions to ecological processes, and the validation of their utility as indicators of environmental condition. Several unique characteristics of nematodes facilitated those developments. In summary, nematodes occur in all soil and aquatic systems: in acidified forest soils, in heavily polluted soil, on heavy clay, in deep sea sediments, in rotting plant material, in compost and in any habitat in which organic material is decomposed. Different nematode taxa exhibit specificities of food sources and changes in the food web are mirrored in shifts among feeding groups. Many families within the Tylenchina feed exclusively on the roots of higher plants but never on bacteria. Cephalobidae and Plectidae feed on bacteria but not on higher plants or fungi. Mononchidae and Anatonchidae are specialist predators of other nematodes and do not feed on higher plants or fungi. The transparent nature of nematodes allows easy observation of mouth and pharyngeal structures, which allows inference of feeding habit. Robust techniques for extracting nematodes from soil and other substrates have been developed and can be applied to all taxonomic and functional groups. Most importantly, nematodes have variable responses to stress factors; some species are extremely sensitive to pollutants and others extremely tolerant (Korthals *et al.*, 1996; Ferris *et al.*, 2004b; Tenuta and Ferris, 2004). They vary in lifespan; some species with a generation time of days, others months or even a year. The matrix of feeding habit, generation time and sensitivity to environmental disturbance allows the designation of functional guilds of different nematode taxa with similar response characteristics (Bongers and Bongers, 1998; Ferris and Bongers, 2006).

Concluding Points

- The nematode body plan and life history, although apparently limiting, are sufficiently adaptable to allow species of the phylum Nematoda to occupy a wide range of habitats, utilize a wide range of resources, become extremely diverse and achieve large populations. They are the most abundant multicellular animals on earth.

- Nematodes have impacts on populations of many organisms and on ecosystem processes. They play key roles in many ecosystem services and processes.
- The abundance and diversity of nematode species vary with ecosystem, substrate and management.
- Nematodes have been successfully used as environmental bioindicators, whether in relation to diverse ecosystem services, plant diseases, management of parasites in grazing mammals, human health, or insect control, etc. Some such indicators have been used for decades.
- There are opportunities for further studies on the use of nematodes as indicators for other ecosystem functions, such as the state of restoration in semi-natural ecosystems and the capacity of soils to sustain diverse plant communities. However, indicators must be tailored for each question and environment, with the response variable(s) and indicator being context-specific.
- Anthelmintic drenches developed for nematodes of livestock relieve suffering when applied to humans. Nematode-trapping fungi, extensively studied in relation to plant-feeding nematodes, have also been successfully deployed to manage gastrointestinal nematodes of livestock. These cases clearly demonstrate the transferability of knowledge of the relationships between nematodes and their total environment.
- In ecosystem terms, root-feeding nematodes influence diversity and succession in natural vegetation; their absence may result in plant invasiveness, especially when plants are introduced in new habitats without their original root feeders. Further, in cropping terms, nematodes and other soil biota influence plant–enemy interactions above ground; a healthy crop, therefore, depends on a healthy soil.

Acknowledgements

Preparation of this chapter has been aided by funding from the Foundation for Research Science and Technology (New Zealand) and by N.S.F. (US) grant DEB-0120169. T.M. is a postdoctoral fellow with the Flemish Fund for Scientific Research F.W.O.

References

Alkemade, R., Wielemaker, A., de Jong, S.A. and Sandee, A.J.J. (1992) Experimental evidence for the role of bioturbation by the marine nematode *Diplolaimella dievengatensis* in stimulating the mineralization of *Spartina anglica* detritus. *Marine Ecology Progress Series* 90, 149–155.

Aller, R.C. and Aller, J.Y. (1992) Meiofauna and solute transport in marine muds. *Limnology and Oceanography* 37, 1018–1033.

Alston, D.G., Bradley, J.R., Coble, H.D. and Schmitt, D.P. (1991) Impact of population density of *Heterodera glycines* on soybean canopy growth and weed competition. *Plant Disease* 75, 1016–1018.

Back, M., Haydock, P. and Jenkinson, P. (2006) Interactions between the potato cyst nematode *Globodera rostochiensis* and diseases caused by *Rhizoctonia solani* AG3 in potatoes under field conditions. *European Journal of Plant Pathology* 114, 215–223.

Bae, Y.S. and Knudsen, G.R. (2001) Influence of a fungus-feeding nematode on growth and biocontrol efficacy of *Trichoderma harzianum*. *Phytopathology* 91, 301–306.

Bardgett, R.D. (2005) *The Biology of Soil: A Community and Ecosystem Approach.* Oxford University Press, Oxford.

Bardgett, R.D., Cook, R., Yeates, G.W. and Denton, C.S. (1999) The influence of nematodes on below-ground processes in grassland ecosystems. *Plant and Soil* 212, 23–33.

Bardgett, R.D., Bowman, W.D, Kaufmann, R. and Schmidt, S.K. (2005) A temporal approach to linking aboveground and belowground ecology. *Trends in Ecology and Evolution* 20, 634–641.

Bargmann, C.I. and Mori, I. (1997) Chemotaxis and thermotaxis. In: Riddle, D.L., Blumenthal, T., Meyer, B.J. and Priess, J.R. (eds) *C. elegans* II. Cold Spring Harbor Laboratory Press, Woodbury, pp. 717–737.

Bedding, R.A. (1993) Biological control of *Sirex noctilio* using the nematode *Deladenus siricidicola*. In: Bedding, R., Akhurst, R. and Kaya, H. (eds) *Nematodes and the Biological Control of Insect Pests*. CSIRO Publications, East Melbourne, pp. 11–20.

Berkelmans, R., Ferris, H., Tenuta, M. and van Bruggen, A.H.C. (2003) Effects of long-term crop management on nematode trophic levels other than plant feeders disappear after one year of disruptive soil management. *Applied Soil Ecology* 23, 223–235.

Beveridge, I., Chiton, N.B. and Spratt, D.M. (2002) The occurrence of species flocks in the nematode genus *Cloacina* (Strongyloidea: Cloacininae), parasitic in the stomach of kangaroos and wallabies. *Australian Journal of Zoology* 50, 597–620.

Bezemer, T.M., De Deyn, G.B., Bossinga, T.M., van Dam, N.M., Harvey, J.A. and Van der Putten, W.H. (2005) Soil community composition drives aboveground plant-herbivore-parasitoid interactions. *Ecology Letters* 8, 652–661.

Bird, A.F. and Ryder, M.H. (1993) Feeding of the nematode *Acrobeloides nanus* on bacteria. *Journal of Nematology* 25, 493–499.

Bird, A.F. and Stynes, B.A. (1977) The morphology of a *Corynebacterium* sp. parasitic on annual ryegrass. *Phytopathology* 67, 828–830.

Bisset, S.A. and Morris, C.A. (1996) Feasibility and implications of breeding sheep for resilience to nematode challenge. *International Journal for Parasitology* 26, 857–868.

Blair, J.E., Ikeo, K., Gojobori, T. and Hedges, S.B. (2002) The evolutionary position of nematodes. *BMC Evolutionary Biology* 2, 7.

Blanchard, G.F. (1990) Overlapping microscale dispersion patterns of meiofauna and microphytobenthos. *Marine Ecology Progress Series* 68, 101–111.

Blaxter, M.L. (1998) *Caenorhabitis elegans* is a nematode. *Science* 282, 2041–2046.

Blaxter, M.L., De Ley, P., Garey, J.R., Liu, L.X., Scheldeman, P., Vierstraete, A., Vanfleteren, J.R., Mackey, L.Y., Dorris, M., Frisse, L.M., Vida, J.T. and Thomas, W.K. (1998) A molecular evolutionary framework for the phylum Nematoda. *Nature (London)* 392, 71–75.

Blouin, M., Zuily-Fodil, Y., Pham-Thi, A.T., Laffray, D., Reversat, G., Pando, A., Tondoh, J. and Lavelle, P. (2005) Belowground organism activities affect plant aboveground phenotype, inducing plant tolerance to parasites. *Ecology Letters* 8, 202–208.

Boag, B., Neilsen, R., Robinson, D., Scrimgeour, C.M. and Handley, L.L. (1998) Wild rabbit host and some parasites show trophic-level relationships for delta C-13 and delta N-15: a first report. *Isotopes in Environmental and Health Studies* 34, 81–85.

Boemare, N. (2002) Biology, taxonomy, and systematics of *Photorhabdus* and *Xenorhabdus*. In: Gaugler, R. (ed.) *Entomopathogenic Nematology*. CABI Publishing, Wallingford, pp. 57–78.

Bongers, T. (1990) The maturity index: an ecological measure of environmental disturbance based on nematode species composition. *Oecologia* 83, 14–19.
Bongers, T. and Bongers, M. (1998) Functional diversity of nematodes. *Applied Soil Ecology* 10, 239–251.
Bongers, T., Ilieva-Makulec, K. and Ekschmitt, K. (2001) Acute sensitivity of nematode taxa to $CuSO_4$ and relationships with feeding-type and life-history classification. *Environmental Toxicology and Chemistry* 20, 1511–1516.
Brinkman, E.P., Duyts, H. and Van der Putten, W.H. (2005) Consequences of variation in species diversity in a community of root-feeding herbivores for nematode dynamics and host plant biomass. *Oikos* 110, 417–427.
Brown, D.J.F., Halbrendt, J.M., Robbins, R.T. and Vrain, T.C. (1993) Transmission of nepoviruses by *Xiphinema americanum*-group nematodes. *Journal of Nematology* 25, 349–354.
Brown, D.J.F., Robertson, W.M. and Trudgill, D.L. (1995) Transmission of viruses by plant nematodes. *Annual Review of Plant Pathology* 33, 223–249.
Bucknell, D., Hoste, H., Gasser, R.B. and Beveridge, I. (1996) The structure of the community of strongyloid nematodes in domestic equids. *Journal of Helminthology* 70, 185–192.
Carman, K.R., Fleeger, J.W. and Pomarico, S.M. (1997) Response of a benthic food web to hydrocarbon contamination. *Limnology and Oceanography* 42, 561–571.
Castillo, P., Navas-Cortes, J.A., Gomar-Tinoco, D., Di Vito, M. and Jimenez-Diaz, R.M. (2003) Interactions between *Meloidogyne artiellia*, the cereal and legume root-knot nematode, and *Fusarium oxysporum* f. sp *ciceris* race 5 in chickpea. *Phytopathology* 93, 1513–1523.
CESC (*C. elegans* Sequencing Consortium) (1998) Genome sequence of the nematode *C. elegans*: a platform for investigating biology. *Science* 282, 2012–2018.
Chen, J. and Ferris, H. (1999) The effects of nematode grazing on nitrogen mineralization during fungal decomposition of organic matter. *Soil Biology and Biochemistry* 31, 1265–1279.
Chia, F.S. and Warwick, R.M. (1969) Assimilation of labelled glucose from seawater by marine nematodes. *Nature (London)* 224, 720–721.
Cobb, N.A. (1915) Nematodes and their relationships. Yearbook of the United States Department of Agriculture 1914, 457–490.
Coleman, D.C., Odum, E.P. and Crossley, D.A. (1992) Soil biology, soil ecology and global change. *Biology and Fertility of Soils* 14, 104–111.
Coull, B.C. (1999) Role of meiofauna in estuarine soft-bottom habitats. *Australian Journal of Ecology* 24, 327–343.
Cupp, E.W. and Cupp, M.A. (2005) Impact of ivermectin community-level treatments on elimination of adult *Onchocerca volvulus* when individuals receive multiple treatments each year. *American Journal of Tropical Medicine and Hygiene* 73, 1159–1161.
De Deyn, G.B. and Van der Putten, W.H. (2005) Linking above- and belowground biodiversity. *Trends in Ecology and Evolution* 20, 625–633.
De Deyn, G.B., Raaijmakers, C.E., Zoomer, H.R., Berg, M.P., de Ruiter, P.C., Verhoef, H.A., Bezemer, T.M. and Van der Putten, W.H. (2003) Soil invertebrate fauna enhances grassland succession and diversity. *Nature (London)* 422, 711–713.
De Deyn, G.B., Raaijmakers, C.E., van Ruijven, J., Berendse, F. and Van der Putten, W.H. (2004) Plant species identity and diversity effects on different trophic levels of nematodes in the soil food web. *Oikos* 106, 576–586.
De Deyn, G.B., van Ruijven, J., Raaijmakers, C.E., de Ruiter, P.C. and Van der Putten, W.H. (2007) Above- and belowground insect herbivores differentially alter nematode communities by plant diversity shifts. *Oikos* 116, 923–930.
de la Peña, E., Rodríguez Echeverría, S., Van der Putten, W.H., Freitas, H. and Moens, M. (2006) Mechanism of control of root-feeding nematodes by mycorrhizal fungi in the dune grass *Ammophila arenaria*. *New Phytologist* 169, 829–840.

de Ley, P. and Blaxter, M. (2002) Systematic position and phylogeny. In: Lee, D.L. (ed.) *The Biology of Nematodes*. Taylor and Francis, London, pp. 1–30.

de Ley, P. and Blaxter, M.L. (2004) A new system for Nematoda: combining morphological characters with molecular trees, and translating clades into ranks and taxa. *Nematology Monographs and Perspectives* 2, 633–653.

De Mesel, I., Derycke, S., Swings, J., Vincx, M. and Moens, T. (2003) Influence of bacterivorous nematodes on the decomposition of cordgrass. *Journal of Experimental Marine Biology and Ecology* 296, 227–242.

De Mesel, I., Derycke, S., Moens, T., Van der Gucht, K., Vincx, M. and Swings, J. (2004) Top-down impact of bacterivorous nematodes on the bacterial community structure: a microcosm study. *Environmental Microbiology* 6, 733–744.

De Mesel, I., Derycke, S., Swings, J., Vincx, M. and Moens, T. (2006) Role of nematodes in decomposition processes: does within-trophic group diversity matter? *Marine Ecology – Progress Series* 321, 157–166.

de Moura, R.M., de Oliveira, I.S. and de Castro Torres, G.R. (2006) [Register of the vinegar nematode in northeastern Brazil.] *Nematologia Brasileira* 30, 67–70.

de Rooij-Van der Goes, P.C.E.M. (1995) The role of plant-parasitic nematodes and soil-borne fungi in the decline of *Ammophila arenaria* (L.) Link. *New Phytologist* 129, 661–669.

de Rosa, R., Grenier, J.K., Andreeva, T., Cook, C.E., Adoutte, A., Akam, M., Carroll, S.B. and Balavoine, G. (1999) Hox genes in brachiopods and priapulids and protostome evolution. *Nature (London)* 399, 772–776.

Debenham, N.J., Lambshead, P.J.D., Ferrero, T.J. and Smith, C.R. (2004) The impact of whale falls on nematode abundance in the deep sea. *Deep-sea Research Part 1 – Oceanographic Research Papers* 51, 701–706.

Donald, P.A., Pierson, P.E., Martin, S.K.S., Sellers, P.R., Noel, G.R., MacGuidwin, A.E., Faghihi, J., Ferris, V.R., Grau, C.R., Jardine, D.J., Melakeberhan, H., Niblack, T.L., Stienstra, W.C., Tylka, G.L., Wheeler, T.A. and Wysong, D.S. (2006) Assessing *Heterodera glycines*-resistant and susceptible cultivar yield response. *Journal of Nematology* 38, 76–82.

dos Santos, G.A.P., Derycke, S., Genevois, V.G.F., Coelho, L.C.B.B., Correia, M.T.S. and Moens, T. (2008) Differential effects of food availability on population development and fitness of three species of estuarine, bacterial-feeding nematodes. *Journal of Experimental Marine Biology and Ecology* 355, 27–40.

Dover, C.L.V. (2000) *The Ecology of Deep-sea Hydrothermal Vents*. Princeton University Press, Princeton, New Jersey.

Eisenback, J.D. (1993) Interactions between nematodes with root-rot fungi. In: Khan, M.W. (ed.) *Nematode Interactions*. Chapman and Hall, London, pp. 134–174.

Ettema, C.H. and Bongers, T. (1993) Characterization of nematode colonization and succession in disturbed soil using the Maturity Index. *Biology and Fertility of Soils* 16, 79–85.

Ettema, C.H. and Wardle, D.A. (2002) Spatial soil ecology. *Trends in Ecology and Evolution* 17, 177–183.

Ettema, C.H. and Yeates, G.W. (2003) Nested spatial biodiversity patterns in a New Zealand forest and pasture soil. *Soil Biology and Biochemistry* 35, 339–342.

Faulkner, L.R. and Darling, H.H. (1961) Pathological histology, hosts and culture of the potato rot nematode. *Phytopathology* 51, 778–786.

Ferris, H. and Bongers, T. (2006) Nematode indicators of organic enrichment. *Journal of Nematology* 38, 3–12.

Ferris, H. and Matute, M. (2003) Structural and functional succession in the nematode fauna of a soil food web. *Applied Soil Ecology* 23, 93–110.

Ferris, H., Mullens, T.A. and Foord, K.E. (1990) Stability and characteristics of spatial description parameters for nematode populations. *Journal of Nematology* 22, 427–439.

Ferris, H., Lau, S. and Venette, R. (1995) Population energetics of bacterial-feeding nematodes: respiration and metabolic rates based on carbon dioxide production. *Soil Biology and Biochemistry* 27, 319–330.

Ferris, H., Venette, R.C. and Lau, S.S. (1996) Dynamics of nematode communities in tomatoes grown in conventional and organic farming systems, and their impact on soil fertility. *Applied Soil Ecology* 3, 161–175.

Ferris, H., Venette, R.C. and Lau, S.S. (1997) Population energetics of bacterial-feeding nematodes: carbon and nitrogen budgets. *Soil Biology and Biochemistry* 29, 1183–1194.

Ferris, H., Venette, R.C., van der Meulen, H.R. and Lau, S.S. (1998) Nitrogen mineralization by bacterial-feeding nematodes: verification and measurement. *Plant and Soil* 203, 159–171.

Ferris, H., Bongers, T. and de Goede, R.G.M. (2001) A framework for soil food web diagnostics: extension of the nematode faunal analysis concept. *Applied Soil Ecology* 18, 13–29.

Ferris, H., Venette, R.C. and Scow, K.M. (2004a) Soil management to enhance bacterivore and fungivore nematode populations and their nitrogen mineralization function. *Applied Soil Ecology* 24, 19–35.

Ferris, H., Bongers, T. and de Goede, R. (2004b) Nematode faunal analyses to assess food web enrichment and connectance. Pp 503–510 in R.C. Cook and D.J. Hunt (eds) Proceedings of the Fourth International Congress of Nematology. Nematology Monographs and Perspectives 2. Brill, Netherlands. 866p.

Fisher, R. (2003) Spatial and temporal variations in nematode assemblages in tropical seagrass sediments. *Hydrobiologia* 493, 43–63.

Fleeger, J.W., Tita, G., Carman, K.R., Millward, R.N., Moser, E.B., Portier, R.J. and Gambrell, R.P. (2006) Does bioturbation by a benthic fish modify the effects of sediment contamination on saltmarsh benthic microalgae and meiofauna? *Journal of Experimental Marine Biology and Ecology* 330, 180–194.

Flint, M.L. (ed.) (1985) *Integrated Pest Management for Tomatoes*, 3rd edn. University of California Statewide IPM Project, Division of Agriculture and Natural Resources, Oakland, Publ. 3274.

Folkertsma, R.T., van Koert, P., Rouppe van der Voort, J.N.A.M., de Groot, K.E., Kammenga, J.E., Helder, J. and Bakker, J. (2001) The effects of founding events and agricultural practices on the genetic structure of three metapopulations of *Globodera pallida*. *Phytopathology* 91, 753–758.

Fritz, K.M., Evans, M.A. and Feminella, J.W. (2004) Factors affecting biomass allocation in the riverine macrophyte *Justicia americana*. *Aquatic Botany* 78, 279–288.

Fu, S., Ferris, H., Brown, D. and Plant, R. (2005) Does the positive feedback effect of nematodes on the biomass and activity of their bacteria prey vary with nematode species and population size? *Soil Biology and Biochemistry* 37, 1979–1987.

Gallucci, F., Steyaert, M. and Moens, T. (2005) Can field distributions of marine predacious nematodes be explained by sediment constraints on their foraging success? *Marine Ecology Progress Series* 304, 167–178.

Gerlach, S.A. (1977) Attraction to decaying organisms as a possible cause for patchy distribution of nematodes in a Bermuda beach. *Ophelia* 16, 151–165.

Ghafouri, S. and McGhee, J.D. (2007) Bacterial residence time in the intestine of *Caenorhabditis elegans*. *Nematology* 9, 87–91.

Giere, O., Windoffer, R. and Southward, E.C. (1995) The bacterial endosymbiosis of the gutless nematode, *Astomonema southwardorum* – structural aspects. *Journal of the Marine Biological Association of the United Kingdom* 75, 153–164.

Goodell, P. and Ferris, H. (1980) Plant-parasitic nematode distributions in an alfalfa field. *Journal of Nematology* 12, 136–141.

Green, C.D. and Plumb, S.C. (1970) The interrelationships of some *Heterodera* spp. indicated by the specificity of the male attractants emitted by their females. *Nematologica* 16, 39–46.

Griffiths, B.S., Ritz, K., Bardgett, R.D., Cook, R., Christensen, S., Ekelund, F., Sørenson, S.J., Bååth, E., Bloem, J., De Ruiter, P.C., Dolfing, J. and Nicolardot, B. (2000) Ecosystem response of pasture soil communities to fumigation-induced microbial diversity reductions: an examination of the biodiversity–ecosystem function relationship. *Oikos* 90, 279–294.

Griffiths, B.S., Neilson, R. and Bengough, A.G. (2003) Soil factors determined nematode community composition in a two year pot experiment. *Nematology* 5, 889–897.

Gunapala, N. and Scow, K.M. (1998) Dynamics of soil microbial biomass and activity in conventional and organic farming systems. *Soil Biology and Biochemistry* 30, 805–816.

Gutierrez, A. and Sommer, R.J. (2004) Evolution of dnmt-2 and mbd2-like genes in the free-living nematodes *Pristionchus pacificus, Caenorhabditis elegans* and *Caenorhabditis briggsae. Nucleic Acids Research* 323, 6388–6396.

Hamels, I., Moens, T., Muylaert, K. and Vyverman, W. (2001) Trophic interactions between ciliates and nematodes from an intertidal flat. *Aquatic Microbial Ecology* 26, 61–72.

Haverkort, A.J., Boerma, M., Velma, R. and Van de Waart, M. (1992) The influence of drought and cyst nematodes on potato growth. 4. Effects on crop growth under field conditions of 4 cultivars differing in tolerance. *Netherlands Journal of Plant Pathology* 98, 179–191.

Heininger, P., Höss, S., Claus, E., Pelzer, J. and Traunspurger, W. (2007) Nematode communities in contaminated river sediments. *Environmental Pollution* 146, 64–76.

Heip, C., Vincx, M. and Vranken, G. (1985) The ecology of marine nematodes. *Oceanography and Marine Biology* 23, 399–489.

Hellwig-Armonies, M., Armonies, W. and Lorenzen, S. (1991) The diet of *Enoplus brevis* (Nematoda) in a supralittoral salt marsh of the North Sea. *Helgoländer Meeresuntersuchungen* 45, 357–372.

Hentschel, U., Berger, E.C., Bright, M., Felbeck, H. and Ott, J.R. (1999) Metabolism of nitrogen and sulfur in ectosymbiotic bacteria of marine nematodes (Nematoda, Stilbonematinae). *Marine Ecology – Progress Series* 183, 149–158.

Hewitt, W.B., Raski, D.J. and Goheen, A.C. (1958) Nematode vector of soil-borne fanleaf virus of grapevines. *Phytopathology* 48, 586–595.

Höckelmann, C., Moens, T. and Jüttner, F. (2004) Odor compounds from cyanobacterial biofilms acting as attractants and repellents for free-living nematodes. *Limnology and Oceanography* 49, 1809–1819.

Hol, W.H.G., de la Peña, E., Moens, M. and Cook, R. (2007) Interaction between a fungal endophyte and root herbivores of *Ammophila arenaria. Basic and Applied Ecology* 8, 500–509.

Höss, S., Traunspurger, W., Severin, G.W., Juttner, I., Pfister, G. and Schramm, K.W. (2004) Influence of 4-nonylphenol on the structure of nematode communities in freshwater microcosms. *Environmental Toxicology and Chemistry* 23, 1268–1275.

Hudson, P.J., Dobson, A.P. and Newborn, D. (1998) Prevention of population cycles by parasite removal. *Science* 282, 2256–2258.

Hussey, R.S. and Roncadori, R.W. (1981) Influence of *Aphelenchus avenae* on vesicular-arbuscular endomycorrhizal growth response in cotton. *Journal of Nematology* 13, 48–52.

Hyman, L.H. (1951) *The invertebrates: Acanthocephala, Aschelminthes, and Entoprocta. The pseudocoelomate Bilateria.* volume III. McGraw-Hill, New York.

Ingham, R.E., Trofymow, J.A., Ingham, E.R. and Coleman, D.C. (1985) Interactions of bacteria, fungi and their nematode grazers on nutrient cycling and plant growth. *Ecological Monographs* 55, 119–140.

Jatala, P., Jensen, H.J. and Russell, S.A. (1974) *Pristionchus lheritieri* as a carrier of *Rhizobium japonicum. Journal of Nematology* 6, 130–131.

Jones, W.J., Won, Y-J., Maas, P.A.Y., Smith, P.J., Lutz, R.A. and Vrijenhoek, R.C. (2006) Evolution of habitat use by deep-sea mussels. *Marine Biology* 148, 841–851.

Jørgensen, S.E. and Mejer, H. (1979) A holistic approach to ecological modelling. *Ecological Modelling* 7, 169–189.
Kaya, H.K. and Stock, S.P. (1997) Techniques in insect nematology. In: Lacey, L.A. (ed.) *Manual of Techniques in Insect Pathology*. Academic Press, London, pp. 281–324.
Kardol, P., Bezemer, T.M., van der Wal, A. and Van der Putten, W.H. (2005) Successional trajectories of soil nematode and plant communities in a chronosequence of ex-arable lands. *Biological Conservation* 126, 317–327.
Klekowski, R.Z., Wasilewska, L. and Paplinska, E. (1972) Oxygen consumption by soil-inhabiting nematodes. *Nematologica* 18, 391–403.
Klingler, J. (1965) On the orientation of plant nematodes and of some other soil animals. *Nematologica* 11, 4–18.
Korthals, G.W., Van de Ende, A., Van Megen, H., Lexmond, T.M., Kammenga, J.E. and Bongers, T. (1996) Short-term effects of cadmium, copper, nickel, and zinc on soil nematodes from different feeding and life-history strategy groups. *Applied Soil Ecology* 4,107–117.
Laakso, J. and Setälä, H. (1999) Sensitivity of primary production to changes in the architecture of belowground food webs. *Oikos* 87, 57–64.
Lacey, L.A., Frutos, R., Kaya, H.K. and Vail, P. (2001) Insect pathogens as biological control agents: Do they have a future? *Biological Control* 21, 230–248.
Lambshead, P.J.D. (1993) Recent developments in marine benthic biodiversity research. *Oceanis* 19, 5–24.
Lambshead, P.J.D., Tietjen, J., Ferrero, T. and Jensen, P. (2000) Latitudinal diversity gradients in the deep sea with special reference to North Atlantic nematodes. *Marine Ecology Progress Series* 194, 159–167.
Lambshead, P.J.D., Brown, C.J., Ferrero, T.J., Mitchell, N.J., Smith, C.R., Hawkins, L.E. and Tietjen, J. (2002) Latitudinal diversity patterns of deep-sea marine nematodes and organic fluxes: a test from the central equatorial Pacific. *Marine Ecology Progress Series* 236, 129–135.
Lee, D.L. (ed.) (2002) *The Biology of Nematodes*. Taylor and Francis, London.
Luc, M., Sikora, R.A. and Bridge, J. (eds) (2005) *Plant Parasitic Nematodes in Subtropical and Tropical Agriculture*. CABI, Wallingford.
Maas, P.W.Th., Oremus, P.A.I. and Otten, H. (1983) Nematodes (*Longidorus* sp. and *Tylenchorhynchus microphasmis* Loof) in growth and nodulation of Sea buckthorn (*Hippophaë rhamnoides* L.). *Plant and Soil* 73, 141–147.
Magurran, A.E. (1988) *Ecological Diversity and its Measurement*. Cambridge University Press, Cambridge.
Mauchline, T.H., Kerry, B.R. and Hirsch, P.R. (2004) The biocontrol fungus *Pochonia chlamydosporia* shows nematode host preference at the intraspecific level. *Mycological Research* 108, 161–169.
McCallum, M.E. and Dusenbery, D.B. (1992) Computer tracking as a behavioral gc detector – nematode responses to vapor of host roots. *Journal of Chemical Ecology* 18, 585–592.
McKenna, P.B. (1997) Checklist of helminth parasites of terrestrial mammals in New Zealand. *New Zealand Journal of Zoology* 24, 277–290.
Michiels, I.C. and Traunspurger, W. (2005) Seasonal variation of biodiversity and assemblage structure in freshwater nematodes. *Archiv für Hydrobiologie* 163, 183–194.
Millennium Ecosystem Assessment (2005) *Ecosystems and human well-being: current state and trends. Findings of the Condition and Trends Working Group*. Island Press, Washington.
Minoshima, H., Jackson, L.E., Cavagnaro, T.R. and Ferris, H. (2007) Short-term fates of ^{13}C–depleted cowpea shoots in no-till and standard tillage soils. *Soil Science Society of America Journal* 71, 1859–1866.
Moens, T., Verbeeck, L., de Maeyer, A., Swings, J. and Vincx, M. (1999) Selective attraction of marine bacterivorous nematodes to their bacterial food. *Marine Ecology Progress Series* 176, 165–178.

Moens, T., Herman, P.M.J., Verbeeck, L., Steyaert, M. and Vincx, M. (2000) Predation rates and prey selectivity in two predacious estuarine nematodes. *Marine Ecology Progress Series* 205, 185–193.

Moens, T., Luyten, C., Middelburg, J.J., Herman, P.M.J. and Vincx, M. (2002) Tracing organic matter sources of estuarine tidal flat nematodes with stable carbon isotopes. *Marine Ecology Progress Series* 234, 127–137.

Moens, T., Yeates, G.W. and De Ley, P. (2004) Use of carbon and energy sources by nematodes. *Nematology Monographs and Perspectives* 2, 529–545.

Moens, T., Bouillon, S. and Gallucci, F. (2005a) Dual stable isotope abundances unravel trophic position of estuarine nematodes. *Journal of the Marine Biological Association of the United Kingdom* 85, 1401–1407.

Moens, T., dos Santos, G.A.P., Thompson, F., Swings, J., Fonsêca-Genevois, V. and De Mesel, I. (2005b) Do nematode mucus secretions affect bacterial growth? *Aquatic Microbial Ecology* 40, 77–83.

Moens, T., Bergtold, M. and Traunspurger, W. (2006) Feeding ecology of free-living benthic nematodes. In: Eyualem, A., Andrássy, I. and Traunspurger, W. (eds) *Freshwater Nematodes: Ecology and Taxonomy*. CAB International Publishing, Cambridge, USA, pp. 105–131.

Montagna, P.A. (1995) Rates of metazoan meiofaunal microbivory: a review. *Vie et Milieu* 45, 1–9.

Morris, C.A., Vlassoff, A., Bisset, S.A., Baker, R.L., Watson, T.G., West, C.J. and Wheeler, M. (2000) Continued selection of Romney sheep for resistance or susceptibility to nematode infection: estimates of direct and correlated responses. *Animal Science* 70, 17–27.

Mulder, C., Schouten, A.J., Hund-Rinke, K. and Breure, A.M. (2005) The use of nematodes in ecological soil classification and assessment concepts. *Ecotoxicology and Environmental Safety* 62, 278–289.

Müller, F. and Lenz, R. (2006) Ecological indicators: theoretical fundamentals of consistent application in environmental management. *Ecological Indicators* 6, 1–5.

Musat, N., Giere, O., Gieseke, A., Thiermann, F., Amann, R. and Dubilier, N. (2007) Molecular and morphological characterization of the association between bacterial endosymbionts and the marine nematode *Astomonema* sp. from the Bahamas. *Environmental Microbiology* 9, 1345–1353.

Needham, T. (1744) A letter concerning chalky tubulous concretions, called malm: with some microscopical observations on the farina of the red lily, and of worms discovered in smutty corn. *Philosophical Transactions of the Royal Society of London* 42, 634–641.

Neher, D. (2001) Role of nematodes in soil health and their use as indicators. *Journal of Nematology* 33,161–168.

Nehring, S. (1993) Tube-dwelling meiofauna in marine sediments. *Internationale Revue der Gesamten Hydriobiologie* 78, 521–534.

Nehring, S., Jensen, P. and Lorenzen, S. (1990) Tube-dwelling nematodes: tube construction and possible ecological effects on sediment–water interfaces. *Marine Ecology Progress Series* 64, 123–128.

Northam, F.E. and Orr, C.C. (1982) Effects of a nematode on biomass and density of silverleaf nightshade. *Journal of Range Management* 35, 536–537.

Okada, H. and Ferris, H. (2001) Temperature effects on growth and nitrogen mineralization of fungi and fungal-feeding nematodes. *Plant and Soil* 234, 253–262.

Okada, H. and Kadota, I. (2003) Host status of 10 fungal isolates for two nematode species, *Filenchus misellus* and *Aphelenchus avenae. Soil Biology and Biochemistry* 35, 1601–1607.

Okada, H., Harada, H., Tsukiboshi, T. and Araki, M. (2005) Characteristics of *Tylencholaimus parvus* (Nematoda: Dorylaimida) as a fungivorus nematode. *Nematology* 7, 843–849.

Oremus, P.A.I. and Otten, H. (1981) Factors affecting growth and nodulation of *Hippophaë rhamnoides* L. ssp. *rhamnoides* in soils from two successional stages of dune formation. *Plant and Soil* 63, 317–331.

Pantone, D.J., Williams, W.A. and Maggenti, A.R. (1989a) An alternative approach for evaluating the efficiacy of potential biocontol agents of weeds. 1. Inverse linear model. *Weed Science* 37, 771–777.

Pantone, D.J., Williams, W.A. and Maggenti, A.R. (1989b) An alternative approach for evaluating the efficiacy of potential biocontol agents of weeds. 2. Path analysis. *Weed Science* 37, 778–783.

Perry, R.N. and Moens, M. (eds) (2006) *Plant Nematology*. CABI, Wallingford.

Peters, L. and Traunspurger, W. (2005) Species distribution of free-living nematodes and other meiofauna in littoral periphyton communities of lakes. *Nematology* 7, 267–280.

Petersen, J.J. (1985) Nematode parasites. *American Mosquito Control Association Bulletin* 6, 110–122.

Phillips, D.A., Ferris, H., Cook, D.R. and Strong, D.R. (2003) Molecular control points in rhizosphere food webs. *Ecology* 84, 816–826.

Pielou, E.C. (1975) *Ecological Diversity*. Wiley, New York.

Piśkiewicz, A.M., Duyts, H., Berg, M.P., Costa, S.R. and Van der Putten, W.H. (2007) Soil microorganisms control plant ectoparasitic nematodes in natural coastal foredunes. *Oecologia* 152, 505–514.

Pline, M. and Dusenbery, D.B. (1987) Responses of the plant-parasitic nematode *Meloidogyne incognita* to carbon dioxide determined by video camera-computer tracking. *Journal of Chemical Ecology* 13, 1617–1624.

Polz, M.F., Ott, J.A., Bright, M. and Cavanaugh, C.M. (2000) When bacteria hitch a ride. *American Society of Microbiology News* 66, 531–539.

Postma-Blaauw, M.B., de Vries, F.T., de Goede, R.G.M., Bloem, J., Faber, J.H. and Brussard, L. (2005) Within-trophic group interactions of bacterivorous nematode species and their effects on the bacterial community and nitrogen mineralization. *Oecologia* 142, 428–439.

Powell, N.T. (1971) Interactions between nematodes and fungi in disease complexes. *Annual Review of Phytopathology* 9, 253–273.

Richard-Lenoble, D., Chandenier, J. and Gaxotte, P. (2003) Ivermectin and filiariasis. *Fundamental and Clinical Pharmacology* 17, 199–203.

Riemann, F. and Helmke, E. (2002) Symbiotic relations of sediment-agglutinating nematodes and bacteria in detrital habitats: The enzyme-sharing concept. *Marine Ecology* 23, 93–113.

Riemann, F. and Schrage, M. (1978) The mucus-trap hypothesis on feeding of aquatic nematodes and implications for biodegradation and sediment texture. *Oecologia* 34, 75–88.

Riemann, F. and Schrage, M. (1988) Carbon dioxide as an attractant for the free-living marine nematode *Adoncholaimus thalassophygas*. *Marine Biology* 98, 81–85.

Riera, P., Richard, P., Grémare, A. and Blanchard, G. (1996) Food source of intertidal nematodes in the Bay of Marennes-Oléron (France), as determined by dual stable isotope analysis. *Marine Ecology Progress Series* 142, 303–309.

Riffle, J.W. (1967) Effect of an *Aphelenchoides* species on the growth of a mycorrhizal and pseudomycorrhizal fungus. *Phytopathology* 57, 541–544.

Riffle, J.W. (1975) Two *Aphelenchoides* species suppress formation of *Suillus granulatus* ectomycorrhizae with *Pinus ponderosa* seedlings. *Plant Disease Reporter* 59, 951–955.

Robertson, G.P. and Freckman, D.W. (1995) The spatial distribution of nematode trophic groups across a cultivated ecosystem. *Ecology* 76, 1425–1433.

Robinson, A.F. (1995) Optimal release rates for attracting *Meloidogyne incognita, Rotylenchulus reniformis*, and other nematodes to carbon-dioxide in sand. *Journal of Nematology* 27, 42–50.

Rodger, S., Griffiths, B.S., McNicol, J.W., Wheatley, R.W. and Young, I.M. (2004) The impact of bacterial diet on the migration and navigation of *Caenorhabditis elegans*. *Microbial Ecology* 28, 358–365.

Roggen, D.R., Raski, D.J. and Jones, N.O. (1966) Cilia in nematode sensory organs. *Science* 152, 515–516.

Ruess, L. and Dighton, J. (1996) Cultural studies on soil nematodes and their fungal hosts. *Nematologica* 42, 330–346.

Rühm, R., Dietsche, E., Harloff, H.J., Lieb, M., Franke, S. and Aumann, J. (2003) Characterization and partial purification of a white mustard kairomone that attracts the beet cyst nematode, *Heterodera schachtii. Nematology* 5, 17–22.

Salinas, K.A., Edenborn, S.L., Sextone, A.J. and Kotcon, J.B. (2007) Bacterial preferences of the bacterivorous soil nematode *Cephalobus brevicauda* (Cephalobidae): effect of bacterial size and type. *Pedobiologia* 51, 55–64.

Sánchez-Moreno, S., Minoshima, H., Ferris, H. and Jackson, L.E. (2006) Linking soil properties and nematode community composition: effects of soil management on soil food webs. *Nematology* 8, 703–715.

Sánchez-Moreno, S., Smukler, S., Ferris, H., O'Green, A.T., and Jackson, L.E. (2008) Nematode diversity, food web condition, and chemical and physical properties in different soil habitats of an organic farm. *Biology and Fertility of Soils* 44, 727–744.

Sankaralingam, A., and McGawley, E.C. (1994) Influence of *Rhizoctonia solani* on egg hatching and infectivity of *Rotylenchus reniformis. Journal of Nematology* 26, 486–491.

Santos, G.A.P. dos, Derycke, S., Genevois, V.G.F., Coelho, L.C.B.B., Correia, M.T.S. and Moens, T. (2008) Differential effects of food availability on population development and fitness of three species of estuarine, bacterial-feeding nematodes. *Journal of Experimental Marine Biology and Ecology* 355, 27–40.

Schratzberger, M., Bolam, S., Whomersley, P. and Warr, K. (2006) Differential response of nematode colonist communities to the intertidal placement of dredged material. *Journal of Experimental Marine Biology and Ecology* 334, 244–255.

Schratzberger, M., Warr, K. and Rogers, S.I. (2007) Functional diversity of nematode communities in the southwestern North Sea. *Marine Environmental Research* 63, 368–389.

Schroeder, J., Thomas, S.H. and Murray, L.W. (2005) Impacts of crop pests on weeds and weed-crop interactions. *Weed Science* 53, 918–922.

Shaw, J.L., Moss, R. and Pike, A.W. (1989) Development and survival of the free-living stages of *Trichostrongylus tenuis*, a caecal parasite of the red grouse *Lagopus lagopus scoticus. Parasitology* 99, 105–113.

Siddiqi, M.R. (2000) *Tylenchida: Parasites of Plants and Insects*, 2nd edn. CABI Publishing, Wallingford.

Sikora, R.W. and Carter, W.W. (1987) Nematode interactions with fungal and bacterial plant pathogens: fact or fantasy. In: Veech, J.A. and Dickson, D.W. (eds) *Vistas on Nematology*. Society of Nematologists, Hyattsville, pp. 307–312.

Small, R.W. (1987) A review of the prey of predatory soil nematodes. *Pedobiologia* 30, 179–206.

Soetaert, K., Vanaverbeke, J., Heip, C., Herman, P.M.J., Middelburg, J.J., Sandee, A. and Duineveld, G. (1997) Nematode distribution in ocean margin sediments of the Goban Spur (northeast Atlantic) in relation to sediment geochemistry. *Deep-Sea Research part I – Oceanographic Research Papers* 44, 1671–1683.

Spiegel, Y., Burrows, P.M. and Bar-Eyal, M. (2003) A chemo attractant in onion root exudates recognized by *Ditylenchus dipsaci* in laboratory bioassay. *Phytopathology* 93, 127–132.

Stanton, N.L. (1988) The underground in grasslands. *Annual Review of Ecology and Systematics* 19, 573–589.

Sterner, R.W. and Elser, J.J. (2002) *Ecological Stochiometry – the biology of elements from molecules to the biosphere*. Princeton University Press, Princeton, New Jersey.

Steyaert, M., Herman, P.M.J., Moens, T., Widdows, J. and Vincx, M. (2001) Tidal migration of nematodes on an estuarine tidal flat (The Molenplaat, Schelde Estuary, SW Netherlands). *Marine Ecology – Progress Series* 224, 299–304.

Sudhaus, W. (1976) Vergleichende Untersuchungen zur Phylogenie, Systematik. Ökologie, Biologie und Ethologie der Rhabditidae (Nematoda). *Zoologia* 125, 1–229.

Suess, E. (1980) Particulate organic carbon flux in the oceans – surface productivity and oxygen utilization. *Nature* 288, 260–263.

Sutherland, J.R. and Fortin, J.A. (1968) Effect of the nematode *Aphelenchus avenae* on some ectotrophic, mycorrhizal fungi and on a red pine mycorrhizal relationship. *Phytopathology* 58, 519–523.

Temple, S.R. (1993) A comparison of conventional, low input and organic farming systems: the transition phase and long term viability. Progress Report on Project LW-89-18.1 to USDA-SARE.

Tenuta, M. and Ferris, H. (2004) Relationship between nematode life-history classification and sensitivity to stressors: ionic and osmotic effects of nitrogenous solutions. *Journal of Nematology* 36, 85–94.

Thiermann, F., Vismann, B. and Giere, O. (2000) Sulphide tolerance of the marine nematode *Oncholaimus campylocercus* – a result of internal sulfur formation. *Marine Ecology – Progress Series* 193, 251–259.

Tietjen, J.H., Deming, J.W., Rowe, G.T., Macko, S. and Wilke, R.J. (1989) Meiobenthos of the Hatteras Abyssal Plain and Puerto Rico Trench: abundance, biomass and associations with bacteria and particulate fluxes. *Deep-Sea Research* 36, 1567–1577.

Togashi, K. and Shigesada, N. (2006) Spread of the pinewood nematode vectored by the Japanese pine sawyer: modeling and analytical approaches. *Population Ecology* 48, 271–283.

Townshend, J.L. and Blackith, R.E. (1975) Fungal diet and the morphometric relationships in *Aphelenchus avenae. Nematologica* 11, 395–408.

Traunspurger, W. (2000) The ecology of lotic nematodes. *Freshwater Biology* 44, 29–45.

Traunspurger, W., Bergtold, M. and Goedkoop, W. (1997) The effects of nematodes on bacterial activity and abundance in a freshwater sediment. *Oecologia* 112, 118–122.

Van der Putten, W.H. (2003) Plant defense belowground and spatio-temporal processes in natural vegetation. *Ecology* 84, 2269–2280.

Van der Putten, W.H. and Van der Stoel, C.D. (1998) Plant parasitic nematodes on spatio-temporal variation in natural vegetation. *Applied Soil Ecology* 10, 253–262.

Van der Putten, W.H., Van Dijk, C. and Peters, B.A.M. (1993) Plant-specific soil-borne diseases contribute to succession in foredune vegetation. *Nature (London)* 362, 53–56.

Van der Putten, W.H., Yeates, G.W., Duyts, H., Schreck Reis, C. and Karssen, G. (2005) Invasive plants and their escape from root herbivory: a worldwide comparison of the root-feeding nematode communities of the dune grass *Ammophila arenaria* in natural and introduced ranges. *Biological Invasions* 7, 733–746.

Van der Putten, W.H., Cook, R., Costa, S., Davies, K.G., Fargette, M., Freitas, H., Hol, W.H.G., Kerry, B.R., Maher, N., Mateille, T., Moens, M., de la Peña, E., Piśkiewicz, A., Raeymaekers, A., Rodríguez-Echeverría, S. and Van der Wurff, A.W.G. (2006) Nematode interactions in nature: models for sustainable control of nematode pests of crop plants? *Advances in Agronomy* 89, 227–260.

Van Gaever, S., Vanreusel, A., Hughes, J.A., Bett, B.J. and Kiriakoulakis, K. (2004) The macro- and micro-scale patchiness of meiobenthos associated with the Darwin Mounds (north-east Atlantic). *Journal of the Marine Biological Association of the United Kingdom* 84, 547–556.

Van Gaever, S., Moodley, L., De Beer, D. and Vanreusel, A. (2006) Meiobenthos at the Arctic Håkon Mosby Mud Volcano, with a parental caring nematode thriving in sulphide-rich sediments. *Marine Ecology Progress Series* 321, 143–155.

Van Gundy, S.D., Kirkpatrick, J.D. and Golden, J. (1977) Nature and role of metabolic leakage from root-knot nematode galls and infection by *Rhizoctonia solani. Journal of Nematology* 9, 113–121.

van Oevelen, D., Moodley, L., Soetaert, K. and Middelburg, J.J. (2006) The trophic significance of bacterial carbon in a marine intertidal sediment: Results of an *in situ* stable isotope labeling study. *Limnology and Oceanography* 51, 2349–2359.

van Ruijven, J., De Deyn, G.B., Raaijmakers, C.E., Berendse, F. and Van der Putten, W.H. (2005) Interactions between spatially separated herbivores indirectly alter plant diversity. *Ecology Letters* 8, 30–37.

Vanfleteren, J.R. (1980) Nematodes as nutritional models. In: Zuckerman, B.M. (ed.) *Nematodes as Biological Models*, volume 2. Academic Press, New York, pp. 47–79.

Venette, R.C. and Ferris, H. (1998) Influence of bacterial type and density on population growth of bacterial-feeding nematodes. *Soil Biology and Biochemistry* 30, 949–960.

Verschoor, B.C. (2002) Carbon and nitrogen budgets of plant-feeding nematodes in grasslands of different productivity. *Applied Soil Ecology* 20, 15–25.

Verschoor, B.C., De Goede, R.G.M., De Vries, F.W. and Brussaard, L. (2001) Changes in the composition of the plant-feeding nematode community in grasslands after cessation of fertiliser application. *Applied Soil Ecology* 17, 1–17.

Viglierchio, D.R. (1991) *The World of Nematodes*. David R. Viglierchio, Davis, California.

Viketoft, M., Palmborg, C., Sohlenius, B., Huss-Danell, K. and Bengtsson, J. (2005) Plant species effects on soil nematode communities in experimental grasslands. *Applied Soil Ecology* 30, 90–103.

Wall, D.H. (ed.) (2004) *Sustaining Biodiversity and Ecosystem Services in Soils and Sediments*. Island Press, Washington.

Waller, P.J., Schwan, O., Ljungström, B-L., Rydzik, A. and Yeates, G.W. (2004) Evaluation of biological control of sheep parasites using *Duddingtonia flagrans* under commercial farming conditions on the island of Gotland, Sweden. *Veterinary Parasitology* 126, 199–315.

Wang, J., Johnson, C.S. and Eisenback, J.D. (1997) Enhanced hatching of *Globodera tabacum* solanacearum juveniles by root exudates of flue-cured tobacco. *Journal of Nematology* 29, 484–490.

Wardle, D.A. (2002) *Communities and Ecosystems: Linking the Aboveground and Belowground Components*. Princeton University Press, Princeton, New Jersey.

Wardle, D.A., Yeates, G.W., Watson, R.N. and Nicholson, K.S. (1995) The detritus food-web and diversity of soil fauna as indicators of disturbance regimes in agro-ecosystems. *Plant and Soil* 170, 35–43.

Wardle, D.A., Bardgett, R.D., Klironomos, J.N., Setala, H., Van der Putten, W.H. and Wall, D.H. (2004) Ecological linkages between aboveground and belowground biota. *Science* 304, 1629–1633.

Wilson, E.O. (ed.) (1988) *Biodiversity*. National Academy of Sciences / Smithsonian Institution, Washington.

Wubben, M.J.E., Su, H., Rodermel, S.R. and Baum, T.J. (2001) Susceptibility to the sugar beet cyst nematode is modulated by ethylene signal transduction in *Arabidopsis thaliana*. *Molecular Plant-Microbe Interactions* 14, 1206–1212.

Yeates, G.W. (1969a) Predation by *Mononchoides potohikus* (Nematoda: Diplogasteridae) in laboratory culture. *Nematologica* 15, 1–9.

Yeates, G.W. (1969b) A note on the function of the prerectum in *Diplenteron potohikus* (Yeates, 1969) (Nematoda: Diplogasteridae). *Nematologica* 15, 430–431.

Yeates, G.W. (1984) Variation in soil nematode diversity under pasture with soil and year. *Soil Biology and Biochemistry* 16, 95–102.

Yeates, G.W. (2003) Nematodes as soil indicators: functional and biodiversity aspects. *Biology and Fertility of Soils* 37, 199–210.

Yeates, G.W. and Boag, B. (2006) Female size shows similar trends in all clades of the Phylum Nematoda. *Nematology* 8, 111–127.

Yeates, G.W., Wardle, D.A. and Watson, R.N. (1999a) Responses of soil nematode populations, community structure and temporal variability to agricultural intensification over a seven-year period. *Soil Biology and Biochemistry* 31, 1721–733.

Yeates, G.W., Saggar, S., Hedley, C.B. and Mercer, C.F. (1999b) Increase in ^{14}C-carbon translocation to the soil microbial biomass when five plant-parasitic nematode infect roots of white clover. *Nematology* 1, 295–300.

Zoon, F.C., Troelstra, S.R. and Maas, P.W.Th. (1993) Ecology of the plant-feeding nematode fauna associated with Sea buckthorn (*Hippophaë rhamnoides* L. ssp. *rhamnoides*) in different stages of dune succession. *Fundamental and Applied Nematology* 16, 247–258.

Appendix 1: Outline Classification of the Phylum Nematoda

Molecular approaches have permitted an integrated view of the phylum. While results are generally congruent with traditional relationships, on one hand they have so far not resolved uncertainty among marine groups, while on the other hand, the Order Rhabditida is a monophyletic group equivalent with all of 'Secernentea' (=Phasmidia) in many previous classifications (Blaxter *et al.*, 1998; De Ley and Blaxter, 2002). The classification below, taken from De Ley and Blaxter (2004), indicates the position of many families of interest to readers of this volume.

Class Enoplea

Subclass Enoplia

Order Enoplida: Ironidae, Alaimidae
Order Triplonchida: Diphtherophoridae, Trichodoridae, Tobrilidae, Prismatolaimidae, Tripylidae

Subclass Dorylaimia

Order Dorylaimida: Dorylaimidae, Aporcelaimidae, Longidoridae, Belondiridae, Leptonchidae, Tylencholaimidae, Nygolaimidae
Order Mononchida: Bathyodontidae, Anatonchidae, Mononchidae, Mylonchulidae
Order Isolaimida: Isolaimiidea
Order Mermithida: Mermithidae, Tetradonematidae
Order Trichinellida: Capillariidae, Trichinellidae, Trichuridae

Class Chromadorea

Subclass Chromadoria

Order Desmoscolecida: Desmoscolecidae
Order Chromadorida: Chromadoridae, Ethmolaimidae, Cyatholaimdae

Order Desmodorida: Desmodoridae, Microlaimidae
Order Monhysterida: Monhysteridea, Sphaerolaimidae
Order Araeolaimida: Axonolaimidae, Diplopeltidae
Order Plectida: Leptolaimidae, Bastianiidae, Rhabdolaimidae, Plectidae, Chronogasteridae, Metateratocephalidae, Haliplectidae, Aulolaimidae
Order Rhabditida:
Suborder Spirurina: Thelastomatidae, Oxyuridae, Rhigonematidae, Hethidae, Camallanidae, Hedruridae, Tetrameridae, Filariidae, Ascarididae, Heterakidae
Suborder Tylenchina: Panagrolaimidae, Steinernematidae, Cephalobidae, Aphelenchidae, Criconematidae, Anguinindea, Hoplolaimidae, Meloidogynidae, Tylenchidae, Pratylenchidae, Drilonematidae
Suborder Rhabditina: Bunonematidae, Diplogastridae, Mesorhabditidae, Peloderidae, Rhabditidae, Heterorhabditidae, Trichostrongylidae, Metastrongylidae

2 Nematode Diversity in Terrestrial, Freshwater Aquatic and Marine Systems

M. Hodda,[1] L. Peters[2] and W. Traunspurger[2]

[1]*Nematode Biosystematics and Ecology, Australian National Insect Collection, CSIRO Entomology, Australia;* [2]*Department of Animal Ecology, University of Bielefeld, Germany*

Introduction

The general concept of diversity as being the range of different types of organisms present in an area of interest over a particular time has general agreement among scientists, policy makers and the general public. At this general level, many share this same general definition, and most at least assume that everyone else shares it too (Williams, 1993; Williams, *et al.*, 1993; Gaston, 1996; Setala *et al.*, 1998). However, more specific definitions are less agreed, particularly with regard to how diversity is measured and the area or time span which should be considered (e.g. Setala *et al.*, 1998). Add to this the possible confounding effects of various different methods of measurement, and the result is that the simple concept of diversity becomes complex (Gaston, 1996). One could say that the concept of diversity itself is quite diverse.

Whatever the measure of diversity, nematodes are certainly one of the most diverse groups of organisms on earth, possibly the most diverse (May, 1988). They are also the most numerous multicellular organisms on earth, in terms of numbers of individuals. Together, these facts mean that studies of nematode diversity are necessarily major undertakings, because sophisticated techniques may be required to obtain samples and huge numbers of individuals of many different types will be involved. Consequently, comprehensive surveys of any habitat are very rare.

Furthermore, nematodes are relatively small, simple animals, so that increasingly scarce expert knowledge is necessary to divide them into species, genera, or trophic groups. The definitions of taxonomic categories vary considerably across the phylum, and ecological categories can be rather fuzzy and overlapping (Yeates *et al.*, 1993; Ferris, 1999).

Much is known about economically important species and groups, such as terrestrial plant-root parasites. However, such groups are not representative of nematodes as a whole: the very ecological and biological properties which mean that they are economic pests also differentiate them from the vast majority of species which are not pests.

The net result of all the characteristics of nematode diversity outlined above is that there are few studies of nematode diversity, and even fewer truly comparable ones from which general patterns can be drawn. Studies are limited in the number of groups identified, the detail to which groups are defined, or the level of evidence supporting the groups. Only very small proportions of an area of interest, or a small subsample of all nematodes, are included in many studies.

Nevertheless, despite the difficulties, nematode diversity is of great interest to a great range of people. The range includes both those with specific interests in nematodes, such as pest managers and ecologists, as well as those with more general interests in other scientific disciplines, including soil, freshwater, marine or deep-sea biologists, environmental scientists, organic agriculturalists, ecologists, evolutionary biologists, global change biologists, pollution monitors and others (Ingham *et al.*, 1985; Yeates, 1987a; Coleman *et al.*, 1992; Wardle *et al.*, 1998).

Diversity has been linked to many other ecosystem properties, such as ecological resilience, evolutionary hot spots, the functioning of various ecological units, and even human aesthetics (Collins and Benning, 1996). While the link between nematode diversity and human aesthetics may be less than that applicable to more iconic biota like butterflies or birds, the importance of nematodes in decomposition, nutrient cycling, substrate food webs and parasitism means that their diversity is potentially critical for many different ecosystem processes (Baldwin *et al.*, 2000; Wall and Virginia, 2000; Lavelle *et al.*, 2006; Van der Putten *et al.*, 2006).

This chapter is an attempt to bring together the studies of nematode diversity in terrestrial, freshwater aquatic and marine environments, and to synthesize these into some sort of general hypotheses. Such hypotheses can then be used as the bases for a range of other scientific activities, including biomonitoring and hypothesis testing. The effects of many of the variables in studies of nematode diversity, such as methods used, resolution or scope of studies, and measures of diversity are discussed first. With this background, studies on nematode diversity are then summarized and compared at a general level. Discussion at a general level only is possible because of the myriad differences between the studies. Finally, hypotheses of the general patterns and processes of nematode diversity are discussed. These general patterns form the natural background for biomonitoring: to evaluate any effect of an environmental change one must know the situation expected without the change, as well as the expected magnitude and direction of the effects of the environmental change.

The Concept of Diversity

Units of measure and method of tally

The concept of diversity has three basic components: some measure of the range of organisms present; a specification of the area of interest; and the time span of interest. All of these components potentially affect the outcomes of any single study of diversity, any comparisons of studies, and any sort of meta-analysis of many studies, such as that attempted in this chapter. Yet there is very little standardization of these components.

In its simplest form, diversity is simply the number of taxa, often species, genera, families or other taxonomic units, and this is often termed richness (note though that there are other uses for the term 'richness' as discussed below). Using species as the unit for diversity studies has been advocated on theoretical grounds, and they are frequently used in studies of nematode diversity (Norton and Ponchilla, 1968; Schmidt and Norton, 1972; Norton and Schmidt, 1978; Bernard, 1992; Ferris, 1999; Yeates, 2003). In field studies, patterns of diversity have been shown in some instances using species which were invisible using other units to measure diversity (e.g. Norton and Ponchilla, 1968; Schmidt and Norton, 1972; Norton and Schmidt, 1978).

In a taxonomically complex group such as nematodes, this immediately raises the issue of whether the definitions of taxa of a particular rank are comparable. Among terrestrial nematodes, for example, a great deal of taxonomic work has been published on plant-parasites (mainly Order Tylenchida, e.g. Siddiqi, 1986, 2000; Fortuner and Luc, 1987; Luc 1987; Luc and Fortuner, 1987; Luc *et al.*, 1987, 1988; Maggenti *et al.*, 1987, 1988; Raski and Luc, 1987), but much less on other orders (e.g. the only comparable work on Aphelenchida is Hunt, 1993). The biological species – defined as the set of potentially interbreeding organisms (Mayr, 1942) – is the basic unit of biology, and has a clear meaning as the unit on which evolution operates. However, very few nematode species have been experimentally verified by breeding studies (e.g. de Guiran and Bruguier, 1989). Recent advances in genetic analysis have allowed inference of non-interbreeding populations by these methods and these studies have revealed the presence of a considerable number of cryptic species, but these genetic species also have not been verified by breeding studies. Studies of genetic structure have also revealed considerable population structure within some nematode species. There may also be substantial biological variation within interbreeding populations, as for example in the genus *Ditylenchus*, where a great number of host races, each with different responses to hosts, are known (Dropkin, 1988; Gentzsch, 1990; Sturhan and Brzeski, 1991; Janssen, 1994). There are also a number of other ways to define species, for example phylogenetic, genetic, or ecological species concepts (Dobzhansky, 1937; Hey, 2001; Pigliucci, 2003). Hence there are theoretical limitations to studies using the species as the basic unit for biodiversity, and species concepts are potentially very important (Ferris, 1999).

The greatest limitation in using species for studies of nematode diversity is logistic. A great deal of expertise and resources are required to identify any substantial number of nematodes to species. Hence, other taxonomic ranks are often used instead. It has even been argued that nematode species will never be used for other than strictly scientific studies of nematode diversity, and that genera or higher taxonomic categories should be studied, because these will be the categories used for routine regulatory or management monitoring (Yeates and Bongers, 1999). Genera are most commonly used, but there is even less biological justification for using genera as a basic unit in studies of nematode diversity than species. Many genera are not distinct evolutionary divergences supported by rigorous cladistic analysis, or strictly comparable to each other. There are also distinct ecological differences within at least some genera (Traunspurger, 1997a; Fiscus and Neher, 2002).

Families have also been used in studies of nematode diversity, but may have even less biological meaning than the lower taxonomic ranks, because of the ecological and biological diversity within them (Ferris *et al.*, 2001; Jones *et al.*, 2006).

Sometimes different taxonomic ranks have been used within one study, with the rank used for each taxon depending on the level of expertise or information available. This approach has the advantage of producing the maximum taxonomic resolution for the minimum resources, but the interpretations of such measures of diversity have to take account of which taxa are being distinguished to a greater or lesser degree. Studies of nematode diversity are mostly internally consistent in the taxonomic units used to measure diversity, but differences between studies can render comparisons dubious. Even within a single study, the use of different taxonomic ranks for different groups may introduce artefacts. For example, the relatively well-known plant-parasitic Tylenchida – identified to species – may be replaced by the much more poorly-known Dorylaimida – identified only to family – so that it may appear that there has been a huge decrease in diversity when in fact there has been no change in the number of species. Even if a single taxonomic rank is used, there may have been different results had a different taxonomic rank been used. Thus caution is needed in comparing different studies.

Some approaches to studying diversity add a measure of relatedness to the numbers of taxa: presence of many widely divergent evolutionary lineages is intuitively more 'diverse' than the same number of closely-related lineages (Faith, 1992; Warwick and Clarke, 1995; Williams, 1995; Williams *et al.*, 1997; Clarke and Warwick, 1998, 2001). This approach has been most used when diversity is being assessed for conservation purposes, and for organisms larger than nematodes, but there have been a number of studies of nematode diversity using these measures (e.g. Clarke and Warwick, 2001; Nicholas and Trueman, 2005).

More complex measurements of diversity include components of the number of individuals of each taxon: presence of similar numbers of individuals of many taxa is more diverse than where most individuals come from a few taxa and many taxa are represented by a very few individuals. Indeed some indices measure only the evenness of abundances, rather than the number of taxa (see Neher and Darby, Chapter 4, this volume). Many indices of diversity combine measures of numbers of taxa and evenness, with some being most influenced by rare species, and others by abundant species (reviewed by Pielou, 1975; Yeates and Bongers, 1999; Begon *et al.*, 2006). The Shannon–Wiener diversity index H′ is frequently used (Norton and Niblack, 1991; Bernard, 1992).

There is even a measure of diversity, which is itself a measure of the differences between different indices of diversity (Tsiafouli *et al.*, 2006). Even more detailed descriptions of nematode diversity are provided by graphical methods, comparing curves of some measure of abundance versus species ranked in order of abundance (e.g. Heip, 1974; Lambshead *et al.*, 1983; Krebs, 1989; Colwell *et al.*, 2004). Absolute, percentage or cumulative abundances may be used, or some transformation of them. While these measures may have advantages over simpler measures of diversity (including scale independence, see below), their complexity makes them less amenable to comparisons.

Rather than using taxonomic groups as the units for measuring diversity, ecological or trophic categories can be used instead with all the methods just discussed. The most common ecological categories used are related to population dynamic 'strategies'. In terrestrial nematodes, this is known as colonizer-persister score: colonizers (colonizers, opportunists, vagrants, r-strategists) have short life cycles, high fecundity and large population fluctuations in response to fluctuating environments or resources; persisters (perpetuators, competitors, stayers, K-strategists) have longer life cycles, low fecundity, and maintain relatively stable populations (Pianka, 1970, 1972; Bongers, 1990). Another approach uses trophic categories related to the types of food, and the ways of consuming it. The categories used differ considerably between terrestrial, marine and freshwater aquatic nematodes. Table 2.1 lists the categories used, their approximate equivalences, and a suggested uniform scheme for all environments. The relative proportions of the various trophic groups are more frequently used to infer changes in ecosystem function, rather than as units for the measurement of diversity, but trophic diversity is often used because trophic groups are easier and quicker to identify than taxonomic groups. There is some theoretical and empirical support for using ecological measures to assess diversity such as population dynamic or trophic groups, some related to lower variance between samples because rare species with highly variable occurrences are aggregated by these sort of measures (Freckman and Caswell, 1985; Ekschmitt, 1998). However, important ecological interactions can also be overlooked by using groups based solely on food or dynamics (Brussaard, 1998).

There are numerous examples of studies where the results were different when different measures of diversity were calculated (e.g. Wright and Coleman, 2002; Viketoft *et al.*, 2005; Tsiafouli *et al.*, 2006).

Area and number of individuals sampled

The area within which diversity of nematodes (or other organisms) should be assessed is as uncertain as the correct measure of diversity. In general ecological theory, different processes seem to operate in controlling diversity at different spatial scales, and a distinction is made between alpha, beta and gamma diversity, representing local, regional and inter-regional scales (Whittaker, 1972). While these terms may be easily defined for some organisms (e.g. plants), they are ill-defined for nematodes. Indeed, nematodes may show almost continuous variation over scales spanning many orders of magnitude (Hodda, 1990; Coleman *et al.*, 1992; Barrett *et al.*, 2004). In practice,

Table 2.1. A uniform trophic categorization of free-living terrestrial and aquatic nematodes, with equivalences in other systems (Wieser, 1953, 1959; Boucher, 1973; Yeates *et al.*, 1993; Traunspurger, 1997b). *Percentage genera from 122 freshwater genera of Germany (Traunspurger, unpublished data).

Proposed uniform category	Secondary category	Habitat category	Notes	Terrestrial category	Description	Freshwater aquatic category	Description	Marine category	Description	Percentage genera terrestrial	Percentage genera freshwater*	Percentage genera marine
Internal plant feeders	Sedentary plant modifiers	Roots	Rare in aquatic substrates, e.g. *Heterodera litoralis, H. spinicaudata, Meloidogyne mersa*	1a	Sedentary endoparasites of plant roots and seeds	Not listed		Not listed		4.9	<0.1	<0.1
		Seeds and above substrate parts	e.g. *Anguina*	1a (pt)	Seed feeders	Not listed		Not listed		0.2	0	<0.1
	Migratory destroyers	Roots	e.g. *Pratylenchus, Radopholus*	1b	Migratory endoparasites of roots	Not listed		Not listed		3.7	<0.1	<0.1
		Above substrate	e.g. *Ditylenchus, Aphelenchoides*	1a (pt), 1b (pt), 1e (pt) or 1f (pt)	Stem and leaf feeders	Not listed		Not listed		0.7	<0.1	<0.1
External plant feeders	Sub-surface feeders	Roots		1c	Semi-endoparasites of roots	Not listed		Not listed		2.5	14.7 (for the four rows from here down)	<0.1
	Surface feeders	Roots		1d	Root ectoparasites	Not listed		Not listed		10.4		<0.1
	Browsers	Higher plants		1e	Epidermal and root hair feeders	4 (part)	Suction feeders	Not listed		3.0		<0.1
		Lower plants	Rare in marine environments, e.g. *Halenchus*	1f	Algal, lichen or moss feeders	4 (part)	Suction feeders	Not listed		0.7		<0.1

Fungal feeders		Hyphae		2	Hyphal feeders	4 (part)	Suction feeders	Not listed		8.1	12.3	0
Microbial feeders	Suspension feeders	Selective	e.g. *Alaimus*	3 (part)	Bacterial feeders	1		1a	Selective deposit feeders	0.4		25.1
		Non-selective	e.g. *Turbatrix*	3 (part)	Bacterial feeders	1		1b (others) (part)	Non-selective deposit feeders (others)	4.6		3.2
	Food processors	Selective	Rare in terrestrial substrates e.g. *Monhystera*	4	Substrate ingestion	2 (part)	Deposit feeders	1b (aggregate feeders)	Non-selective deposit feeders (aggregate feeders)	1.7	36.1	6.6
		Piercers	e.g. Rhabditidae	3 (part)	Bacterial feeders	1 (part)		1b (others) (part), 2a (part)	Non-selective deposit feeders	16.0		6.8
		Crushers	e.g. Cephalobidae	3 (part)	Bacterial feeders	1 (part)		1b (others) (part), 2a (part)	Non-selective deposit feeders	17.4		7.2
Algae feeders	Scrapers		Rare in terrestrial substrates e.g. *Achromadora*	6	Unicellular eukaryote feeders	4 (part)	Epistrate feeders	2a	Epistrate feeders	4.9	7.4	23.3
Omnivores			Includes less specialized predators, e.g. *Tobrilus* and some Diplogasteridae	3 (part), 8	Omnivores	4 (part)	Suction feeders	2b (part)	Omnivore predators	3.5	9.8	19.6
Specialized predators	Ingesters		e.g. Mononchida	5a	Predators – ingesters	3 (part)	Chewers	2b (part)	Omnivore predators	7.4	3.3	5.7
	Piercers		e.g. Dorylaimida	5b	Predators – piercers	4 (part)	Suction feeders	2b	Omnivore predators	7.4	16.4	2.0
Non-feeding dispersal or infective stages			Rare in aquatic substrates	7	Dispersal or infective stages	Not listed		Not listed		2.5	<0.1	<0.1

the scale on which nematode diversity has been measured has varied from individual samples covering 1-10 cm^2 surface area, through groups of samples covering about 10-100 m^2 surface area less intensively, to large areas covering 10-100 km^2 even less intensively. The reason for using these vastly different scales is often related to the reasons for a particular study, and the sort of scale that the ecological process or environmental change occurs on. For example, a particular vegetation type may occur over a wide area, pollution may affect a moderate area near a source, and other organisms (plants or burrowers) may affect a purely local area.

There is little evidence of whether nematodes respond or operate ecologically on all or any particular combination of spatial scales (Hodda, 1990; Ettema, 1998). Consequently, there may be substantial effects on results from measuring nematode diversity on different spatial scales. For example, a change to greater small-scale patchiness in nematode diversity may be detected at the middle scale, but not at the others, and large-scale homogenization of nematode diversity may be detectable at the small scale, but not at larger scales (Lambshead and Hodda, 1994). The nature of spatial correlation in data for nematodes has been investigated very infrequently, despite a considerable potential influence on both the measurement and analysis of diversity (Lambshead and Hodda, 1994; Wallace and Hawkins, 1994; Liang *et al.*, 2005; Simmons *et al.*, 2008).

A suggested response to difficulties in defining the appropriate scale to measure nematode diversity is measures which are, to a greater or lesser extent, independent of the area sampled (Renyi, 1961; Ricotta, 2000; Coleman and Whitman, 2006). Such measures, while avoiding having to justify a particular scale of sampling, are often so complex that they are difficult to compare. In any case, they also measure a fundamentally different property of diversity than the number of taxa in a given area. Various formulations of the relationship between the number of taxa or groups identified and the number of individuals, are also common (e.g. Bunge, 1993; Gotelli and Colwell, 2001; Brose *et al.*, 2003).

Related, but not necessarily equivalent to the area and time span of a study, is the number of specimens examined. The abundance and diversity of nematodes in many habitats is so high that sub-sampling of large numbers of nematodes is necessary for detailed studies of diversity. If one sub-samples, the number of individuals identified to species level needs to be large enough to represent rare as well as abundant species (Fig. 2.1). Larger numbers of individuals examined increases the chances of greater diversity, particularly for the measures of diversity including numbers of taxa. Unfortunately, the relationship between number of taxa found and number of individuals identified seems neither linear, nor precisely predictable for nematodes. Typically, a relatively few taxa account for a majority of specimens identified in many studies, and there are a large number of taxa represented by one or a few specimens (Fig. 2.2). This seems to apply in freshwater, marine or estuarine sediments, periphyton and terrestrial soils, but does not apply to internal parasites (Poinar, 1983; Hodda and Nicholas, 1985; Hodda *et al.*, 1997; Nicholas and Hodda, 1999; Peters and Traunspurger, 2005; Traunspurger *et al.*, 2006). This seems to apply whether species are identified morphologically

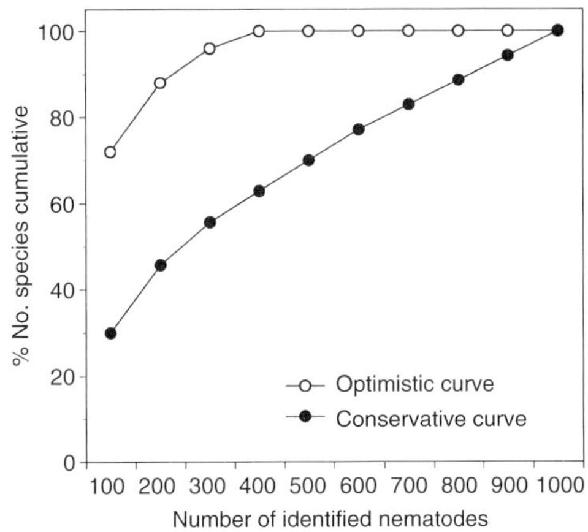

Fig. 2.1. Optimistic and conservative ranking of the percentage of species number of 10 replicate samples of 11 alpine, oligotrophic lakes in Germany (sample size of each lake >1000 individuals). The number of new species found per new sample is plotted against cumulative number of identified individuals. Open circles represent optimistic estimates and closed circles conservative estimates of rarefaction curves. Data from Michiels and Traunspurger (2005).

or by molecular methods (Bloemers *et al.*, 1997; Ekschmitt, 1998; Mullin *et al.*, 2004), but is not universal for nematodes (e.g. Nicholas, 2001). The converse also seems to apply: for a given species most samples contain few or no individuals, but a few will contain many individuals (Norton and Niblack, 1991; Bernard, 1992) (Fig. 2.3).

Time span

The final component of the concept of diversity is the time span over which it is measured. This may be a single time, a season, a year or multiple years. The effect on measures of diversity, beyond the simple effect of adding more individuals, has not been studied.

While it may be possible in some cases to recalculate data on diversity to make it comparable between different studies, in many cases, data that are strictly comparable are impossible to obtain. The myriad aspects of diversity may in fact mean that only very extensive surveys using large numbers of samples and individuals identified to species over multiple spatial scales, and over a period of time, may be sufficient to fully describe nematode diversity. Furthermore, very extensive analysis of the huge data set would also be required (Lawton *et al.*, 1996, 1998). The resources necessary to accomplish such studies are seldom available, but the results of less intensive or extensive studies must be interpreted with regard to the potential artefacts introduced by the concept of diversity being referred to.

Many of the measures of diversity used in studies of nematodes are listed elsewhere (see Neher and Darby, Chapter 4, this volume).

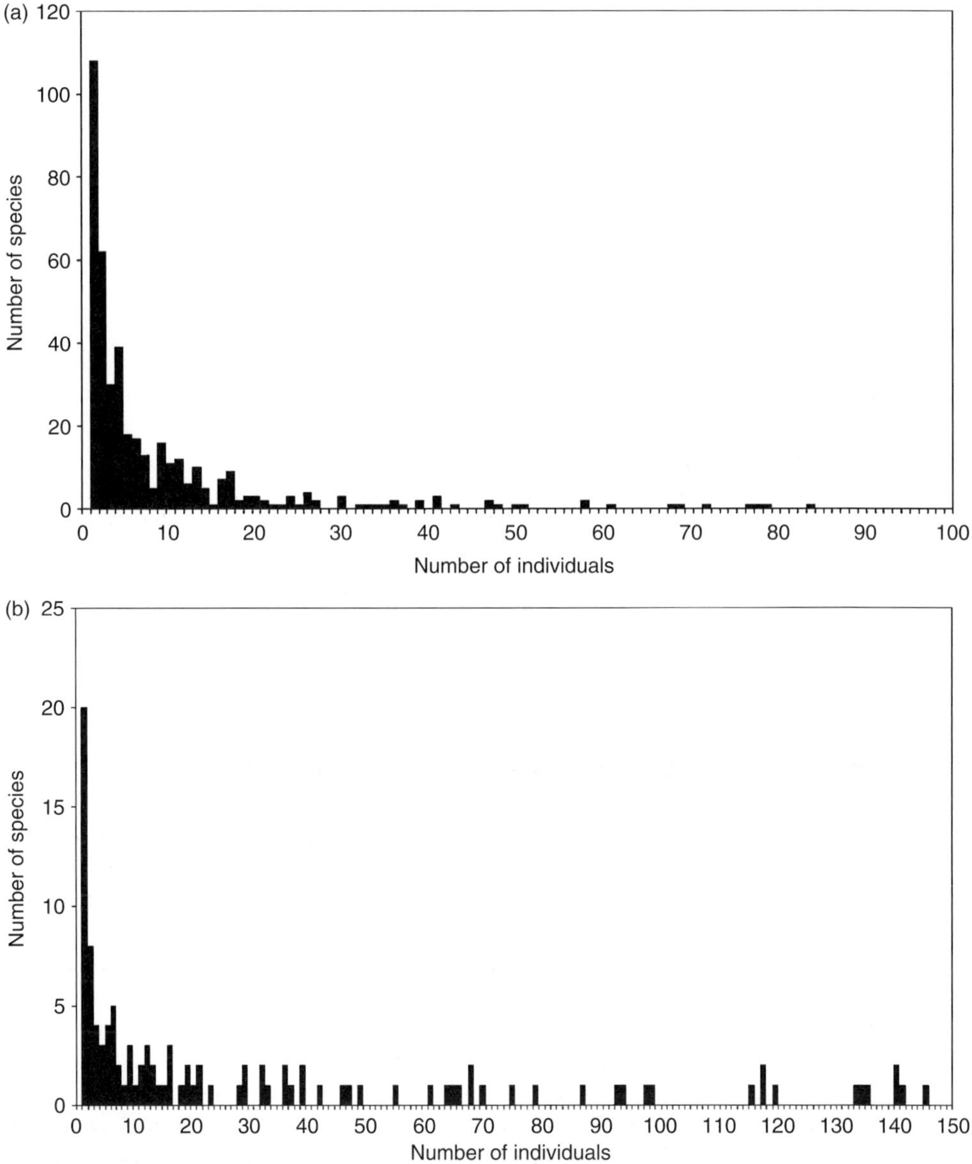

Fig. 2.2. Number of species with particular numbers of individuals in large nematode surveys. (a) 4923 nematodes from 24 samples of tropical forest soils in Cameroon (data from Hodda *et al.*, 1997). Species with 143, 143, 205, 208, 232 and 234 individuals not shown. (b) 12,074 nematodes from 136 samples of grain-growing agricultural soil in southeastern Australia (Harden and Cowra). Data from Hodda *et al.* (1999). Species with 151, 154, 154, 220, 232, 268, 355, 381, 393, 409, 428, 446, 458, 536, 625, 774, 1012 and 1286 individuals not shown.

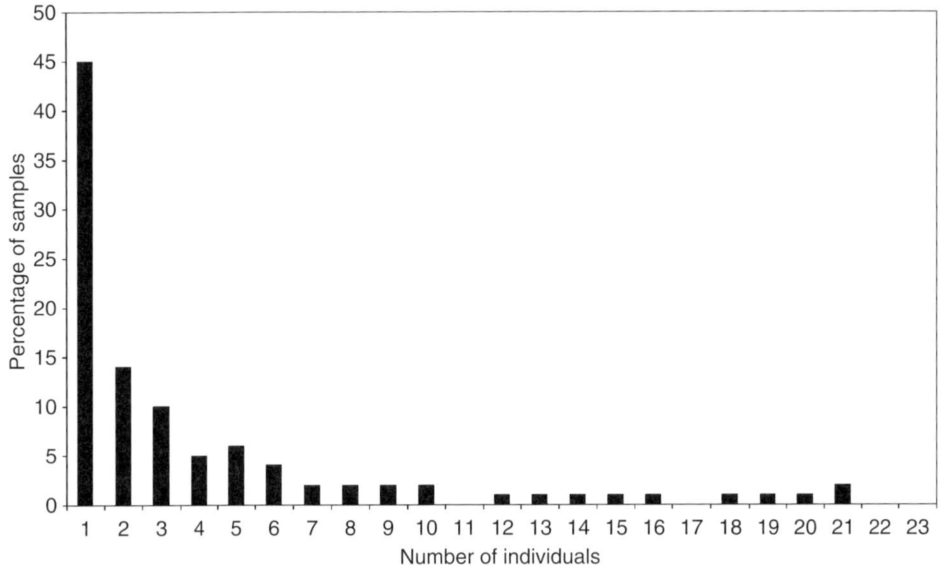

Fig. 2.3. Percentage of samples containing particular numbers of individuals of *Helicotylenchus robustus* in maize fields. Data from Norton and Niblack (1991).

The Effects of Methods on Measurement of Diversity

The methods used in a study of nematode diversity can also have important effects on the results. Among the most important influences are the depth of sampling, the treatment of samples, and the methods used for extraction from substrate.

For nematodes from particulate substrates (soils or sediments), the depth of sampling is important because some nematodes occur only near the surface, and some only at depth (e.g. *Paratylenchus* 40 cm – Geraert, 1965; *Radopholus similis* 366 cm – Wallace, 1963; *Longidorus* 30 cm – Evans, 1978; *Paratrichodorus teres* 45 cm depth on some crops – Kuiper and Loof, 1964; *Meloidogyne* in some soils 150 cm – Koen, 1966; and *Meloidodera*, *Meloidogyne* and *Xiphinema* 11–12 m on deep-rooted desert plants – Freckman and Virginia, 1989). Not only can there be significant populations of certain nematodes at depths such as these, which are below normal sampling, but the depth distribution can vary in different habitats or circumstances (Pitcher, 1967; Freckman and Virginia, 1989). Some apparent changes in diversity are nothing more than some nematodes moving deeper in the substrate. This may occur in terrestrial soils in response to drying or root distribution, as well as in marine and aquatic sediments in response to oxygen availability (Traunspurger and Drews, 1996).

Shaking or disturbance of samples before extraction of nematodes can also result in loss of certain species (Maas and Brinckman, 1980), as can disturbance when soils are very dry and some nematodes are in a dry, brittle anhydrobiotic state (Slack *et al.*, 1972; Simons, 1973). The opposite may also occur if soils are stored for some time, and the conditions are favourable for some nematodes (Franklin *et al.*, 1971). In particular, the temperature of storage can substantially change nematode populations within samples in different ways: some nematodes are favoured by cold and others by heat (Barker *et al.*, 1969; Whyte and Gowen, 1974; Hooper, 1986). Fixing before extraction can also greatly affect the nematodes extracted (Elmiligy and De Grisse, 1970; Freckman *et al.*, 1977).

The use of different mesh sizes for extraction of nematodes in the many methods with a sieving stage also has an important effect on the diversity and species composition obtained as a result. Larger mesh sizes are sometimes suggested, particularly in marine and freshwater aquatic samples, as a way of obtaining samples with less extraneous matter. Mesh sizes of 100 μm, 64 μm and 42 μm have been suggested by Mare (1942), Giere (1993) and Fenchel (1978), respectively. 'Clean' samples obtained with larger meshes may allow more samples to be processed, thus improving spatial coverage of nematode diversity. However, small nematodes, both the adults of smaller species and small juveniles, may never appear in results, thus causing results which are serious underestimates. To avoid missing small nematodes mesh sizes of less than 35 μm, and preferably 20 μm are often suggested (Hodda and Eyualem, 2006; Traunspurger *et al.*, 2006).

The arrangement of samples can affect the probability of finding species, based on the scale at which aggregations occur within a substrate (Been and Schomaker, 2006). The size of aggregations can differ for species and change in response to environmental conditions.

Extraction methods have the largest potential to influence results of nematode diversity. Extraction methods may be ineffective for certain types of nematodes because of their size, activity or attachment to the substrate (Bloemers and Hodda, 1995; Yeates and Bongers, 1999; Bell *et al.*, 2005; Hodda and Eyualem, 2006). The chemical and physical properties of a substrate may also affect extraction efficiency, although it is not known whether all types of nematodes are affected equally or whether there are differential effects on different groups (Yeates and Bongers, 1999).

Molecular Methods

Molecular methods have been proposed as ways of avoiding some of the issues described in the preceding sections, particularly regarding availability of expertise and comparability of the units used to measure biodiversity (Floyd *et al.*, 2002; Griffiths *et al.*, 2006; Nielson *et al.*, Chapter 8, this volume). However, many of the issues remain: genetic markers can vary enormously in their efficiency between different groups of taxa (Powers *et al.*, 1997); different species may have different amounts of genetic variation which may be related to dispersal patterns (Zauner *et al.*, 2007); the resource requirements for molecular studies may be even greater than for conventional ones

(Floyd *et al.*, 2002); and methods have not yet advanced to a level where sampling methods are unimportant in the results.

There have been no studies directly comparing estimates of numbers of nematode species by morphological and molecular means. Perhaps the closest approach to comparative data is studies of *Festuca* grasslands. Using morphological methods, 23 species were found in single samples of about 300 individuals in New Zealand (Yeates, 1974). Molecular analysis of 74 individuals from a comparable grassland in southern Scotland yielded 19 taxa (Floyd *et al.*, 2002). However, the composition of the fauna in the molecular study was different to that in most other studies of grasslands (Orr and Dickerson, 1966; Yeates, 1974; Hodda and Wanless, 1994). The Dorylaimida, which are among the terrestrial groups least sampled or characterized molecularly, were vastly underrepresented in the molecular study. Chalk grasslands in southern England gave values of between 13 and 29 species per sample of 200 individuals, with an average of 21 species in 180 samples (Hodda and Wanless, 1994). The similarity in the numbers of species in these completely independent studies is reassuring, however the variance in numbers of species among replicate cores in the study of chalk grasslands also emphasizes how inadequate most data on diversity are, and that this similarity is far from conclusive.

Molecular methods have been very successful in apparently detecting previously unknown population structure within species, and the presence of cryptic species (e.g. Giblin-Davis *et al.*, 2004; Derycke *et al.*, 2005; Barrett *et al.*, 2006; Oliviera *et al.*, 2006; De Ley *et al.*, 2007).

Total Nematode Diversity

A total of about 14,000 free-living, invertebrate- and plant-associated nematode species are known, described and accepted (Table 2.2) (Platt and Warwick, 1983, 1988; Andrassy, 1992; Hugot *et al.*, 2001; Warwick *et al.*, 1998). This total excludes parasites of vertebrates for which about another 12,000 species are accepted (Hugot *et al.*, 2001). A similar number of published species names are regarded as synonyms of valid species.

About 42% of species described are terrestrial, 39% are marine, 12% are entomophilic, and 5% are freshwater. The numbers of species described reflect a bias towards description of terrestrial nematodes: of all nematode species, an estimated 50% are marine, 35% are terrestrial, and 15% are entomophilic (Ayoub, 1980). The described species represent an estimated 3–10% of the total number of species which exist (Kaestner, 1965; Poinar, 1983; Hugot *et al.*, 2001). This implies 140,000 to 500,000 species of free-living, invertebrate- and plant-associated nematodes in total.

In terms of species described, Panagrolaimida is the largest order, with Tylenchina the largest suborder including about three-quarters of the species in the order. Dorylaimida, Monhysterida, Chromadorida, Enoplida and Desmodorida also contain substantial numbers of species (Table 2.2). Several orders are quite small, and the effort accorded the various orders differs

Table 2.2. Numbers of nematode species and genera in all orders and suborders of non-vertebrate-parasitic nematodes under the classification of Hodda (2007).

Order	Number of valid species described	Number of valid genera	Habitats	References
Benthimermithida	10	4	Deep sea	Miljutin, 2004, Tsesunov, 1997
Rhaptothyreida	2	1	Deep sea	Tsesunov, 1997
Enoplida	800	60	Marine, freshwater and a few terrestrial substrates	Gerlach and Riemann, 1973
Ironida	400	40	Marine, freshwater, terrestrial substrates	Gerlach and Riemann, 1973
Rhabdolaimida	20	5	Terrestrial substrates	Gerlach and Riemann, 1973
Tripyloidida	50	10	Marine substrates	Gerlach and Riemann, 1973
Campydorida	1	1	Terrestrial substrates	Mullin *et al.*, 2003
Alaimida	120	15	Terrestrial substrates	
Trefusiida	130	25	Terrestrial substrates	Gerlach and Riemann, 1973
Oncholaimida	500	60	Marine substrates	Gerlach and Riemann, 1973
Tripylida	230	40	Freshwater, terrestrial substrates	Gerlach and Riemann, 1973
Diphtherophorida	110	10	Terrestrial substrates	Jairajpuri and Ahmad, 1992, Decraemer, 1995
Tobrilida	150	20	Freshwater, terrestrial substrates	Gerlach and Riemann, 1973
Dorylaimida	1900	240	Freshwater, terrestrial substrates	Jairajpuri and Ahmad, 1992
Nygolaimida	100	20	Freshwater, terrestrial substrates	Jairajpuri and Ahmad, 1992
Bathyodontida	10	5	Terrestrial substrates	Zullini and Peneva, 2006
Mononchida	300	40	Freshwater, terrestrial substrates	Zullini and Peneva, 2006; Jairajpuri and Khan, 1982
Mermithida	500	70	Terrestrial substrates, entomophilic	Gafurov, 1997, Holovachov and De Ley, 2006b

Isolaimida	10	1	Terrestrial substrates	
Marimermithida	10	5	Deep sea	Tsesunov, 1997
Chromadorida	1200	100	Most marine substrates, few freshwater or terrestrial	Gerlach and Riemann, 1973
Selachinematida	120	20	Marine substrates	Gerlach and Riemann, 1973
Desmoscolecida	110	25	Marine substrates	Gerlach and Riemann, 1973
Desmodorida	800	60	Marine substrates	Gerlach and Riemann, 1973
Monhysterida	1200	120	Most marine substrates, few freshwater or terrestrial	Gerlach and Riemann, 1973
Aulolaimida	10	3	Marine, freshwater, terrestrial substrates	Gerlach and Riemann, 1973
Plectida	500	60	Marine, freshwater, terrestrial substrates, entomophilic	Gerlach and Riemann, 1973, Holovachov and De Ley, 2006a
Teratocephalida	25	5	Terrestrial substrates	Goodey, 1963
Diplogasterida	310	30	Terrestrial substrates, entomophilic	Sudhaus and Fuerst von Lieven, 2003
Rhabditida	600	80	Terrestrial substrates, entomophilic	Andrassy, 1983, Sudhaus and Fitch, 2001
Panagrolaimida: Panagrolaimina	550	50	Terrestrial substrates, entomophilic	Goodey, 1963; Hodda, 2003a, 2003b
Panagrolaimida: Tylenchina	2650	200	Terrestrial substrates, plants, entomophilic, very few freshwater or marine substrates	Siddiqi, 2000; Hodda, 2003a
Panagrolaimida: Cephalobina	350	30	Terrestrial substrates, entomophilic	Nadler *et al.*, 2006
Drilonematida	60	15	Invertebrates	
Total	13,838	1,470		

enormously. Most described Tylenchina are associated with plants, and it may be speculated that their radiation is associated with that of plants. If so, the question arises as to whether the lesser number of species associated with the even more speciose insects is an artefact of sampling intensity, the greater difficulty in isolating appropriate material, the cryptic nature of many entomophilic nematode species (e.g. Davies and Giblin-Davis, 2004), a different degree of specificity in the relationships with hosts, or the evolutionary potential of the various groups. (Although most nematode associates of insects are in the orders Mermithida, Plectida, Diplogasterida and Rhabditida, there are some Tylenchida, too.)

Diversity in Natural Environments

In a single region, the relationships between diversity of nematodes in soils and vegetation show no consistent patterns. As in agricultural situations, vegetation does not appear to influence nematode diversity directly. In Poland, grasslands had a higher diversity (H′) than forest (Wasilewska, 1979). In Scotland, forest had a higher diversity (measured as number of genera) than moorland (Keith *et al.*, 2006). In Utah, there was no difference in diversity (measured by H′) between meadows and a range of forest types (Bennet and Vetter, 1980). Likewise in western Canada, there were no differences in numbers of genera among sports field, forest and orchard sites (Chevalier and Webster, 2006).

Soils under native vegetation almost always have higher diversity than adjacent cropland soils (Korenko and Schmidt, 2006). Urban and non-urban forest soils have similar nematode diversities (Pavao-Zuckerman and Coleman, 2007).

In terrestrial substrates, the total nematode diversity seems to be similar directly under and between plants (Bell *et al.*, 2005).

Comparisons of total numbers of species found in different habitats must be circumspect for the reasons discussed above, even in comprehensive, detailed studies over multiple sampling periods. However, the relationships of the largest numbers of species found in each habitat may be the best measure with which to compare different habitats (Table 2.3). Minimum values and mean values are not presented because they are too influenced by sampling intensity, numbers of specimens examined etc.

The various studies conducted by Hodda and colleagues, probably represent the most comparable set of detailed studies of nematode species diversity. Although sampling effort, core size and depth differed among the studies, the methods and taxonomic concepts remained very similar. These studies show a steady decrease in diversity from tropical forest to temperate forest to temperate grassland to cultivated fields (Table 2.3). This pattern perhaps represents decreasing soil heterogeneity and/or disturbance, rather than a direct effect of latitude or vegetation. Studies by the same authors of mangroves and sandy beaches, which span a similar latitudinal range, show that the far more heterogeneous mangrove sediments (with fallen leaves,

Table 2.3. Highest numbers of species recorded from terrestrial, freshwater aquatic and marine habitats. Sample sizes vary.

Habitat	Maximum	Maximum single sample/core	Number of specimens	Notes	References
Terrestrial:					
Tropical forest	431	89	6500	194 genera, Mbalmayo, Cameroon	Bloemers *et al.*, 1997
	153			Korup, Cameroon	Price and Siddiqi, 1994
Temperate deciduous forest	175	29		Mixed species, Indiana, USA	Johnson *et al.*, 1972
	125			Beech forest, Slovakia	Saly, 1985
	112			Beech forest, Carpathians, Romania	Popovici, 1989
Temperate evergreen forest	254	52		Eucalypt forest and woodlands, Southeast Australia	Hodda unpub.; Reay unpub.
Coniferous forest	106			Spruce forest, Germany	Ruess, 1995
Tallgrass prairie	375		8400	Kansas, USA	Mullin *et al.*, 2004
	228			Kansas, USA	Orr and Dickerson, 1966
	213			Kansas, USA	Todd *et al.*, 2006
Shrubland	34			Negev Desert, Israel	Pen-Mouratov and Steinberger, 2005
	32			Heathland, northern Sweden	Ruess *et al.*, 1998
Desert, hot	25			Chihuahuan Desert, New Mexico, USA	Wall and Virginia, 1999
Temperate grassland	226	35		Denmark	Overgaard Nielsen, 1949
	154	29	3600	England	Hodda and Wanless, 1994
	112			Meadow, Czech Republic	Hanel, 2003
	108			Czech Republic	Hanel, 1995
Polar grassland (tundra)	160			Arctic tundra, Sweden	Sohlenius, 1980
	89	45		Spitzbergen	Loof, 1971
	75	40		Ellesmere Is, Canada	Mulvey, 1963, 1969a, b, c; Das, 1964; Anderson, 1969; Wu, 1969
	30	19		Signy Is, Antarctic	Spaull, 1973

Continued

Table 2.3. Continued

Habitat	Maximum	Maximum single sample/core	Number of specimens	Notes	References
	26			Macquarie Is, Australia	Bunt, 1954
Cultivated	137			Poland	Wasilewska, 1967
	133	38	12,704	Grain crops, southeastern Australia	Hodda *et al.*, 1999
	100			Tennessee, USA	Baird and Bernard, 1984
Polar	45			Maritime Antarctic total	Maslen and Convey, 2006
	14			Continental Antarctica total	Maslen and Convey, 2006
Freshwater aquatic:					
Sediment:					
Eutrophic lake	152		12,000	Lake Obersee, Germany	Michiels and Traunspurger, 2004b
	63		2,042	Lake Hopfensee, Germany	Traunspurger, 2002
Mesotrophic lake	67		2,832	Lake Spitzingsee, Germany	Traunspurger, 2002
	55		3,116	Lake Sulzberger See, Germany	Traunspurger, 2002
Oligotrophic lake	116		45,000	Lake Konigssee, Germany	Traunspurger, 1996a, b
	95		6,184	Loch Ness, Scotland	David, 1998
	47		3,544	Lake Lustsee, Germany	Traunspurger, 2002
	42		20,000	Lake Brunnsee, Germany	Bergtold and Traunspurger, 2004
River	113		12,000	River Krähenbach, Germany	Beier and Traunspurger, 2003b
	71		1,200	River Krähenbach, Germany	Beier and Traunspurger, 2003a
	51			Danube, Austria	Danielopol, 1979; Eder, 1983
Hard substrates:					
Mixed	34	1–19	3,675	17 lakes along a trophic gradient, Sweden	Peters and Traunspurger, 2005
Oligotrophic	29		2,000	Excl. Dorylaimidae, Lake Königssee, Germany	Traunspurger, 1992
	19	6–13	715	Lake Constance, Austria, Germany and Switzerland	Peters *et al.*, 2005
Mesotrophic-eutrophic	22		8,387	Stones, reed, rush, eastern Holstein, Germany	Schneider, 1922

Marine aquatic:					
Estuary	175			River Elbe, Germany	Riemann, 1966
Seagrass	75	62		Western Australia	Hourston *et al.*, 2005
Mangrove	54			Hunter River, Australia	Hodda and Nicholas, 1985
	32			Darwin, Australia	Hodda and Nicholas, 1987; Hodda and Nicholas, unpublished
Sandy beach	113		9,755	Low energy fine sand beach, Clyde Estuary, Scotland	Lambshead, 1986
	67			High energy coarse sand beach, NSW, Australia	Nicholas and Hodda, 1999; Nicholas, 2001
Shallow sea bed	178		2,253	Irish Sea	Boucher and Lambshead, 1995
	110			83 genera, Norwegian fjord	Somerfield *et al.*, 2006
Sea bed (*c.* 1000 m depth)	277			113 genera, Greenland	Fonseca and Soltwedel, 2007
	223		2,630	145 genera, South China Sea	Huang *et al.*, 2007
	189		808	Western Mediterranean Sea	Soetaert *et al.*, 1991
	160		741	Western Mediterranean Sea	Soetaert *et al.*, 1991
Deep sea (> 4000 m depth)	205[a]		2,467	Hebble, North Atlantic Ocean	Thistle *et al.*, 1995
	128[a]		477	Equatorial central Pacific abyssal plain	Brown, 1998
	125[a]		1,401	North Atlantic abyssal plain (Porcupine)	Rice and Lambshead, 1994

[a]These numbers may be a considerable underestimate. In studies of deep sea nematode diversity, specimens are often identified only to genus. Maximum numbers of genera found in studies of the deep sea range from 113 to 136 (Raes and Vanreusel, 2006; Fonseca and Soltwedel, 2007). Detailed studies of some of these genera have found almost 10 species per genus (29 species in 3 genera – Brandt *et al.*, 2006), so actual numbers of species may be up to 1000.

crab holes, roots, and a redox potential discontinuity of varying depth) have similar numbers of species to the homogenous but frequently disturbed sandy beaches. Terrestrial soils appear far more diverse than either freshwater or marine sediments. Note, however, that some marine environments may have a very high species turnover (Nicholas and Hodda, 1999).

Although most freshwater nematode studies share similar methods of sampling and analysis, there are few consistent patterns of nematode species diversity in freshwater habitats including rivers and lakes of various depths. Lakes especially seem very variable in the number of nematode species present (Table 2.3). In lakes, the distribution and abundance of nematodes within a sediment is strongly influenced by depth (Traunspurger, 1996a,b; Eyualem *et al.*, 2006), but there is no clear evidence for any relationship between nematode diversity and sediment depth or water depth. There are still too few studies of freshwater nematodes for worthwhile hypotheses on the effects of trophic status of a water body or flow velocity on nematode species diversity (Table 2.3).

There does not seem to be any unequivocal evidence for any of the proposed global latitudinal gradients of nematode diversity (Procter, 1990; Vanhove *et al.*, 1999, Lambshead *et al.*, 2000, 2002; Gobin and Warwick, 2006; Brandt *et al.*, 2007). There have been confounding factors in most tests of these hypotheses. For example, an apparent gradient in species numbers with latitude in sandy beaches (Nicholas and Trueman, 2005), is confounded with a gradient of disturbance: the energy of the surf – and hence homogenization through disturbance of the sand – also decreases from high latitudes to low latitudes. An experimental approach using colonization of artificial substrates in marine habitats also produced no evidence of latitudinal gradients in nematode diversity (Gobin and Warwick, 2006).

In general, substrates disturbed by human activity have fewer nematode species than 'natural' situations. Thus 30% more species were observed in regenerating prairie than in adjacent agricultural land (Todd *et al.*, 2006). In a managed meadow, nearly twice as many species were found as in cultivated fields, with an intermediate number in fields where cultivation ceased two years previously (Hanel, 2003). More species were found in undisturbed tropical forest than after clearing, cultivation or during regeneration (Bloemers *et al.*, 1997). In deserts, too, diversity of nematodes has been observed to be greater where there is less disturbance and soil crusts are well-developed (Darby *et al.*, 2007). Similar results were obtained in soybean fields (Okada and Harada, 2007), and a range of other habitats (Baujard *et al.*, 1979a, b; Pate *et al.*, 2000; Cadet *et al.*, 2003a,b; Hanel, 2003).

Soil disturbance appears to reduce the number of genera as well, a result repeated in a considerable number of situations and studies (Yeates and Hughes, 1990; Wardle *et al.*, 1995; Hanel, 2003; Kardol *et al.*, 2005; Korenko and Schmidt, 2006; Tsiafouli *et al.*, 2006). Similar results have been obtained using H′ as the measure of diversity (Wasilewska, 1979; Okada and Harada, 2007).

Cropping of long-lived, woody plants without soil cultivation may have little effect on soil nematode diversity: tea plantations even had a slightly higher generic diversity than adjacent grassland and shrublands (Li *et al.*, 2007). Changing the plant species under cultivation may change some measures of diversity, but not others (Viketoft *et al.*, 2005). The age of the crop may also have no effect on diversity (Jiang *et al.*, 2007). At one location, there was no difference in diversity of fen and pasture (Yuen, 1966).

The decrease in soil nematode diversity following soil disturbance may continue for some time, at least 2 years or longer in some circumstances (Bloemers *et al.*, 1997; Hanel, 2003; Kardol *et al.*, 2005). How soon the changes in diversity occur after disturbance has not been investigated.

Although some patterns of nematode diversity are apparent on large spatial scales, the majority of the variation in nematodes may occur on scales of less than a few kilometres (Hodda, 1990).

Influences on Nematode Diversity

Substrate texture, packing and chemistry

Substrate texture, packing and chemistry can have a substantial effect on the species and genera present in terrestrial soils, but the effects on diversity seem to be small and inconsistent (Yeates and Bird, 1994; Yeates *et al.*, 1997; Alphei, 1998; Althoff and Thein, 2005; Rantalainen *et al.*, 2005) The effects of sediments on diversity may be much greater in freshwater aquatic and marine sediments (Beier and Traunspurger, 2003a,b,c).

Organic matter

Addition of organic matter often results in large increases in microbial populations, with a consequent increase in at least some microbial feeding nematodes. Although the number of taxa present may be the same, the increase in the proportion of the favoured microbial feeders may result in a decrease in many diversity indices including an evenness component (see above). This effect has been observed in both agricultural and forest habitats (Yeates, 1978a,b, 1995; Weiss and Larink, 1991). Composting the organic matter being added may lessen this phenomenon (Nahar *et al.*, 2006). The amount of natural input via litter may also have no effect on diversity (Bengtsson *et al.*, 1998).

As with substrate texture, packing and chemistry, the influence of organic matter may be greater in freshwater and marine nematodes.

Organic management

The interplay of different influences on nematode diversity may be seen in the observed responses to organic management of cropping soils (Table 2.4). Organic management involves a decrease in tillage, an increase in the use of naturally-derived organic matter in various forms and decrease in

Table 2.4. Effects of various forms of agricultural management on terrestrial soil nematodes.

Management	Effect on	Effect	Crop, Location	Reference
Form of chemical in fertilizer	Trophic diversity	–	Wheat, China	Liang *et al.*, 2005
Weed control	No. taxa	–	Pasture, NZ	Yeates *et al.*, 1976
N and lime fertilizer	No. taxa	+	Grassland, UK	Murray *et al.*, 2006
Irrigation	Trophic diversity	Inconsistent	Pasture, Netherlands	Bouwman and Zwart, 1994
Composting of organic input	No. taxa, indices	Inconsistent	Tomatoes, Ohio	Nahar *et al.*, 2006
Cultivar diversity	Evenness	+	Sugarcane, West Indies	Cadet *et al.*, 2007
Grazing, stocking intensity	No. taxa	0	Pasture, Florida	McSorley and Tanner, 2007
	No. taxa	0	Natural grassland, Austria	Zolda, 2006
	No. taxa	0	Shortgrass steppe, Colorado	Freckman and Huang, 1998
Fallowing	No. taxa plant parasites	0	Various agricultural fields, Senegal	Cadet *et al.*, 2003a

the use of inorganic chemical fertilizers and biocides. The exact combination of management practices termed 'organic' varies considerably, and so do the apparent effects on nematode diversity. Some studies have shown either an increase or a decrease in diversity with organic management, depending on which measure of diversity is used (Tsiafouli *et al.*, 2006). Other studies have found no difference irrespective of which measure is used (Yeates *et al.*, 1997). Yet other studies have found inconsistent results over time (Okada and Harada, 2007). Still other studies have found consistent increases in diversity with organic management, but only in some measures of diversity (Van Diepeningen *et al.*, 2006).

Pollution

Pollution by inorganic contaminants generally decreases nematode diversity measured in most ways. There is some discussion about the relative sensitivities of different parameters, but most seem to show similar trends (Table 2.5).

Pollution of many kinds reduces diversity in many different natural systems, but this is not always the case. Pollution by different heavy metals may have different effects (Sanchez-Moreno *et al.*, 2006). Also, the effect of pollution may be modified considerably by other factors. For example, where no effect of heavy metal pollution was reported from soybean fields in China, the considerable disturbance from the agricultural management may have masked the effect of the heavy metals (Li *et al.*, 2006). Prior exposure may also

Table 2.5. Effects observed from a range of pollutants in terrestrial soils.

Pollutant	Effect	Diversity measure(s)	Environment	Reference
Heavy metals	–	No. genera	Riverine forest, Spain	Sanchez-Moreno *et al.*, 2006
	–	No. species		Korthals *et al.*, 1996b
	–	No. species	Moss, Italy	Zullini and Perretti, 1986
	–	H'	Beech forest, Romania	Popovici, 1989
Cu	–	No. species	Maize, Netherlands	Korthals *et al.*, 1996a
Ammonium	– (Small)	Simpson Index	Forest, Netherlands	Tamis, 1986
Ash	–	No. species, no. genera	Grassland, Poland	Dmowska and Ilieva-Makulec, 2006
Acid	0	No. species	Spruce forest, Germany	Ruess *et al.*, 1996
	–	No. species	Spruce forest	Ruess *et al.*, 1998
	0	No. genera	Spruce forest, Germany	Yeates *et al.*, 1994
Oil	–	Trophic diversity	Tallgrass prairie, Oklahoma	Sublette *et al.*, 2007

influence the effect of heavy metals on nematode diversity. In one study, prior exposure over long periods (hundreds of years), may have meant that experimental additions of zinc caused no reductions in soil nematode diversity (Van der Wurff *et al.*, 2007). However, the presence of heavy metals influenced the effects of other stressors on the nematodes: in the same study, where zinc was added, excessive heat reduced diversity compared with treatments of heat without zinc. In another study, experimental addition of fertilizer also modified the effect of oil pollution (Sublette *et al.*, 2007). The effect of pollution of various sorts on all the different aspects of nematode populations are discussed in elsewhere (see Donavaro *et al.*, Chapter 6, this volume and Nagy, Chapter 7, this volume).

Generation and Maintenance of Nematode Diversity

A large range of environmental and biological interactions have been implicated in controlling nematode diversity, including the physical and chemical environment, spatial and temporal heterogeneity, competition and predators or pathogens (Freckman and Caswell, 1985; Ettema, 1998). One review listed 12 abiotic and 15 biotic influences on terrestrial plant-parasitic nematodes (Norton and Niblack, 1991). Diversity may also be controlled by different forces at any of a great range of spatial and temporal scales (Hodda, 1990; Ettema, 1998; Virginia and Wall, 1999; Ettema and Wardle, 2002; Dillon *et al.*, 2008; Simmons *et al.*, 2008). Because of the number and range of influences on nematode diversity, understanding the processes behind the effects is important for predicting the results of the various influences on nematode diversity, so some of the processes apparently controlling nematode diversity are discussed below. This discussion shows that there has been a great deal of speculation but few experimental studies of these processes.

Dispersal

Widespread dispersal by wind and water are thought to be important in generating and maintaining nematode diversity by deposition of species into areas where they are not present, but there have been relatively few demonstrations that the process occurs in the field, and that it occurs frequently (Orr and Newton, 1971; Carrol and Viglierchio, 1981; Whitford and Freckman, 1988).

Up to 28 genera in all feeding groups were found transported by air and deposited in dust traps in one study (Orr and Newton, 1971). However, in other studies only nematodes in an anhydrobiotic state were transported by wind (Carrol and Viglierchio, 1981; Janiek, 1996).

In high energy surf beaches, nematodes may be suspended and transported regularly by the water flow (Hodda, 1981; Nicholas, 2001). Likewise, nematodes may be transported by rapidly-flowing water during floods, or even slower-flowing irrigation water, (Faulkner and Bolander, 1967, 1970a,b; Tobar-Jimenez and Palacios-Mejia, 1975; Odihirin, 1977; Waliullah, 1984, 1989; Roccuzzo and Ciancio, 1991; American Public Health Association, 1998).

The few studies on dispersal and colonization of nematodes in other freshwater and marine habitats provide evidence that nematodes are primarily transported through the water column to new substrates in running freshwater and marine systems. In lakes with stony hard substrates, nematodes are able to quickly colonize new or disturbed substrates mainly by transport through the water column after they passively or actively enter the water (Peters *et al.*, 2005, 2007).

Many terrestrial nematodes are transported by insects and other mobile invertebrates (Timper and Davies, 2004).

The nature and importance of dispersal in generating and maintaining nematode diversity may differ in different environments. Constant movement may be the norm in exposed sandy beaches where sediment and the nematodes within it are constantly remixed (Hodda, 1981; Nicholas and Hodda, 1999; Nicholas, 2001). As a result, competitive exclusion may seldom operate, and more similar species will coexist than elsewhere, resulting in higher diversity. In nematode species feeding opportunistically on bacterial colonies which are highly variable in both space and time, there may be intense competition within each patch of suitable resources resulting in local populations dominated by relatively few species or genera (Cutter, 2006; Phillips, 2006). However, there may be a great number of such patches, each dominated by different nematodes, resulting in a high diversity on a different spatial and temporal scale to that in sandy beaches (Cutter *et al.*, 2006). In many parasitic species, however, dispersal is even more infrequent, but the difficulty in finding hosts may mean that more than one species is seldom present, and so diversity may stem from a strong founder effect and divergence in sub-populations creating the conditions for many species to evolve. This may occur in parasites of vertebrates, invertebrates and plants, (e.g. Jagdale *et al.*, 2006, Webster *et al.*, 2007; Zhou *et al.*, 2007). In some environments, like the dry valleys of Antarctica, gene flow between isolated populations may be sufficient to prevent divergence (Courtright *et al.*, 2000; Adams *et al.*, 2007).

Over very long time spans, plate tectonics may move and isolate nematode populations from each other and thus facilitate allopatric evolutionary divergence (Ferris, 1976; Gerlach, 1977a).

Reproductive rate and generation time

Reproductive rates and generation times vary enormously in nematodes, and have been implicated in successional sequences after disturbance (Anderson and Coleman, 1982; Schiemer, 1983; Yeates *et al.*, 1985; Ettema and Bongers, 1993; Ferris *et al.*, 1995, 1996, 1997; Venette and Ferris 1997). The higher the reproductive rate and shorter the generation time, the more likely that one or a few taxa will dominate a low-diversity fauna. Although demonstrated in microcosms, the importance to nematode diversity in field situations is uncertain.

Plants

In terrestrial environments, the soil nematode diversity is often directly related to the diversity of plants. Such relationships have been observed in natural grasslands, and in experimental mesocosms (Yeates *et al.*, 1983; Wasilewska, 1995; De Deyn *et al.*, 2004). In some studies diversity measured by the number of taxa was related to plant diversity, but in others it was one of the diversity or evenness indices related to plant diversity, not the number of taxa (Viketoft *et al.*, 2005; Cadet *et al.*, 2007). Soil nematode diversity is often higher in areas of natural vegetation with many plant species compared to managed areas with fewer plant taxa, especially in terms of nematodes feeding on roots (Van Gundy and Freckman, 1977; Wallace, 1987). The mechanism behind this has been hypothesized as lack of competition resulting from division of roots into very precisely defined niches in the natural vegetation (Van Gundy and Freckman, 1977), but there is little evidence and the observations may be equally due to other confounding factors, such as level of disturbance (see above).

At a very small scale, horizontal and vertical changes in the amount of roots and organic matter in terrestrial soils can have a strong influence on diversity, but this is not universal (Yeates, 1980, 1981; Alphey, 1985; Freckman and Virginia, 1989; De Goede *et al.*, 1993b; Sohlenius, 1997; Bell *et al.*, 2005). Patchy food sources are also important at small scales in marine environments (Meyers and Hopper, 1966; Gerlach, 1977b).

Other nematodes

Other nematodes may influence diversity either positively or negatively. There may be a positive effect on diversity from the presence of nematodes which are a food source for predatory nematodes. Maintaining high abundances of predatory nematodes may prevent competitive exclusion by reducing populations of dominant species. This may cause a positive feedback loop and result in even greater differences between different situations than would otherwise be the case. For example, the diversity of the omnivorous or predatory family Mononchidae closely follows that of the Cephalobidae, which are their likely prey in terrestrial soils (Yeates and Bongers, 1999). In 13 alpine, freshwater lakes, more nematode species were found where there were more predatory nematodes (Michiels *et al.*, 2004). The number of nematode species increased from 24 to 75 when the ratio of predatory to non-predatory nematodes increased from 0.02 to 0.26 (Fig. 2.4). Note though that in these studies that it was predator abundance rather than diversity which increased the diversity of other species.

Interspecific competition may limit diversity in entomophilic and root-feeding nematodes in soil (Sikora *et al.*, 1979; Freckman and Caswell, 1985; Eisenback and Griffin, 1987; Neumann and Shields, 2006). However there are few observations within other groups of nematodes (Ettema, 1998). Furthermore, entomophilic nematodes may at least temporarily reduce the diversity of other groups of nematodes (omnivorous Dorylaimida) through ecological processes that remain obscure (Chevalier and Webster, 2006). Observations of lower soil nematode diversity in mesocosms containing grasshoppers may be due to the

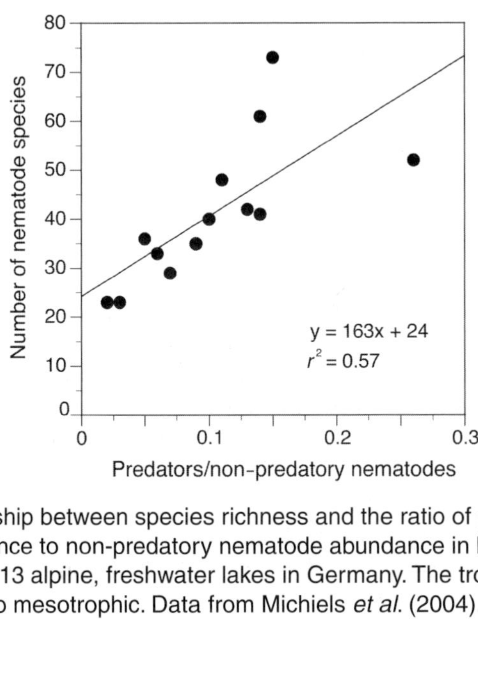

Fig. 2.4. Relationship between species richness and the ratio of predatory nematode abundance to non-predatory nematode abundance in benthic nematode communities from 13 alpine, freshwater lakes in Germany. The trophic status ranged from oligotrophic to mesotrophic. Data from Michiels *et al.* (2004).

same effect (De Deyn *et al.*, 2007). Observations purported as competition between root-feeding nematodes are discussed above.

Pathogens and predators

Although pathogens and predators have been hypothesized as important influences on diversity, there have been few experimental or even observational tests in terrestrial, freshwater and marine substrates (Van der Putten *et al.*, 2006; Galluci *et al.*, 2008). It has been noted that the rate of predation on nematodes is probably density dependent, so predation should increase diversity by increasing the evenness of taxon abundances and facilitate the survival of taxa which would otherwise be excluded by competitive dominants (Hyvonen and Persson, 1996; Yeates and Wardle, 1996). Furthermore, which taxa are susceptible to predation has been observed to differ according to the particular set of conditions (Bilgrami, 1993). So the presence of predators or pathogens may further increase diversity in this way.

Microorganisms as food

Microbial-feeding nematodes have been demonstrated on many occasions to consume, to grow or to reproduce differently according to the particular microorganisms available as food in microcosms and under experimental conditions (e.g. Grewal and Wright, 1992; Ruess and Dighton, 1996; Newsham *et al.*, 2004; Tahseen *et al.*, 2004). Hence, microbial diversity in substrates should exert a strong influence on the diversity of microbial-feeding nematodes. Correlations between microbial diversity and nematode diversity have been observed in a number of situations (e.g. in decomposing litter in terrestrial soils and estuarine mangrove sediments – Hodda and Nicholas, 1985; Ilieva-Makulec *et al.*, 2006). In very simple communities in the Antarctic, however, there seems to be no link between bacterial and nematode diversity (Barrett *et al.*, 2006).

Ecosystem productivity

Despite considerable theoretical speculation that there is a general causal link between ecosystem productivity and biological diversity (e.g. Tilman and Pacala, 1993; Olff and Ritchie, 1998; Hillebrand, 2003; Rajaniemi, 2003), no relationship between nematode diversity and either productivity or total abundance of nematodes has been demonstrated unequivocally in any ecosystem (Yuen, 1966; Yeates, 1984; Hodda and Nicholas, 1985; Sohlenius *et al.*, 1987; Yeates and Hughes, 1990; Wasilewska, 1991; Yeates *et al.*, 1997; Michiels and Traunspurger, 2004a, b).

Many studies have been conducted in freshwater lakes on the relationship between productivity and diversity of nematodes, because this is a crucial question for management of these lakes. Within freshwater sediments and on hard substrates, the number of species coexisting seems independent of the trophic status of their habitats (Fig. 2.5). The eutrophic lakes with the highest productivity ranged from 22 to 152 species; mesotrophic lakes with intermediate productivity ranged from 16 to 70 species; and oligotrophic lakes with the lowest productivity had 17 to 116 species (Traunspurger *et al.*, 2006). Similar absence of patterns were found for hard substrate and periphyton communities (Peters and Traunspurger, 2005, Table 2). It may be that resource availability increases affects nematode species diversity only over long periods under field conditions (Michiels and Traunspurger, 2004a). In mesocosm experiments, however, depletion of nutrients caused species richness and diversity to decline considerably compared to treatments where nutrients were added over two years.

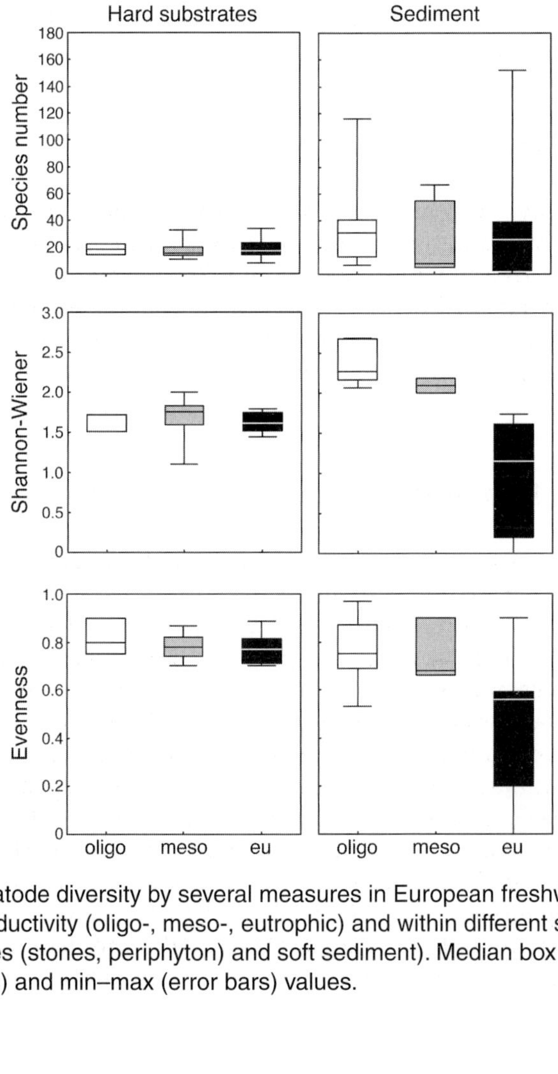

Fig. 2.5. Nematode diversity by several measures in European freshwater lakes of different productivity (oligo-, meso-, eutrophic) and within different substrate types (hard substrates (stones, periphyton) and soft sediment). Median box plot with 25-75% (boxes) and min–max (error bars) values.

Water

In fully terrestrial substrates and those subject to periodic inundation and exposure, the amount of water in a substrate and its chemistry can be important influences on nematode diversity. Water is essential to nematode movement (Nicholas, 1984), so substrates with very limited water may have very limited nematode faunas (Freckman and Mankau, 1986; Virginia and Wall, 1999). The adaptations to dehydration and rates of survival vary between species, and this may allow the coexistence of many taxa in fluctuating environments, such as deserts and marshes (Hodda *et al.*, 2006). Through its influence on movement and other soil properties, soil water has been observed to effect nematode diversity in many environments from Antarctica to natural prairies to agricultural fields and forests (Yeates, 1980, 1981; Alphey, 1985; Sohlenius, 1985, 1997; Freckman and Virginia, 1989; De Goede *et al.*, 1993a; Todd *et al.*, 1999; Virginia and Wall, 1999). However, the way that diversity was influenced was mostly complex rather than straightforward (Yeates, 1980, 1981; Alphey, 1985; Freckman and Virginia, 1989; De Goede *et al.*, 1993b; Sohlenius, 1997; Todd *et al.*, 1999).

Substrate particle size

There have been many casual observations that the body lengths and diameters of nematodes correlates with the particle size of the substrate in which they live (Norton, 1979; Goodell and Ferris, 1980; Wallace *et al.*, 1993; De Goede and Bongers, 1994; Yeates *et al.*, 1997). There is some experimental evidence that terrestrial soil habitats may be partitioned between nematode taxa on this basis, and so it can make a contribution to maintaining diversity (Ettema, 1998).

Influence of Nematode Diversity on Ecosystem Processes

The influence of biological diversity on ecosystem processes was originally seen as simple: higher diversity enhanced most ecosystem processes (Chapin *et al.*, 1997). More recently, the links between ecological processes and diversity have been seen as complex and idiosyncratic, a result of non-linear dynamics and the coupling of semi-independent systems (De Mesel *et al.*, 2006; Vandermeer, 2006). There has also been the suggestion that diversity does not matter, and that many taxa are redundant to ecosystem function (Walker, 1992; Andren *et al.*, 1995). Others have suggested that all species are essential in at least some systems (Chapin *et al.*, 1997), or that rare species may be most important for ecosystem function and stability (Freckman and Virginia, 1997; Wall and Virginia, 1999). Any examinations or experimental investigations of the relationships between diversity and ecosystem processes are subject to considerable difficulties in defining and measuring not only diversity – as discussed above – but also the ecosystem processes. These issues have been discussed in some detail (e.g. Lawton, 1994; Lamont, 1995; Martinez, 1996; Bengtsson *et al.*, 1998).

It has been suggested that soil nematodes drive soil nutrient cycling, soil carbon fluxes, the structure of soil food webs, primary production, vegetation succession and plant species diversity, and other processes (Ingham *et al.*, 1985; Yeates, 1987a; Freckman, 1988; Freckman and Virginia, 1989; Griffiths, 1990; Anderson, 1995; Wall and Moore, 1999; Mikola and Sulkava, 2001; Lavelle *et al.*, 2006; Van der Putten *et al.*, 2006). Correlations between nematode diversity and these ecosystem processes have been observed, but seldom has the nature or direction of the linkages been unequivocally demonstrated, and seldom do observed relationships seem to hold in every situation (Covich *et al.*, 1999; Groffman and Bohlen, 1999; Snelgrove, 1999; Wall, 1999; Wall and Moore, 1999).

Relationships between nematode diversity and bacterial productivity in substrates of many types have been observed frequently (Freckman, 1988; Griffiths *et al.*, 1994; Ferris *et al.*, 1997), but are not found everywhere (Barrett *et al.*, 2006). Relationships between nematode diversity and plant species diversity have also been observed (Wasilewska, 1995; Brinkman *et al.*, 2005), but the particular species involved rather than diversity per se may be the driving force. There is some evidence for an effect of nematodes on the abundance and activity of sediment bacteria in a freshwater lake, but the link between nematode diversity and bacterial productivity or diversity remains unproven in freshwater sediments (Traunspurger *et al.*, 1997). Likewise, observations of a relationship between soil nematode diversity and plant productivity may be more to do with the presence of particular species of plant parasites, rather than nematode diversity (Evans, 1978; Brinkman *et al.*, 2005). Relationships have also been observed between diversity of biota and ecosystem function in freshwater sediments, but the relationship seems very subject to disturbance (Palmer *et al.*, 1997). The presence or absence of soil nematodes may affect bacterial diversity (Brussaard *et al.*, 1997; Swift *et al.*, 1998; Kaffe-Abramovich and Steinberger, 2006), and terrestrial plant biomass (Ingham *et al.*, 1985; Alphei *et al.*, 1996; Brussaard *et al.*, 1997) (but in one study root nitrogen content and nitrogen leaching were inversely related to nematode diversity (Alphei *et al.*, 1996)).

One of the most idiosyncratic of effects of nematode diversity reported is of an indirect influence of entomophilic nematode diversity on bird diversity (Jiang *et al.*, 2005). The mechanism of such an influence remains obscure to say the least!

Conclusions

Nematode diversity seems to lack any convincing consistent patterns to either particular environmental causes or ecological processes. This observation may be a result of one of three situations. First, nematodes may be influenced by basic forces controlling diversity, but subject to such frequent changes in conditions that the equilibrium condition is seldom reached (Wasilewska, 1994). Second, nematode species may all behave individualistically, so that there are a multitude of forces controlling diversity on many different spatial and temporal scales, rather than only a few (Todd *et al.*, 2006). Third, the small proportion of nematode diversity which can feasibly be sampled, coupled with the influence of differences in methods and measurements, a potentially very complex pattern of nematode diversity, and a lack of good systematics combined with the expertise to use it, may all obscure underlying patterns of nematode diversity (Bernard, 1992; Lambshead, 1993; Boucher and Lambshead, 1995; Lawton *et al.*, 1998).

The only consistency in patterns of nematode diversity is the apparent inconsistency. Variation may be regional (Yeates, 1994; Yeates and Bird, 1994; Neher *et al.*, 1995; Yeates and Bongers, 1999) or local (Norton and Oard, 1981; Zolda 2006) and/or seasonal (Yeates and Bird, 1994; McSorley and Frederick, 1996; Stamou *et al.*, 2005), depending on the measure of diversity used (Bernard, 1992; Viketoft *et al.*, 2005), or whether a particular part of the nematode fauna is considered (Van Diepeningen *et al.*, 2006). History and prior events may be important or not (Virginia and Wall, 1999; Maslen and Convey, 2006). Disturbance may be a force controlling diversity, or it may be a force modifying or disrupting other influences on diversity (Palmer *et al.*, 1997).

There seems little doubt that nematode diversity is complicated. But far from this being a reason to abandon studies in the area, it should serve as a reason for further studies. There is little doubt that nematodes and their diversity – in terrestrial soils, in aquatic substrates and marine sediments – are important in influencing and being influenced by a whole raft of ecosystem processes and properties. Only through further modelling, field and microcosm studies will the body of knowledge be expanded to eventually allow elucidation of the general patterns and relationships underlying nematode diversity.

References

Adams, B.J., Wall, D.H., Gozel, U., Dillman, A.R., Chaston, J.M. and Hogg, I.D. (2007) The southernmost worm, *Scottnema lindsayae* (Nematoda): diversity, dispersal and ecological stability. *Polar Biology* 30, 809–815.

Alphei, J. (1998) Differences in soil nematode community structure of beech forests: Comparison between a mull and a moder soil. *Applied Soil Ecology* 9, 9–15.

Alphei, J., Bonkowski, M. and Scheu, S. (1996) Protozoa, Nematoda and Lumbricidae in the rhizosphere of *Hordelymus europaeus* (Poaceae): faunal interactions, response of microorganisms and effects on plant growth. *Oecologia* 106, 111–126.

Alphey, T.J.W. (1985) A study of spatial-distribution and population-dynamics of 2 sympatric species of trichodorid nematodes. *Annals of Applied Biology* 107, 497–509.

Althoff, P.S. and Thien, S.J. (2005) Impact of M1A1 main battle tank disturbance on soil quality, invertebrates, and vegetation characteristics. *Journal of Terramechanics* 42, 159–176.

American Public Health Association (1998) Nematological examination: introduction. In: Clesceri, L.S., Greenberg, A.E. and Eaton, A.D. (eds) *Standard Methods for the Examination of Water and Wastewater*, 20th edn. APHA, Washington, DC, pp. 10–74 to 10–76.

Anderson, J.M. (1995) Soil organisms as engineers: Microsite modulation of macroscale processes. In: Jones, C.G. and Lawton, J.H. (eds) *Linking Species and Ecosystems*. Chapman and Hall, New York, pp. 94–106.

Anderson, R.V. (1969) Comparative morphology and descriptions of 3 new species of *Teratocephalus* from Canada. *Canadian Journal of Zoology* 47, 829–840.

Anderson, R.V. and Coleman, D.C. (1982) Nematode temperature responses – a niche dimension in populations of bacterial-feeding nematodes. *Journal of Nematology* 14, 69–76.

Andrassy, I. (1983) *A Taxonomic Review of the Suborder Rhabditina (Nematoda: Secernentia).* ORSTOM, Paris.

Andrassy, I. (1992) A short census of free-living nematodes. *Fundamental and Applied Nematology* 15, 187–188.

Andren, O., Bengtsson, J. and Clarholm, M. (1995) Biodiversity and species redundancy among litter decomposers. In: Collins, H.P., Robertson, G.P. and Klug, M.J. (eds) *The Significance and Regulation of Soil Biodiversity*. Kluwer Academic Publishers, Dordrecht, pp. 141–151.

Ayoub, S.M. (1980) *Plant Nematology: an Agricultural Training Aid.* (Revised). CDFA, Sacramento, California.

Baird, S.M. and Bernard, E.C. (1984) Nematode population and community dynamics in soybean-wheat cropping and tillage regimes. *Journal of Nematology* 16, 379–386.

Baldwin, J.G., Nadler, S.A. and Freckman Wall, D. (2000) Nematodes: pervading the earth and linking all life. In: Raven, P.R. and Williams, T. (eds) *Nature and Human Society: The Quest for a Sustainable World*. National Academy Press, Washington, DC, pp. 176–191.

Barker, K.R., Nusbaum, C.J. and Nelson, L.A. (1969) Effects of storage temperature and extraction procedure on recovery of plant-parasitic nematodes from field soils. *Journal of Nematology* 1, 240–247.

Barrett, J.E., Virginia, R.A., Wall, D.H., Parsons, A.N., Powers, L.E. and Burkins, M.B. (2004) Variation in biogeochemistry and soil biodiversity across spatial scales in a polar desert ecosystem. *Ecology* 85, 3105–3118.

Barrett, J.E., Virginia, R.A., Wall, D.H., Cary, S.C., Adams, B.J., Hacker, A.L. and Aislabie, J.M. (2006) Co-variation in soil biodiversity and biogeochemistry in northern and southern Victoria Land, Antarctica. *Antarctic Science* 18, 535–548.

Baujard, P., Comps, B. and Scotto La Massese, C. (1979a) Introduction à l'étude écologique de la nematofaune tellurique du massif landais (France). *Revue d'Ecologie et de Biologie du Sol* 16, 61–78.

Baujard, P., Boulbria, A., Ham, R., Laumond, C. and Scotto La Massese, C. (1979b) Premieres données sur la nematofaune associée aux dépérissements du pin maritime dans l'Ouest de la France. *Annales des Sciences Forestières* 36, 331–339.

Been, T.H. and Schomaker, C. (2006) Distribution patterns and sampling. In: Perry, R.N. and Moens, M. (eds) *Plant Nematology*. CABI Publishing, Wallingford, pp 302–326.

Begon, M., Harper, J.L. and Townsend, C.R. (2006) *Ecology: from Individuals to Ecosystems*, 4th edn. Blackwell Science, Oxford.

Beier, S. and Traunspurger, W. (2003a) Seasonal distribution of free-living nematodes in the Krahenbach, a fine-grained submountain carbonate stream in southwest Germany. *Nematology* 5, 113–136.

Beier, S. and Traunspurger, W. (2003b) Seasonal distribution of free-living nematodes in the stream Körsch: a coarse-grained submountain carbonate stream in Southwest Germany. *Nematology* 5, 481–504.

Beier, S. and Traunspurger, W. (2003c) Temporal dynamics of meiofauna communities in two small submountain carbonate streams with different grain size. *Hydrobiologia* 498, 107–131.

Bell, N.L., Davis, L.T., Sarathchandra, S.U., Barratt, B.I.P., Ferguson, C.M. and Townsend, R.J. (2005) Biodiversity of indigenous tussock grassland sites in Otago, Canterbury and the central North Island of New Zealand II. Nematodes. *Journal of the Royal Society of New Zealand* 35, 303–319.

Bengtsson, J. (1998) Which species? What kind of diversity? Which ecosystem function? Some problems in studies of relations between biodiversity and ecosystem function. *Applied Soil Ecology* 10, 191–199.

Bengtsson, J., Lunqvist, H., Saetre, P., Sohlenius, B. and Solbreck, B. (1998) Effects of organic matter removal on the soil food web: forestry practices meet ecological theory. *Applied Soil Ecology* 9, 137–143.

Bennet, L.W. and Vetter, R.T. (1980) Diversity trends among the invertebrate litter decomposers of a subalpine spruce-fir sere: a test of Odum's hypotheses. *Proceedings of the International Soil Zoology Colloquium* 7, 785–94.

Bergtold, M. and Traunspurger, W. (2004) The benthic community in the profundal of Lake Brunnsee: seasonal and spatial patterns. *Archives of Hydrobiology* 160, 527–554.

Bernard, E.C. (1992) Soil nematode biodiversity. *Biology and Fertility of Soils* 14, 99–103.

Bilgrami, A.L. (1993) Analysis of the predation by *Aporcelaimellus nivalis* on prey nematodes from different prey trophic categories. *Nematologica* 39, 356–365.

Bloemers, G.F. and Hodda, M. (1995) A method for extracting nematodes from a tropical forest soil. *Pedobiologia* 39, 331–343.

Bloemers, G.F., Hodda, M., Lambshead, P.J.D., Lawton, J.H. and Wanless, F.R. (1997) The effects of forest disturbance on diversity of tropical soil nematodes. *Oecologia* 111, 575–582.

Boag, B. and Robertson, L. (1983) A technique for studying the micro distribution of nematodes in undisturbed soil. *Revue de Nematologie* 6, 146–148.

Bongers, T. (1990) The maturity index: an ecological measure of environmental disturbance based on nematode species composition. *Oecologia* 83, 14–19.

Boucher, G. (1972) First ecological data on marine free living nematodes from a coastal mud station of Banyuls. *Vie et Milieu* 23B, 69–100.

Boucher, G. and Lambshead, P.J.D. (1995) Ecological biodiversity of marine nematodes in samples from temperate, tropical, and deep sea regions. *Conservation Biology* 9, 1594–1604.

Bouwman, L.A. and Zwart, K.B. (1994) The ecology of bacterivorous protozoans and nematodes in arable soil. *Agriculture, Ecosystems and Environment* 51, 145–160.

Brandt, A., De Broyer, C., De Mesel, I., Ellingsen, K.E., Gooday, A.J., Hilbig, B., Linse, K., Thomson, M.R.A. and Tyler, P.A. (2006) The biodiversity of the deep southern Ocean benthos. *Philosophical Transactions of the Royal Society B-Biological Sciences* 362, 39–66.

Brinkman, E.P., Duyts, H. and Van der Putten, W.H. (2005) Consequences of variation in species diversity in a community of root-feeding herbivores for nematode dynamics and host plant biomass. *Oikos* 110, 417–427.

Brose, U., Martinez, N.D. and Williams, R.J. (2003) Estimating species richness: Sensitivity to sample coverage and insensitivity to spatial patterns. *Ecology* 84, 2364–2377.

Brown, C.J. (1998) Effects of a phytodetrital input on nematode communities of the abyssal, equatorial Pacific. PhD thesis, Southampton University, Southampton.

Brussaard, L. (1998) Soil fauna, guilds, functional groups and ecosystem processes. *Applied Soil Ecology* 9, 123–135.

Brussaard, L., Behan-Pelletier, V.M., Bignell, D.E., Brownm, V.K., Didden, W.A.M., Folgarait, P.I., Fragoso, C., Freckman, D.W., Gupta, V.V.S.R, Hattori, T., Hawksworth, D., Klopatek, C., Lavelle, P., Malloch, D., Rusek, J., Soderstrom, B., Tiedje, J.M. and Virginia, R.A. (1997) Biodiversity and ecosystem functioning in soil. *Ambio* 26, 563–570.

Bunge, J., Fitzpatrick, M. (1993) Estimating the number of species – a review. *Journal of the American Statistical Association* 88, 364–371.

Bunt, J.S. (1954) The soil-inhabiting nematodes of Macquarie Island. *Australian Journal of Zoology* 2, 264–274.

Cadet, P., Pate, E. and Ndiaye-Faye, N. (2003a) Nematode community changes and survival rates under natural fallow in the Sudano-Sahelian area of Senegal. *Pedobiologia* 47, 149–160.

Cadet, P., Pate, E. and Thioulouse, J. (2003b) Relationship of nematode communities to human demographics and environment in agricultural fields and fallow lands in Senegal. *Journal of Tropical Ecology* 19, 279–290.

Cadet, P., Berry, S.D., Leslie, G.W. and Spaull, V.W. (2007) Management of nematodes and a stalk borer by increasing within-field sugarcane cultivar diversity. *Plant Pathology* 56, 526–535.

Carroll, J.J. and Viglierchio, D.R. (1981) On the transport of nematodes by the wind. *Journal of Nematology* 13, 476–482.

Chapin, F.S. III, Walker, B.H., Hobbs, R.J., Hooper, D.U., Lawton, J.H., Sala, O.E. and Tilman, D. (1997) Biotic control over the functioning of ecosystems. *Science* 277, 500–504.

Chevalier, R. and Webster, J.M. (2006) Temporary changes in populations of soil organisms after field application of entomopathogenic nematodes. *Journal of Nematology* 38, 466–473.

Clarke, K.R. and Warwick, R.M. (1998) A taxonomic distinctness index and its statistical properties. *Journal of Applied Ecology* 35, 523–531.

Clarke, K.R. and Warwick, R.M. (2001) Practical measures of marine biodiversity based on relatedness of species. *Oceanography and Marine Biology: an Annual Review* 39, 207–231.

Coleman, D.C. and Whitman, W.B. (2006) Linking species richness, biodiversity and ecosystem function in soil systems. *Pedobiologia* 49, 479–497.

Coleman, D.C., Odum, E.P. and Crossley, D.A. Jr (1992) Soil biology, soil ecology, and global change. *Biology and Fertility of Soils* 14, 104–111.

Collins, S.L. and Benning, T.L. (1996) Spatial and temporal patterns in functional diversity. In: Gaston, K.J. (ed.) *Biodiversity: a Biology of Numbers and Difference*. Blackwell, Oxford, pp. 253–280.

Colwell, R.K., Mao, C.X. and Chang, J. (2004) Interpolating, extrapolating, and comparing incidence-based species accumulation curves. *Ecology* 85, 2717–2727.

Courtright, E.M., Wall, D.H., Virginia, R.A., Frisse, L.M., Vida, J.T. and Thomas, W.K. (2000) Nuclear and mitochondrial DNA sequence diversity in the Antarctic nematode *Scottnema lindsayae*. *Journal of Nematology* 32, 143–153.

Covich, A.P., Palmer, M.A. and Crowl, T.A. (1999) The role of benthic invertebrate species in freshwater ecosystems – Zoobenthic species influence energy flows and nutrient cycling. *BioScience* 49, 119–127.

Cross, M.A., Collins, C., Campbell, N., Watts, P.C., Chubb, J.C., Cunningham, C.O., Hatfield, E.M.C. and MacKenzie, K. (2007) Levels of intra-host and temporal sequence variation in a large CO1 sub-units from *Anisakis simplex* sensu stricto (Rudolphi 1809) (Nematoda: Anisakisdae): implications for fisheries management. *Marine Biology* 151, 695–702.

Cutter, A.D. (2006) Nucleotide polymorphism and linkage disequilibrium in wild populations of the partial selfer *Caenorhabditis elegans*. *Genetics* 172, 171–184.

Cutter, A.D., Felix, M-A., Barriere, A. and Charlesworth, D. (2006) Patterns of nucleotide polymorphism distinguish temperate and tropical wild isolates of *Caenorhabditis briggsae*. *Genetics* 173, 2021–2031.

Danielopol, D.L. (1979) Sampling techniques for groundwater organisms. *Stygo News* 2, 5–7.

Das, V.M. (1964) *Hexatylus mulveyi* n sp and *Deladenus durus* (Cobb 1922) Thorne 1941 (Nematoda – Neotylenchidae) from the Canadian Arctic. *Canadian Journal of Zoology* 42, 649–653.

Darby, B.J., Neher, D.A. and Belnap, J. (2007) Soil nematode communities are ecologically more mature beneath late- than early-successional stage biological soil crusts. *Applied Soil Ecology* 35, 203–212.

David, R. (1998) An investigation of biodiversity patterns and processes in nematode populations of Loch Ness. PhD thesis, Southampton University, Southampton.

Davies, K.A. and Giblin-Davis, R.M. (2004) The biology and associations of *Fergusobia* (Nematoda) from the *Melaleuca leucadendra*-complex in eastern Australia. *Invertebrate Systematics* 18, 291–319.

De Deyn, G.B., Raaijmakers, C.E., Van Ruijven, J., Berendse, F. and Van der Putten, W.H. (2004) Plant species identity and diversity effects on different trophic levels of nematodes in the soil food web. *Oikos* 106, 576–586.

De Deyn, G.B., van Ruijven, J., Raaijmakers, C.E., de Ruiter, P.C. and Van der Putten, W.H. (2007) Above- and belowground insect herbivores differentially affect soil nematode communities in species-rich plant communities. Oikos 116, 923–930.

De Goede, R.G.M. and Bongers, T. (1994) Nematode community structure in relation to soil and vegetation characteristics. *Applied Soil Ecology* 1, 29–44.

De Goede, R.G.M., Georgieva, S.S., Verschoor, B.C. and Kamerman, J.W. (1993a) Changes in nematode community structure in a primary succession of blown-out areas in a drift sand landscape. *Fundamental and Applied Nematology* 16, 501–513.

De Goede, R.G.M., Verschoor, B.C. and Georgieva, S.S. (1993b) Nematode distribution, trophic structure and biomass in a primary succession of blown-out areas in a drift sand landscape. *Fundamental and Applied Nematology* 16, 525–538.

de Guiran, G. and Bruguier, N. (1989) Hybridization and phylogeny of the pine wood nematode (*Bursaphelenchus* spp.). *Nematologica* 35, 321–330.

De Ley I.T., Mundo-Ocampo, M., Yoder, M. and De Ley, P. (2007) Nematodes from vernal pools in the Santa Rosa Plateau Ecological Reserve, California I. *Hirschmanniella santarosae* sp. n. (Nematoda: Pratylenchidae), a cryptic sibling species of *H. pomponiensis* Abdel-Rahman and Maggenti, 1987. *Nematology* 9, 405–429.

De Mesel, I., Derycke, S., Swings, J., Vincx, M. and Moens, T. (2006) Role of nematodes in decomposition processes: Does within-trophic group diversity matter? *Marine Ecology Progress Series* 321, 157–166.

Decraemer, W. (1995) The family Trichodoridae: stubby root and virus vector nematodes. *Developments in Plant Pathology* 6, 1–360.

Derycke, S., Remerie, T., Vierstraete, A., Backeljau, T., Vanfleteren, J., Vincx, M. and Moens, T. (2005) Mitochondrial DNA variation and cryptic speciation within the free-living marine nematode *Pellioditis marina*. *Marine Ecology Progress Series* 300, 91–103.

Dillon, A.B., Rolston, A.N., Meade, C.V., Downes, M.J. and Griffin, C.T. (2008) Establishment, persistence, and introgression of entomopathogenic nematodes in a forest ecosystem. *Ecological Applications* 18, 735–747.

Dmowska, E. and Ilieva-Makulec, K. (2006) Secondary succession of nematodes in power plant ash dumps reclaimed by covering with turf. *European Journal of Soil Biology* 42 (Suppl 1), 164–170.

Dobzhansky, T. (1937) *Genetics and the Origin of Species*. Columbia University Press, New York.

Dropkin, V.H. (1988) The concept of race in phytonematology. *Annual Review of Phytopathology* 26, 145–161.

Eder, R. (1983) Interstitial nematodes from the Danube at Fischamend (Lower Austria). *Archives of Hydrobiology* 7, 100–113.

Eisenback, J.D. and Griffin, G.D. (1987) Interactions with other nematodes. In: Veech, J.A. and Dickson, D.W. (eds) *Vistas on Nematology*. Society of Nematologists, Hyattsville, Maryland, pp. 313–320.

Ekschmitt, K. (1998) Population assessments of soil fauna: general criteria for the planning of sampling schemes. *Applied Soil Ecology* 9, 439–445.

Elmiligy, I.A. and De Grisse, A. (1970) Effect of extraction technique and adding fixative to soil before storing on recovery of plant-parasitic nematodes. *Nematologica* 3, 353–358.

Ettema, C.H. (1998) Soil nematode diversity: species coexistence and ecosystem function. *Journal of Nematology* 30, 159–169.

Ettema, C. and Bongers, T. (1993) Characterization of nematode colonization and succession in disturbed soil using the maturity index. *Biology and Fertility of Soils* 16, 79–85.
Ettema, C.H. and Wardle, D.A. (2002) Spatial soil ecology. *Trends in Ecology and Evolution* 117, 177–183.
Evans, K. (1978) Rothamsted Experimental Station Annual Report 2, 27–45.
Eyualem, A., Traunspurger, W. and Michiels, I.C. (2006) Dynamics of freshwater nematodes: abundance, biomass and diversity. In: Eyualem, A., Andrassy, I. and Traunspurger, W. (eds) *Freshwater Nematodes: Ecology and Taxonomy*. CABI Publishing, Wallingford, pp. 77–93.
Faith, D.P. (1992) Conservation evaluation and phylogenetic diversity. *Biological Conservation* 61, 1–10.
Faulkner, L.R. and Bolander, W.J. (1967) Occurrence of large nematode populations in irrigation canals of South Central Washington. *Nematologica* 12, 591–600.
Faulkner, L.R. and Bolander, W.J. (1970a) Acquistition and distribution of nematodes in irrigation waterways of Columbia-basin in eastern Washington. *Journal of Nematology* 2, 362–367.
Faulkner, L.R. and Bolander, W.J. (1970b) Agriculturally-polluted irrigation water as a source of plant-parasitic nematode infestation. *Journal of Nematology* 2, 368–374.
Fenchel, T. (1978) Ecology of microbenthos and meiobenthos. *Annual Reviews in Ecology and Systematics* 9, 99–121.
Ferris, H., Lau, S.S. and Venette, R., (1995) Population energetics of bacterial-feeding nematodes – respiration and metabolic rates based on CO_2 production. *Soil Biology and Biochemistry* 27, 319–330.
Ferris, H., Eyre, M., Venette, R.C. and Lau, S.S. (1996) Population energetics of bacterial-feeding nematodes: Stage-specific development and fecundity rates. *Soil Biology and Biochemistry* 28, 271–280.
Ferris, H., Venette, R.C. and Lau, S.S. (1997) Population energetics of bacterial-feeding nematodes: carbon and nitrogen budgets. *Soil Biology and Biochemistry* 29, 1183–1194.
Ferris, H., Bongers, T. and de Goede, R.G.M. (2001) A framework for soil food web diagnostics: extension of the nematode faunal analysis concept. *Applied Soil Ecology* 18, 13–29.
Ferris, V.R. (1999) Species concepts do matter in nematology. *Journal of Nematology* 31, 93–94.
Ferris, V.R., Goseco, C.G. and Ferris, J.M. (1976) Biogeography of free-living soil nematodes from perspective of plate tectonics. *Science* 193, 508–510.
Fiscus, D.A. and Neher, D.A. (2002) Distinguishing sensitivity of free-living soil nematode genera to physical and chemical disturbances. *Ecological Applications* 12, 565–575.
Floyd, R., Abebe, E., Papert, A. and Blaxter, M. (2002) Molecular barcodes for soil nematode identification. *Molecular Ecology* 11, 839–850.
Fonseca, G. and Soltwedel, T. (2007) Deep-sea meiobenthic communities underneath the marginal ice zone off Eastern Greenland. *Polar Biology* 30, 607–618.
Fortuner, R. and Luc, M. (1987) A reappraisal of Tylenchina (Nemata). 6. The family Belonolaimidae Whitehead, 1960. *Revue de Nematologie* 10, 183–202.
Franklin, M.T., Clark, S.A. and Course, J.A. (1971) Population changes and development of *Meloidogyne naasi* in the field. *Nematologica* 17, 575–590.
Freckman, D.W. (1988) Bacterivorous nematodes and organic-matter decomposition. *Agriculture, Ecosystems and Environment* 24, 195–217.
Freckman, D.W. and Caswell, E.P. (1985) The ecology of nematodes in agroecosystems. *Annual Review of Phytopathology* 23, 275–296.
Freckman, D.W. and Huang, S.P. (1998) Response of the soil nematode community in a shortgrass steppe to long-term and short-term grazing. *Applied Soil Ecology* 9, 39–44.
Freckman, D.W. and Mankau, R. (1986) Abundance, distribution, biomass and energetics of soil nematodes in a northern Mojave Desert ecosystem. *Pedobiologia* 29, 129–142.
Freckman, D.W. and Virginia, R.A. (1989) Plant-feeding nematodes in deep-rooting desert ecosystems. *Ecology* 70, 1665–1678.

Freckman, D.W. and Virginia, R.A. (1997) Low-diversity Antarctic soil nematode communities: distribution and response to disturbance. *Ecology* 78, 363–369.

Freckman, D.W., Kaplan, D.T. and Van Gundy, S.D. (1977) A comparison of techniques for extraction and study of anhydrobiotic nematodes from dry soils. *Journal of Nematology* 9, 176–181.

Gafurov, A.K. (1997) *Mermithids (systematics, biology, phylogeny, practical significance)*. Donish, Dushanbe, Tajikistan.

Gallucci, F., Fonseca, G. and Soltwedel, T. (2008) Effects of megafauna exclusion on nematode assemblages at a deep-sea site. *Deep-sea Research part I – Oceanography Research Papers* 55, 332–349.

Gaston, K.J. (1996) What is biodiversity? In: Gaston, K.J. (ed.) *Biodiversity: a Biology of Numbers and Difference*. Blackwell, Oxford, pp. 1–9.

Gentzsch, D. (1990) Races of *Ditylenchus dipsaci* including practical hints. *Gesunde Pflanzen* 42, 432–435.

Geraert, E. (1965) The genus *Paratylenchus*. *Nematologica* 11, 301–334.

Gerlach, S.A. (1977a) Means of meiofauna dispersal. *Mikrofauna des Meeresbodens* 61, 89–103.

Gerlach, S.A. (1977b) Attraction to decaying organisms as a possible cause for patchy distribution of nematodes in a Bermuda Beach. *Ophelia* 16, 151–165.

Gerlach, S.A. and Riemann, F. (1973) The Bremerhaven checklist of aquatic nematodes. *Veroeffentlichen des Instituts für Meeresforscchung Bremerhaven*, Suppl 4, 1–736.

Giblin-Davis, R.M., Davies, K.A., Taylor, G.S. and Thomas, W.K. (2004) Entomophilic nematode models for studying biodiversity and cospeciation. In: Chen, Z.X., Chen, S.Y. and Dickson, D.W. (eds) *Nematology: Advances and Perspectives, vol 1: Nematode Morphology, Physiology and Ecology*. CABI Publishing, Wallingford, pp. 493–540.

Giere, O. (1993) *Meiobenthology: The Microscopic Fauna in Aquatic Sediments*. Springer-Verlag, Berlin.

Gobin, J.F. and Warwick, R.M. (2006) Geographical variation in species diversity: a comparison of marine polychaetes and nematodes. *Journal of Experimental Marine Biology and Ecology* 330, 234–244.

Goodell, P. and Ferris, H. (1980) Plant-parasitic nematode distributions in an alfalfa field. *Journal of Nematology* 12, 136–141.

Goodey, J.B. (1963) *Soil and Freshwater Nematodes*. Methuen, London.

Gotelli, N.J. and Colwell, R.K. (2001) Quantifying biodiversity: procedures and pitfalls in the measurement and comparison of species richness. *Ecology Letters* 4, 379–391.

Grewal, P.S. and Wright, D. (1992) Migration of *Caenorhabditis elegans* (Nematoda Rhabditidae) larvae towards bacteria and the nature of the bacterial stimulus. *Fundamental and Applied Nematology* 15, 159–166.

Griffiths, B. (1990) Approaches to measuring the contribution of nematodes and protozoa to nitrogen mineralization in the rhizosphere. *Soil Use and Management* 6, 88–90.

Griffiths, B.S., Ritz, K., and Wheatley, R.E. (1994) Nematodes as indicators of enhanced microbiological activity in a Scottish organic farming system. *Soil Use and Management* 10, 20–24.

Griffiths, B.S., Donn, S., Neilson, R. and Daniell, T.J. (2006) Molecular sequencing and morphological analysis of a nematode community. *Applied Soil Ecology* 32, 325–337.

Grillo, V., Jackson, F., Cabaret, J. and Gilleard, J.S. (2007) Population genetic analysis of the ovine parasitic nematode *Teladorsagia circumcincta* and evidence for a cryptic species. *International Journal for Parasitology* 37, 435–447.

Groffman, P.M. and Bohlen, P.J. (1999) Soil and sediment biodiversity. *BioScience* 49, 139–148.

Hanel, L. (1995) Secondary successional stages of soil nematodes in cambisols of south Bohemia. *Nematologica* 41, 197–218.

Hanel, L. (2003) Recovery of soil nematode populations from cropping stress by natural secondary succession to meadow land. *Applied Soil Ecology* 22, 255–270.

Heip, C. (1974) New index measuring evenness. *Journal of the Marine Biological Association of the UK* 54, 555–557.

Hey, J. (2001) The mind of the species problem. *Trends in Ecology and Evolution* 16, 326–329.

Hillebrand, H. (2003) Opposing effects of grazing and nutrients on diversity. *Oikos* 100, 592–600.

Hodda, M. (1981) Behavioural adaptions of nematodes in an exposed sandy beach. Honours thesis, Australian National University, Canberra.

Hodda, M. (1986) Sandworms. *Australian Natural History* 21, 520–523.

Hodda, M. (1990) Variation in estuarine littoral nematodes over 3 spatial scales. *Estuarine and Coastal Shelf Science* 30, 325–340.

Hodda, M. (2003a) *Checklist of Aphelenchida*, CSIRO, Canberra, 138pp. Available from: http://www.ento.csiro.au/science/nematodes/checklist_dec2003.rtf (accessed 31 October 2007).

Hodda, M. (2003b) *Nematodes of Australia: Free-living, plant- and insect-associated*. CSIRO, Canberra. Available from: http://ento.csiro.au/science/nematodes (accessed 31 October 2007).

Hodda, M. (2007) Phylum nematoda. *Zootaxa* 1668, 265–293.

Hodda, M. and Eyualem, A. (2006) Methods for processing freshwater nematodes. In: Eyualem, A., Andrassy, I. and Traunspurger, W. (eds) *Freshwater Nematodes: Ecology and Taxonomy*. CAB International, Wallingford, UK, pp. 31–45.

Hodda, M. and Nicholas, W.L. (1985) Meiofauna associated with mangroves in the Hunter-River-estuary and Fullerton Cove, southeastern Australia. *Australian Journal of Marine and Freshwater Resources* 36, 41–50.

Hodda, M. and Nicholas, W.L. (1987) Free-living nematodes from Darwin mangroves. *Beagle* 4, 7–10.

Hodda, M. and Wanless, F.R. (1994) Nematodes from an English chalk grassland: species distributions. *Nematologica* 40, 116–32.

Hodda, M., Bloemers, G.F., Lawton, J.H. and Lambshead, P.J.D. (1997) Effects of clearing and subsequent land-use on diversity of free-living soil nematodes in tropical forest. *Oecologia* 111, 575–582.

Hodda, M., Stewart, E., FitzGibbon, F., Reid, I., Longstaff, B.C. and Packer, I. (1999) *Nematodes – Useful Indicators of Soil Conditions*. RIRDC Publication No 98/141, Canberra.

Hodda, M., Ocana, A. and Traunspurger, W. (2006) Nematodes from extreme freshwater habitats. In: Eyualem, A., Andrassy, I. and Traunspurger, W. (eds) *Freshwater Nematodes: Ecology and Taxonomy*. CAB International, Wallingford, UK, pp. 179–210.

Holovachov, O. and De Ley, P. (2006a) Order Plectida. In: Eyualem, A., Traunspurger, W. and Andrassy, I. (eds) *Freshwater Nematodes: Taxonomy and Ecology*. CAB International, Wallingford, UK, pp. 611–647.

Holovachov, O. and De Ley, P. (2006b) Order Mermithida and families incertae sedis (Bastianidae and Rhabdolaimidae). In: Eyualem, A., Traunspurger, W. and Andrassy, I. (eds) *Freshwater Nematodes: Taxonomy and Ecology*. CABI Publishing, Wallingford, pp. 722–735.

Hooper, D.J. (1986) Extraction of free-living stages from soil. In: Southey, J.F. (ed.) *Laboratory Methods for Work with Plant and Soil Nematodes*. HMSO, Norwich, pp. 5–30.

Hourston, M., Warwick, R.M., Valesini, F.J. and Potter, I.C. (2005) To what extent are the characteristics of nematode assemblages in nearshore sediments on the west Australian coast related to habitat type, season and zone? *Estuarine and Coastal Shelf Science* 64, 601–612.

Huang, Y., Zhang, Z.N. and Liu, X.S. (2007) The marine nematodes community in winter southern Yellow Sea. *Oceanologia et Limnologia Sinica* 38, 199–205.

Hugot, J.P., Baujard, P. and Morand, S. (2001) Biodiversity in helminths and nematodes as a field of study: an overview. *Nematology*, 3, 199–208.

Hunt, D.J. (1993) *Aphelenchida, Longidoridae and Trichodoridae: their Systematics and Bionomics*. CAB International, Wallingford, UK.

Hyvonen, R. and Persson, T. (1996) Effects of fungivorous and predatory arthropods on nematodes and tardigrades in microcosms with coniferous forest soil. *Biology and Fertility of Soils* 21, 121–127.

Ilieva-Makulec, K., Olejniczak, I. and Szanser, M. (2006) Response of soil micro- and mesofauna to diversity and quality of plant litter. *European Journal of Soil Biology* 42 (Suppl 1), 244–249.

Ingham, R.E., Trofymow, J.A., Ingham, E.R. and Coleman, D.C. (1985) Interactions of bacteria, fungi, and their nematode grazers: Effects on nutrient cycling and plant growth. *Ecological Monographs* 55, 119–140.

Jagdale, G.B., Saeb, A.T.M., Somasekhar, N. and Grewal, P.S. (2006) Genetic variation and relationships between isolates and species of the entomopathogenic nematode genus *Heterorhabditis* deciphered through isozyme profiles. *Journal of Parasitology* 92, 509–516.

Jairajpuri, M.S. and Ahmad, W. (1992) *Dorylaimida: Free-living, Predaceous and Plant-parasitic Nematodes*. E.J. Brill, Oxford.

Jairajpuri, M.S. and Khan, W.U. (1982) *Predatory Nematodes (Mononchida)*. Associated Publishing, New Delhi.

Janiec, K. (1996) Short distance wind transport of microfauna in Maritime Antarctic (King George Island, South Shetland Islands). *Polish Polar Research* 17, 203–211.

Janssen, G.J.W. (1994) The relevance of races in *Ditylenchus dipsaci* (Kuhn) Filipjev, the stem nematode. *Fundamental and Applied Nematology* 17, 469–473.

Jiang, D., Li, Q., Liu, F., Jiang, Y. and Liang, W. (2007) Vertical distribution of soil nematodes in an age sequence of *Caragana microphylla* plantations in the Horqin Sandy Land, Northeast China. *Ecological Research* 22, 49–56.

Jiang, K., Yu, M., Ding, P., Xu, X., Jiang, P., Zhou, C. and Lu, G. (2005) Avian community response to vegetation succession caused by the pine wood nematode in Zhejiang, China. *Shengwu Duoyangxing* 13, 496–506.

Johnson, S.R., Ferris, V.R. and Ferris, J.M. (1972) Nematode community structure of forest woodlots. I. Relationships based on similarity coefficients of nematode species. *Journal of Nematology* 4, 175–183.

Jones, K.L., Todd, T.C., Wall-Beam, J.L., Coolon, J.D., Blair, J.M. and Herman, M.A. (2006) Molecular approach for assessing responses of microbial-feeding nematodes to burning and chronic nitrogen enrichment in a native grassland. *Molecular Ecology* 15, 2601–2609.

Kaestner, A. (1965) *Lehrbuch der Speziellen Zoologie. vol. I. Wirbellose. I. Protozoa, Mesozoa, Parazoa, Coelenterata*, Prostomia, Jena.

Kaffe-Abramovich, T. and Steinberger, Y. (2006) Soil microbial functional diversity response following nematocide and biocide amendments in a desert ecosystem. *Soil Biology and Biochemistry* 38, 1966–1976.

Kardol, P., Bezemer, T.M., Van der Wal, A. and Van der Putten, W.H. (2005) Successional trajectories of soil nematode and plant communities in a chronosequence of ex-arable lands. *Biological Conservation* 126, 317–327.

Keith, A.M., Van der Wal, R., Brooker, R.W., Osler, G.H.R., Chapman, S.J. and Burslem, D.F.R.P. (2006) Birch invasion of heather moorland increases nematode diversity and trophic complexity. *Soil Biology and Biochemistry* 38, 3421–3430.

Koen, H. (1966) The influence of seasonal variations on the vertical distribution of *Meloidogyne javanica* in sandy soils. *Nematologica* 12, 297–301.

Korenko, V. and Schmidt, C. (2006) Effects of agricultural practices in the rice crop system on nematode communities in Uruguay. *Nematologia Mediterranea* 34, 151–159.

Korthals, G.W., Alexiev, A.D., Lexmond, T.M., Kammenga, J.E. and Bongers, T. (1996a) Long-term effects of copper and pH on the nematode community in an agroecosystem. *Environmental Toxicology and Chemistry* 15, 979–985.

Korthals, G.W., Ende, A.V.D., Megen, H.V., Lexmond, T.M., Kammenga, J.E. and Bongers, T. (1996b) Short-term effects of cadmium, copper, nickel and zinc on soil nematodes from different feeding and life-history strategy groups. *Applied Soil Ecology* 4, 107–117.

Krebs, C. (1989) *Ecological Methodology*. Harper and Row, New York.
Kuiper, K. and Loof, P.A.A. (1964) Observations on some *Paratrichodorus* species. *Nematologica* 10, 77.
Lambshead, P.J.D. (1986) Sub-catastrophic sewage and industrial waste contamination as revealed by marine nematode faunal analysis. *Marine Ecology Progress Series* 29, 247–260.
Lambshead, P.J.D. (1993) Recent developments in marine benthic biodiversity research. *Oceanis* 19, 5–24.
Lambshead, P.J.D. (2004) Marine nematode biodiversity. In: Chen, Z.X., Chen, S.Y. and Dickson, D.W. (eds) *Nematology Advances and Perspectives*. CABI, Wallingford. pp. 438–468.
Lambshead, P.J.D. and Hodda, M. (1994) The impact of disturbance on measurements of variability in marine nematode populations. *Viet Milieu* 44, 21–27.
Lambshead, P.J.D., Platt, H.M. and Shaw, K.M. (1983) The detection of differences among assemblages of marine benthic species based on an assessment of dominance and diversity. *Journal of Natural History* 17, 859–874.
Lambshead, P.J.D., Tietjen, J., Ferrero, T. and Jensen, P. (2000) Latitudinal diversity gradients in the deep sea with special reference to North Atlantic nematodes. *Marine Ecology Progress Series* 194, 159–167.
Lambshead, P.J.D., Brown, C.J., Ferrero, T.J., Mitchell, N.J., Smith, C.R., Hawkins, L.E. and Tietjen, J. (2002) Latitudinal diversity patterns of deep-sea marine nematodes and organic fluxes: a test from the central equatorial Pacific. *Marine Ecology Progress Series* 236, 129–135.
Lamont, B.B. (1995) Testing the effect of ecosystem composition structure on its functioning. *Oikos* 74, 283–295.
Lavelle, P., Decaens, T., Aubert, M., Barot, S., Blouin, M., Bureau, F., Margerie, P., Mora, P. and Rossi, J.P. (2006) Soil invertebrates and ecosystem services. *European Journal of Soil Biology* 42 (Suppl), 3–15.
Lawton, J.H. (1994) What do species do in ecosystems? *Oikos* 71, 367–374.
Lawton, J.H., Bignell, D.E., Bloemers, G.F., Eggleton, P. and Hodda, M. (1996) Carbon flux and diversity of nematodes and termites in Cameroon forest soils. *Biodiversity and Conservation* 5, 261–276.
Lawton, J.H., Bignell, D.E., Bolton, B., Bloemers, G.F., Eggleton, P., Hammond, P.M., Hodda, M., Holt, R.D., Larsen, T.B., Mawdsley, N.A., Stork, N.E., Srivastava, D.S. and Watt, A.D. (1998) Biodiversity inventories, indicator taxa and effects of habitat modification in tropical forest. *Nature* 391, 72–76.
Li, Q., Jiang, Y. and Liang, W. (2006) Effect of heavy metals on soil nematode communities in the vicinity of a metallurgical factory. *Journal of Environmental Science* (China) 18, 323–328.
Li, Y., Feng, J., Chen, J. and Wu, J. (2007) Original vegetation type affects soil nematode communities. *Applied Soil Ecology* 35, 68–78.
Liang, W., Li, Q., Jiang, Y. and Neher, D.A. (2005) Nematode faunal analysis in an aquic brown soil fertilised with slow-release urea, northeast China. *Applied Soil Ecology* 29, 185–192.
Loof, P.A.A. (1971) Free living and plant parasitic nematodes from Spitzbergen collected by H Van Rossen. *Mededelingen Landbouwhogeschool, Wageningen* 71, 1–86.
Luc, M. (1987) A reappraisal of Tylenchina (Nemata). 7. The family Pratylenchidae Thorne, 1949. *Revue de Nematologie* 10, 203–218.
Luc, M. and Fortuner, R. (1987) A reappraisal of Tylenchina (Nemata). 5. The family Dolichodoridae Chitwood, 1950. *Revue de Nematologie* 10, 177–181.
Luc, M., Maggenti, A.R., Fortuner, R., Raski, D.J. and Geraert, E. (1987) A reappraisal of Tylenchina (Nemata) 1. For a new approach to the taxonomy of Tylenchina. *Revue de Nematologie* 10, 127–134.
Luc, M., Maggenti, A.R. and Fortuner, R. (1988) A reappraisal of Tylenchina (Nemata). 9. The family Heteroderidae Filip'ev and Schuurmans Stekhoven, 1941. *Revue de Nematologie* 11, 159–176.

Maas, P.W.T. and Brinkman, H. (1980) Sampling of soil for nematode vectors of plant viruses in the Netherlands. *Mededelingen van de Faculteit Landbouwwetenschappen Gent* 45, 769–773.

Maggenti, A.R., Luc, M., Raski, D.J., Fortuner, R. and Geraert, E. (1987) A reappraisal of Tylenchina (Nemata). 2. Classification of the suborder Tylenchina (Nemata: Diplogasteria). *Revue de Nematologie* 10, 135–142.

Maggenti, A.R., Luc, M., Raski, D.J., Fortuner, R. and Geraert, E. (1988) A reappraisal of Tylenchina (Nemata). 11. List of generic and supra-generic taxa, with their junior synonyms. *Revue de Nematologie* 11, 177–188.

Mare, M.F. (1942) A study of a marine benthic community with special reference to the micro-organisms. *Journal of the Marine Biological Association* 25, 517–554.

Martinez, D. (1996) Defining and measuring functional aspects of biodiversity. In: Gaston, K.J. (ed.) *Biodiversity: a Biology of Numbers and Difference*. Blackwell, Oxford, pp 114–148.

Maslen, N.R. and Convey, P. (2006) Nematode diversity and distribution in the southern maritime Antarctic – clues to history? *Soil Biology and Biochemistry* 38, 3141–3151.

May, R.M. (1988) How many species are there on earth? *Science* 241, 1441.

Mayr, E. (1942) *Systematics and the Origin of Species*. Columbia University Press, New York.

McSorley, R. and Frederick, J.J. (1996) Nematode community structure in rows and between rows of a soybean field. *Fundamental and Applied Nematology* 19, 251–261.

McSorley, R. and Tanner, G.W. (2007) Effects of cattle stocking rates on nematode communities in south Florida. *Rangeland Ecology and Management* 60, 31–35.

Meyers, S.P. and Hopper, B.E. (1966) Attraction of the marine nematode *Metoncholaimus* sp. to fungal substrates. *Bulletin of Marine Science* 16, 142–150.

Michiels, I.C. and Traunspurger, W. (2004a) Impact of resource availability on species composition and diversity in freshwater nematodes. *Oecologia* 142, 98–103.

Michiels, I.C. and Traunspurger, W. (2004b) A three year study of seasonal dynamics of a zoobenthos community in a eutrophic lake. *Nematology* 6, 655–669.

Michiels, I.C. and Traunspurger, W. (2005) Benthic community patterns and the composition of feeding types and reproductive modes in freshwater nematodes. *Nematology* 7, 21–36.

Michiels, I.C., Matzak, S. and Traunspurger, W. (2004) Maintenance of biodiversity through predation in freshwater nematodes? *Nematology Monographs and Perspectives* 2, 723–737.

Mikola, J. and Sulkava, P. (2001) Responses of microbial-feeding nematodes to organic matter distribution and predation in experimental soil habitat. *Soil Biology and Biochemistry* 33, 811–817.

Miljutin, D.M. (2004) New findings of deep-sea nematodes of genus *Benthimermis* Petter, 1980 (Nematoda, Benthimermithidae) with description of seven new species. *Zoosystematics* 26, 21–48.

Mullin, P.G., Harris, T.S. and Powers, T.O. (2003) Systematic status of *Campydora* Cobb, 1920 (Nematoda: Campydorina). *Nematology* 5, 699–711.

Mullin, P.G., Harris, T.S., Higgins, R.S. and Powers, T.O. (2004) An 18S DNA barcode applied to nematodes from the Konza Tallgrass Prairie. *Journal of Nematology* 36, 336.

Mulvey, R.H. (1963) Some soil inhabiting, fresh-water, and plant-parasitic nematodes from the Canadian Arctic and Alaska. *Arctic* 16, 203–204.

Mulvey, R.H. (1969a) Soil-inhabiting nematodes of orders Araeolaimida, Chromadorida, Enoplida and Monhysterida from Canadian high Arctic. *Canadian Journal of Zoology* 47, 365–382.

Mulvey, R.H. (1969b) Nematodes of genus *Tylenchorhynchus* (Tylenchoidea – Nematoda) from Canadian high Arctic. *Canadian Journal of Zoology* 47, 1245–1248.

Mulvey, R.H. (1969c) Nematodes of family Neotylenchidae (Tylenchida – Nematoda) from Canadian high Arctic. *Canadian Journal of Zoology* 47, 1261–1268.

Murray, P.J., Cook, R., Currie, A.F., Dawson, L.A., Gange, A.C., Grayston, S.J. and Treonis, A.M. (2006) Interactions between fertilizer addition, plants and the soil environment: Implications for soil faunal structure and diversity. *Applied Soil Ecology* 33, 199–207.

Nadler, S.A., De Ley, P., Mundo-Ocampo, M., Smythe, A.B., Stock, S.P., Bumbarger, D., Adams, B.J., De Ley, I.T., Holovachov, O. and Baldwin, J.G. (2006) Phylogeny of Cephalobina (Nematoda): Molecular evidence for recurrent evolution of probolae and incongruence with traditional classifications. *Molecular Phylogenetics and Evolution* 40, 696–711.

Nahar, M.S., Grewal, P.S., Miller, S.A., Stinner, D., Stinner, B.R., Kleinhenz, M.D., Wszelaki, A. and Doohan, D. (2006) Differential effects of raw and composted manure on nematode community, and its indicative value for soil microbial, physical and chemical properties. *Applied Soil Ecology* 34, 140–151.

Neher, D.A. and Darby, B.J. (2006) Computation and application of nematode community indices: general guidelines. In: Eyualem, A., Traunspurger, W. and Andrassy, I. (eds) *Freshwater Nematodes: Taxonomy and Ecology*. CAB International, Wallingford, UK, pp. 211–222.

Neher, D.A., Peck, S.L., Rawlings, J.O. and Campbell, C.L. (1995) Measures of nematode community structure and sources of variability among and within agricultural fields. *Plant and Soil* 170, 167–181.

Neumann, G. and Shields, E.J. (2006) Interspecific interactions among three entomopathogenic nematodes, *Steinernema carpocapsae* Weiser, *S. feltiae* Filipjev, and *Heterorhabditis bacteriophora* Poinar, with different foraging strategies for hosts in multipiece sand columns. *Environmental Entomology* 35, 1578–1583.

Newsham, K.K., Rolf, J., Pearce, D.A. and Strachan, R.J. (2004) Differing preferences of Antarctic soil nematodes for microbial prey. *European Journal of Soil Biology* 40, 1–8.

Nicholas, W.L. (1984) *Biology of Free-living Nematodes*. 2nd edn. Clarendon Press, Oxford.

Nicholas, W.L. (2001) Seasonal variations in nematode assemblages on an Australian temperate ocean beach; the effect of heavy seas and unusually high tides. *Hydrobiologia* 464, 17–26.

Nicholas, W.L. and Hodda, M. (1999) Free-living nematodes of a temperate, high-energy sandy beach: faunal composition and variation over space and time. *Hydrobiologia* 394, 113–127.

Nicholas, W.L. and Trueman, J.W.H. (2005) Biodiversity of marine nematodes in Australian sandy beaches from tropical and temperate regions. *Biodiversity and Conservation* 14, 823–839.

Norton, D.C. (1979) Relationship of physical and chemical factors to populations of plant-parasitic nematodes. *Annual Review of Phytopathology* 17, 279–299.

Norton, D.C. and Niblack, T.L. (1991) Biology and ecology of nematodes. In: Nickle, W.R. (ed.) *Manual of Agricultural Nematology*. Marcel Dekker, New York, pp. 47–72.

Norton, D.C. and Oard, M. (1981) Plant-parasitic nematodes in loess toposequences planted with corn. *Journal of Nematology* 13, 314–321.

Norton, D.C. and Ponchilla, P.E. (1968) Stylet-bearing nematodes associated with plants in Iowa prairie. *Proceedings of the Iowa Academy of Sciences* 75, 32–35.

Norton, D.C. and Schmitt, D.P. (1978) Community analyses of plant-parasitic nematodes in the Kalsow Prairie, Iowa. *Journal of Nematology* 10, 171–176.

Odihirin, R.A. (1977) Irrigation water as a means for the dissemination of the sugar cane cyst nematode *Heterodera sacchari* at Bacita Sugar Estate. *Occasional Publications of the Nigerian Society of Plant Protection* 2, 58.

Okada, H. and Harada, H. (2007) Effects of tillage and fertilizer on nematode communities in a Japanese soybean field. *Applied Soil Ecology* 35, 582–598.

Olff, H. and Ritchie, M.E. (1998) Effects of herbivores on grassland plant diversity. *Trends in Ecology and Evolution* 13, 261–265.

Oliveira, C.M.G., Ferraz, L.C.C.B. and Neilson, R. (2006) *Xiphinema krugi*, species complex or complex of cryptic species? *Journal of Nematology* 38, 418–428.

Orr, C.C. and Dickerson, O.J. (1966) Nematodes in true prairie soils of Kansas. *Kansas Academy of Sciences Transcripts* 69, 317–334.

Orr, C.C. and Newton, O.H. (1971) Distribution of nematodes by wind. *Plant Disease* 55, 61–63.

Overgaard Nielsen, C. (1949) Studies on the soil microfauna. II. The soil inhabiting nematodes. *Natura Jutlandica* 2, 1–131.

Palmer, M., Covich, A.P., Finlay, B.J., Gibert, J., Hyde, K.D., Johnson, R.K., Kairesalo, T., Lake, S., Lovell, C.R., Naiman, R.J., Ricci, C., Sabater, F. and Strayer, D. (1997) Biodiversity and ecosystem processes in freshwater sediments. *Ambio* 26, 571–577.

Pate, E., Ndiaye-Faye, N., Thioulouse, J., Villenave, C., Bongers, T., Cadet, P. and Debouzie, D. (2000) Successional trends in the characteristics of soil nematode communities in cropped and fallow lands in Senegal (Sonkorong). *Applied Soil Ecology* 14, 5–15.

Pavao-Zuckerman, M.A. and Coleman, D.C. (2007) Urbanization alters the functional composition, but not taxonomic diversity, of the soil nematode community. *Applied Soil Ecology* 35, 329–339.

Pen-Mouratov, S. and Steinberger, Y. (2005) Spatio-temporal dynamic heterogeneity of nematode abundance in a desert ecosystem. *Journal of Nematology* 37, 26–36.

Peters, L. and Traunspurger, W. (2005) Species distribution of free-living nematodes and other meiofauna in littoral periphyton communities of lakes. *Nematology* 7, 267–280.

Peters, L., Traunspurger, W., Wetzel, M.A. and Rothhaupt, K.O. (2005) Community development of free-living aquatic nematodes in littoral periphyton communities. *Nematology* 7, 901–916.

Peters, L., Wetzel, M.A., Traunspurger, W. and Rothhaupt, K.O. (2007) Epilithic communities in a lake littoral zone: the role of water-column transport and habitat development for dispersal and colonization of meiofauna. *Journal of the North American Benthological Society* 26, 232–243.

Phillips, P.C. (2006) One perfect worm. *Trends in Genetics* 22, 405–407.

Pianka, E.R. (1970) On r and K-selection. *American Naturalist* 104, 592–597.

Pianka, E.R. (1972) r and K selection or b and d selection? *American Naturalist* 106, 581–588.

Pielou, E.C. (1975) *Ecological Diversity*. Wiley, New York.

Pigliucci, M. (2003) Species as family resemblance concepts: The (dis-)solution of the species problem? *Bioessays* 25, 596–602.

Pitcher, R.S. (1967) The host–parasite relations and ecology of *Trichodorus viruliferus* on apple roots, as observed from an underground laboratory. *Nematologica* 13, 547–557.

Platt, H.M. and Warwick, R.M. (1983) Free-living marine nematodes. Part 1. British Enoplids. Pictorial key to world genera and notes for the identification of British species. *Synopses of the British Fauna New Series* No. 28, 1–307.

Platt, H.M. and Warwick, R.M. (1988) Free-living marine nematodes. Part 2. British Chromadorids. Pictorial key to world genera and notes for the identification of British species. *Synopses of the British Fauna New Series* 38, 1–502.

Poinar, G.O. Jr. (1983) *The Natural History of Nematodes*. Prentice-Hall, Englewood Cliffs, New Jersey.

Popovici, I. (1989) Soil nematode communities in the Carpathian beech forests of Romania. *Studia Universitatis Babes-Bolyai Biologia* 34, 38–44.

Powers, T.O., Todd, T.C., Burnell, A.M., Murray, P.C.B., Fleming, C.C., Szalanski, A.L., Adams, B.A. and Harris, T.S. (1997) The rDNA internal transcribed spacer region as a taxonomic marker for nematodes. *Journal of Nematology* 29, 441–450.

Price, N.S. and Siddiqi, M.R. (1994) Rainforest nematodes with particular reference to the Korup National Park, Cameroon. *Afro-Asian Journal of Nematology* 4, 117–128.

Procter, D.L.C. (1990) Global overview of the functional roles of soil-living nematodes in terrestrial communities and ecosystems. *Journal of Nematology* 22, 1–7.

Putten, W.H. Van der, Cook, R., Costa, S., Davies, K.G., Fargette, M., Freitas, H., Hol, W.H.G., Kerry, B.R., Maher, N. and Mateille, T. (2006) Nematode interactions in nature: Models for sustainable control of nematode pests of crop plants? *Advances in Agronomy*, 89, 227–260.

Raes, M. and Vanreusel, A. (2005) The metazoan meiofauna associated with a cold-water coral degradation zone in the Porcupine Seabight (NE Atlantic). In: Freiwald, A. and Roberts, J.M. (eds) *Cold-water Corals and Ecosystems*. Springer, Berlin. pp. 821–847.

Raes, M. and Vanreusel, A. (2006) Microhabitat type determines the composition of nematode communities associated with sediment-clogged cold water coral framework in the Porcupine Seabight (NE Atlantic). *Deep-sea Research Part 1, Oceanographic Research Papers* 53, 1880–1894.

Rajaniemi, T.K. (2003) Explaining productivity–diversity relationships in plants. *Oikos* 101, 449–457.

Rantalainen, M-L., Fritze, H., Haimi, J., Pennanen, T. and Setaelae, H. (2005) Species richness and food web structure of soil decomposer community as affected by the size of habitat fragment and habitat corridors. *Global Change Biology* 11, 1614–1627.

Raski, D.J. and Luc, M. (1987) A reappraisal of Tylenchina (Nemata) 10. The superfamily Criconematoidea Taylor, 1936. *Revue de Nematologie* 10, 409–444.

Renyi, A. (1961) On measures of entropy and information. In: Neyman, J. (ed.) *Proceedings 4th Berkeley Symposium on Mmathematical Statistics and Probability*. University of California Press, Berkeley, pp. 547–561.

Rice, A.L. and Lambshead, P.J.D. (1994) Patch dynamics in the deep-sea benthos: the role of a heterogeneous supply of organic matter. In: Giller, P.S., Hildrew, A.G.. and Raffaelli, D.G. (eds) *Aquatic Ecology: Scale, Pattern and Process*. Blackwell, Oxford. pp. 469–497.

Ricotta, C. (2000) From theoretical ecology to statistical physics and back: self similar landscape metrics as a synthesis of ecological diversity and geometrical complexity. *Ecological Modelling* 125, 245–253.

Riemann, F. (1966) Die interstitielle Fauna im Elbe-Aestuar Verbreitung und Systematic. *Archives of Hydrobiology* Supplement 31, 1–279.

Roccuzzo, G. and Ciancio, A. (1991) Notes on nematodes found in irrigation water in southern Italy. *Nematologia Mediterranea* 19, 105–108.

Ruess, L. (1995) Nematode fauna in spruce forest soils – a qualitative/quantitative comparison. *Nematologica* 41, 106–124.

Ruess, L. and Dighton, J. (1996) Cultural studies on soil nematodes and their fungal hosts. *Nematologica* 42, 330–346.

Ruess, L., Sandbach, P., Cudlin, P., Dighton, J., Crossley, A. (1996) Acid deposition in a spruce forest soil: effects on nematodes, mycorrhizas and fungal biomass. *Pedobiologia* 40, 51–66.

Ruess, L., Michelsen, A., Schmidt, I.K., Jonasson, S. and Dighton, J. (1998) Soil nematode fauna of a subarctic heath: potential nematicidal action of plant leaf extracts. *Applied Soil Ecology* 7, 111–124.

Saly, A. (1985) Production of free living nematodes in the protected landscape area of the Slovak Paradise. *Ekologia* (CSSR) 4, 185–209.

Sanchez-Moreno, S., Camargo, J.A. and Navas, A. (2006) Ecotoxicological assessment of the impact of residual heavy metals on soil nematodes in the Guadiamar River basin (southern Spain). *Environmental Monitoring and Assessment* 116, 245–262.

Schiemer, F. (1983) Comparative aspects of food dependence and energetics of free-living nematodes. *Oikos* 41, 32–42.

Schmidt, D.P. and Norton, D.C. (1972) Relationships of plant parasitic nematodes to sites in native Iowa prairies. *Journal of Nematology* 4, 200–205.

Schneider, W. (1922) Freilebende Susswassernematoden aus ostholsteinischen Seen. *Archiv für Hydrobiologie* 13, 696–753.

Setala, H., Laakso, J., Mikola, J. and Huhta, V. (1998) Functional diversity of decomposer organisms in relation to primary production. *Applied Soil Ecology* 9, 25–31.

Siddiqi, M.R. (1986) *Tylenchida: Parasites of Plants and Insects*. CABI Publishing Wallingford.

Siddiqi, M.R. (2000) *Tylenchida: parasites of plants and insects*, 2nd edition. CAB International, Wallingford, UK.

Sikora, R.A., Malek, R.B., Taylor, D.P. and Edwards, D.I. (1979) Reduction of *Meloidogyne naasi* infection of creeping bentgrass by *Tylenchorhynchus agri* and *Paratrichodorus minor*. *Nematologica* 25, 179–183.

Simmons, B.L., Niles, R.K. and Wall, D.H. (2008) Distribution and abundance of alfalfa-field nematodes at various spatial scales. *Applied Soil Ecology* 38, 211–222.

Simons, W.R. (1973) Nematode survival in relation to soil moisture. *Mededelingen Landbouwhogeschool, Wageningen* 73(3), 1–85.

Slack, D.A., Riggs, R.D. and Hamblen, M.L. (1972) The effect of temperature and moisture on the survival of *Heterodera glycines* in the absence of a host. *Journal of Nematology* 4, 263–266.

Snelgrove, P.V.R. (1999) Getting to the bottom of marine biodiversity: sedimentary habitats. *BioScience* 49, 129–138.

Soetaert, K., Heip, C. and Vincx, M. (1991) Diversity of nematode assemblages along a Mediterranean deep sea transect. *Marine Ecology Progress Series* 75, 275–282.

Sohlenius, B. (1980) Abundance, biomass and contribution to energy-flow by soil nematodes in terrestrial ecosystems. *Oikos* 34, 186–194.

Sohlenius, B. (1985) Influence of climatic conditions on nematode coexistence – a laboratory experiment with a coniferous forest soil. *Oikos* 44, 430–438.

Sohlenius, B. (1997) Fluctuations of nematode populations in pine forest soil. Influence by clear-cutting. *Fundamental and Applied Nematology* 20, 103–114.

Sohlenius, B., Bostrom, S. and Sandor, A. (1987) Long-term dynamics of nematode communities in arable soil under 4 cropping systems. *Journal of Applied Ecology* 42, 131–144.

Somerfield, P.J., Cochrane, S.J., Dahle, S. and Pearson, T.H. (2006) Free-living nematodes and macrobenthos in a high-latitude glacial fjord. *Journal of Experimental Marine Biology and Ecology* 330, 284–296.

Spaull, V.W. (1973) Qualitative and quantitative distribution of soil nematodes of Signy Island, South Orkney Islands. *British Antarctic Survey Bulletin* 33–35, 177–184.

Stamou, G.P., Papatheodorou, E.M., Hovardas, A. and Argyropoulou, M.D. (2005) Some structural and functional characteristics of a soil nematode community from a Mediterranean grassland. *Belgian Journal of Zoology* 135, 253–259.

Sturhan, D. and Brzeski, M. (1991) Stem and bulb nematodes: *Ditylenchus* spp. In: Nickle, W.R. (ed.) *Manual of Agricultural Nematology*. Marcel Dekker, New York, pp 423–464.

Sublette, K., Jennings, E., Mehta, C., Duncan, K., Brokaw, J., Todd, T. and Thoma, G. (2007) Monitoring soil ecosystem recovery following bioremediation of a terrestrial crude oil spill with and without a fertilizer amendment. *Soil and Sediment Contamination* 16, 181–208.

Sudhaus, W. and Fitch, D. (2001) Comparative studies on the phylogeny and systematics of the Rhabditidae (Nematoda). *Journal of Nematology* 33, 1–70.

Sudhaus, W. and Fuerst von Lieven, A. (2003) A phylogenetic classification and catalogue of the Diplogastridae (Secernentea, Nematoda). *Journal of Nematode Morphology and Systematics* 6, 43–89.

Swift, M.J., Andren, O., Brussaard, L., Briones, M., Couteaus, M.M., Ekschmitt, K., Kjoller, A., Loiseau, P. and Smith, P. (1998) Global change, soil biodiversity, and nitrogen cycling in terrestrial ecosystems: three case studies. *Global Change Biology* 4, 729–744.

Tahseen, Q., Sheridan, J. and Perry, R.N. (2004) Electropharyngeograms of pharyngeal pumping activity in six species of free-living nematodes. *Nematology* 6, 49–54.

Tamis, W. (1986) Nematoden in een ammoniumdepositiegradient in een groevdennenbos. *Hydrobiologisch Adviesbureau Klink bv., Wageningen, Rapporten en Mededelingen* 23, 1–27.

Thistle, D., Lambshead, P.J.D. and Sherman, K.M. (1995) Changes in proportional abundance of nematode buccal cavity and tail-shape groups in three different regions of the North Atlantic deep sea. *Vie et Milieu* 45, 107–115.

Tilman, D., Pacala, S. (1993) The maintenance of species richness in plant communities. In: Ricklefs, R.E. and Schluter, D. (eds) *Species Diversity in Ecological Communities*. University of Chicago Press, Chicago, pp. 13–25.

Timper, P. and Davies, K.G. (2004) Biotic interactions. In: Gaugler, R. and Bilgrami, A.L. (eds) *Nematode Behaviour*. CAB International, Wallingford, UK, pp. 277–307.

Tobar-Jimenez, A. and Palacios-Mejia, F. (1975) Water as a vehicle for dispersing phytoparasitic nematodes. *Revista Iberica de Parasitologia* 35, 223–259.

Todd, T.C., Blair, J.M. and Milliken, G.A. (1999) Effects of altered soil-water availability on a tallgrass prairie nematode community. *Applied Soil Ecology* 13, 45–55.

Todd, T.C., Powers, T.O. and Mullin, P.G. (2006) Sentinel nematodes of land-use change and restoration in tallgrass prairie. *Journal of Nematology* 38, 20–27.

Traunspurger, W. (1992) A study of the free-living freshwater nematodes of hard substrates in the littoral of the oligotrophic Koenigssee (National Park Berchtesgaden, Germany). *Spixana* 15, 233–238.

Traunspurger, W. (1996a) Distribution of benthic nematodes in the littoral of an oligotrophic lake (Königssee, National Park Berchtesgaden, FRG). *Archives of Hydrobiology* 135, 393–412.

Traunspurger, W. (1996b) Distribution of benthic nematodes in the littoriprofundal and profundal of an oligotrophic lake (Königssee, National Park Berchtesgaden, FRG). *Archives of Hydrobiology* 135, 557–575.

Traunspurger, W. (1997a) Distribution, seasonal occurence, and vertical patterns of *Tobrilus gracilis* and *Tobrilus medius* (Nematoda). *Nematologica* 43, 59–81.

Traunspurger, W. (1997b) Bathymetric, seasonal and vertical distribution of the feeding-types of nematodes in an oligotrophic lake. *Vie et Milieu* 47, 1–7.

Traunspurger, W. (2002) Nematoda. In: Rundle, S.D., Robertson, A.L. and Schmid Araya, J.M. (eds) *Freshwater Meiofauna: Biology and Ecology*. Backhuys, Leiden, pp. 63–104.

Traunspurger, W. and Drews, C. (1996) Vertical distribution of benthic nematodes in an oligotrophic lake: seasonality, species and age segregation. *Hydrobiologia* 331, 33–42.

Traunspurger, W., Bergtold, M. and Goedkoop, W. (1997) The effects of nematodes on bacterial activity and abundance in a freshwater sediment. *Oecolgia* 112, 118–122.

Traunspurger, W., Michiels, I.C. and Eyualem, A. (2006) Composition and distribution of free-living freshwater nematodes: global and local perspectives. In: Eyualem, A., Traunspurger, W. and Andrassy, I. (eds) *Freshwater Nematodes: Taxonomy and Ecology*. CAB International, Wallingford, UK, pp. 46–76.

Tsesunov, A.V. (1997) Marimermithid nematodes: anatomy, position in the nematode system, phylogeny. *Zoologicheskii Zhurnal* 76, 1283–1299.

Tsiafouli, M.A., Argyropoulou, M.D., Stamou, G.P. and Sgardelis, S.P. (2006) Soil nematode biodiversity in organic and conventional agroecosystems of Northern Greece. *Russian Journal of Nematology* 14, 159–169.

Van der Putten. W.H., Cook, R., Costa, S., Davies, K.G., Fargette, M., Freitas, H., Hol, W.H.G., Kerry, B.R., Maher, N. and Mateille, T. (2006) Nematode interactions in nature: models for sustainable control of nematode pests of crop plants? *Advances in Agronomy* 89, 227–260.

Van der Wurff, A.W.G., Kools, S.A.E., Boivin, M.E.Y., Van den Brink. P.J., Van Megen H.H.M., Riksen, J.A.G., Doroszuk, A. and Kammenga, J.E. (2007) Type of disturbance and ecological history determine structural stability. *Ecological Applications* 17, 190–202.

Van Diepeningen, A.D., De Vos, O.J., Korthals, G.W. and Van Bruggen, A.H.C. (2006) Effects of organic versus conventional management on chemical and biological parameters in agricultural soils. *Applied Soil Ecology* 31, 120–135.

Van Gundy, S.D. and Freckman, D.W. (1977) Phytoparasitic nematodes in belowground agroecosystems. *Ecology Bulletin* 25, 320–329.

Vandermeer, J. (2006) Oscillating populations and biodiversity maintenance. *Bioscience* 56, 967–975.

Vanhove, S., Arntz, W. and Vincx, M. (1999) Comparative study of the nematode communities on the southeastern Weddell Sea shelf and slope (Antarctica). *Marine Ecology Progress Series* 181, 237–256.

Venette, R.C. and Ferris, H. (1997) Thermal constraints to population growth of bacterial-feeding nematodes. *Soil Biology and Biochemistry* 29, 63–74.

Viketoft, M., Palmborg, C., Sohlenius, B., Huss-Danell, K. and Bengtsson, J. (2005) Plant species effects on soil nematode communities in experimental grasslands. *Applied Soil Ecology* 30, 90–103.

Virginia, R.A. and Wall, D.W. (1999) How soils structure communities in the Antarctic dry valleys. *BioScience* 49, 973–983.

Waliullah, M.I.S. (1984) Nematodes in irrigated water. *Nematologia Mediterranea* 12, 243–245.

Waliullah, M.I.S. (1989) Nematodes in irrigation canals of the Kashmir Valley, India. *Nematologia Mediterranea* 17, 55–56.

Walker, B.H. (1992) Biodiversity and ecological redundancy. *Conservation Biology* 6, 18–23.

Wall, D.H. (1999) Biodiversity and Ecosystem Functioning. *BioScience* 49, 107–108.

Wall, D.H. and Moore, J.C. (1999) Interactions underground: soil biodiversity, mutualism and ecosystem processes. *BioScience* 49, 109–117.

Wall, D.H. and Virginia, R.A. (1999) Controls on soil biodiversity: insights from extreme environments. *Applied Soil Ecology* 13, 137–150.

Wall, D.H. and Virginia, R.A. (2000) The world beneath our feet: soil biodiversity and ecosystem functioning. In: Raven, P.R. and Williams, T. (eds) *Nature and Human Society: the Quest for a Sustainable World*. National Academy Press, Washington, DC, pp. 225–241.

Wallace, H.R. (1963) *The Biology of Plant Parasitic Nematodes*. Arnold, London.

Wallace, H.R. (1987) A perception of tolerance. *Nematologica* 33, 419–432.

Wallace, M.K. and Hawkins, D.M. (1994) Applications of geostatistics in plant nematology. *Journal of Nematology* (Supp) 26, 626–634.

Wallace, M.K., Rust, R.H., Hawkins, D.M. and MacDonald, D.H. (1993) Correlation of edaphic factors with plant-parasitic nematode population-densities in a forage field. *Journal of Nematology* 25, 642–653.

Wardle, D.A., Yeates, G.W., Watson, R.N. and Nicholson, K.S. (1995) The detritus food-web and the diversity of soil fauna as indicator of disturbance regimes in agro-ecosystems. *Plant and Soil* 170, 35–43.

Wardle, D.A., Verhoef, H.A. and Clarholm, M. (1998) Trophic relationships in the soil microfood-web: predicting the responses to a changing global environment. Soil biota and global change. *Global Change Biology* 4, 713–727.

Warwick, R.M. and Clarke, K.R. (1995) New 'biodiversity' measures reveal a decrease in taxonomic distinctness with increasing stress. *Marine Ecology Progress Series* 129, 301–305.

Warwick, R.M., Platt, H.M. and Somerfield, P.J. (1998) Free-living marine nematodes part 3. Monhysterids. Pictorial key to world genera and notes for the identification of British species. *Synopses of the British Fauna New Series* 53, 1–296.

Wasilewska, L. (1967) Analysis of the occurrence of nematodes in alfalfa crops: I Species composition of nematodes in two alfalfa crops of different age and penetration of species from soil to plants. *Ekologia Polska, Series A,* 15, 31–74.

Wasilewska, L. (1979) The structure and function of soil nematode communities in natural ecosystems and agrocenoses. *Polish Ecological Studies*. 5, 97–145.

Wasilewska, L. (1991) Long-term changes in communities of soil nematodes on fen peat meadows due to the time since their drainage. *Ekologia Polska* 39, 59–104.

Wasilewska, L. (1994) The effect of age of meadows on succession and diversity in soil nematode communities. *Pedobiologia* 38, 1–11.

Wasilewska, L. (1995) Differences in development of soil nematode communities in single-species and multi-species grass experimental treatments. *Applied Soil Ecology* 2, 53–64.

Webster, L.M.I., Johnson, P.C.D., Adam, A., Mable, B.K. and Keller, L.F. (2007) Macrogeographic population structure in a parasitic nematode with avian hosts. *Veterinary Parasitology* 144, 93–103.

Weiss, B. and Larink, O. (1991) Influence of sewage sludge and heavy metals on nematodes in an arable soil. *Biology and Fertility of Soils* 12, 5–9.

Whitford, W.G. and Freckman, D.W. (1988) The role of soil biota in soil processes in the Chihuahuan desert. In: Whitehead, E.E., Hutchinson, C.F., Timmerman, B.N. and Varady, R.G. (eds) *Arid Lands: Today and Tomorrow*. University of Arizona Press, Tucson, pp. 1063–1073.

Whittaker, R.H. (1972) Evolution and measurement of species diversity. *Taxon*, 21, 213–251.

Whyte, E.B. and Gowen, S.R. (1974) Recovery of nematodes from banana roots and soil samples. *Nematropica* 2, 27–31.

Wieser, W. (1953) Die Beziehung zwischen Mundhoehlen-gestalt, Ernahrungsweise und Vorkommen bei freilebenden marinen Nematoden. Eine okologisch-morphologische Studie. *Arkiv foer Zoologi* 4, 439–484.

Wieser, W. (1959) *Free-living Nematodes and Other Small Invertebrates from Puget Sound Beaches*. University of Washington Press, Seattle.

Williams, P., Humphries, C., Vane-Wright, D. and Gaston, K. (1997) Descriptive and predictive approaches to biodiversity measurement. *Trends in Ecology and Evolution* 12, 444–445.

Williams, P.H. (1993) Choosing conservation areas: using taxonomy to measure more of biodiversity. In: Moon, T-Y. (ed.) *Manuscript Collection, International Society for Biological Conservation*. Korean Entomological Institute, Seoul.

Williams, P.H. (1995) Biodiversity value and taxonomic relatedness. In: Hochberg, M.E., Clobert, J. and Barbault, R. (eds) *Genesis and Maintenance of Biological Diversity*. OUP, Oxford, pp. 261–277.

Williams, P.H., Vane-Wright, R.I. and Humphries, C.J. (1993) Measuring biodiversity for choosing conservation areas. In: LaSalle, J. and Gauld, I.D. (eds) *Hymenoptera and Biodiversity*. CABI Publishing, Wallingford, pp. 309–328.

Wright, C.J. and Coleman, D.C. (2002) Responses of soil microbial biomass, nematode trophic groups, N-mineralization, and litter decomposition to disturbance events in the southern Appalachians. *Soil Biology and Biochemistry*, 34, 13–25.

Wu, L.Y. (1969) Five new species of *Tylenchus* Bastian, 1865 (Nematoda-Tylenchidae) from Canadian high Arctic. *Canadian Journal of Zoology* 47, 1005–1010.

Yeates, G.W. (1974) Studies on a climosequence of soils in tussock grasslands. II. Nematodes. *New Zealand Journal of Zoology* 1, 171–177.

Yeates, G.W. (1978a) Populations of nematode genera in soils under pasture. I. Seasonal dynamics in dryland and irrigated pastures on a southern yellow-grey earth. *New Zealand Journal of Agricultural Research* 21, 321–330.

Yeates, G.W. (1978b) Populations of nematode genera in soils under pasture. II. Seasonal dynamics in dryland and effluent-irrigated pastures on a central yellow-grey earth. *New Zealand Journal of Agricultural Research* 21, 331–340.

Yeates, G.W. (1980) Relation of generic nematode populations to soil and plant-parameters. *Journal of Nematology* 12, 242–243.

Yeates, G.W. (1981) Nematode populations in relation to soil environmental-factors – a review. *Pedobiologia* 22, 312–338.

Yeates, G.W. (1984) Variation in soil nematode diversity under pasture with soil and year. *Soil Biology and Biochemistry* 16, 95–102.

Yeates, G.W. (1987a) How plants affect nematodes. *Advances in Ecological Research* 17, 61–113.

Yeates, G.W. (1987b) Significance of developmental stages in the coexistence of 3 species of Mononchoidea (Nematoda) in a pasture soil. *Biology and Fertility of Soils* 5, 225–229.

Yeates, G.W. (1994) Modification and quantification of the nematode maturity index. *Pedobiologia* 38, 97–101.

Yeates, G.W. (1995) Effect of sewage effluent on soil fauna in a *Pinus radiata* plantation. *Australian Journal of Soil Research* 33, 555–564.

Yeates, G.W. (2003) Nematodes as soil indicators: Functional and biodiversity aspects. *Biology and Fertility of Soils* 37, 199–210.

Yeates, G.W. and Bird, A.F. (1994) Some observations on the influence of agricultural practices on the nematode faunae of some South Australian soils. *Fundamental and Applied Nematology* 17, 133–145.

Yeates, G.W. and Bongers, T. (1999) Nematode diversity in agroecosystems. *Agriculture Ecosystems and Environment* 74, 113–135.

Yeates, G.W. and Hughes, K.A. (1990) Effect of three tillages regimes on plant and soil nematodes in an oats/maize rotation. *Pedobiologia* 34, 379–387.

Yeates, G.W. and Wardle, D.A. (1996) Nematodes as predators and prey: relationships to biological control and soil processes. *Pedobiologia* 40, 43–50.

Yeates, G.W., Stout, J.D., Ross, D.J., Dutch, M.E. and Thomas, R.F. (1976) Long-term effects of paraquat-diquat and additional weed control treatments on some physical, biological, and respiratory properties of a soil previously under grass. *New Zealand Journal of Agricultural Research* 19, 51–61.

Yeates, G.W., Stannard, R.E. and Barker, J.M. (1983) Vertical distribution of nematode populations in Horotui soils. *New Zealand Soil Bureau Science Reports* 60, 1–14.

Yeates, G.W., Watson, R. and Steele, K.W. (1985) Complementary distribution of *Meloidogyne, Heterodera* and *Pratylenchus* (Nematoda: Tylenchida) in roots of white clover. *Proceedings of the Australasian Conference on Grassland Invertebrate Ecology* 4, 71–79.

Yeates, G.W., Bongers, T., Goede, R.G.M. de, Freckman, D.W. and Georgieva, S.S. (1993) Feeding habitats in soil nematode families and genera – an outline for soil ecologists. *Journal of Nematology* 25, 315–331.

Yeates, G.W., Orchard, G.W., Speir, T.W., Hunt, J.L. and Hermans, M.C.C. (1994) Impact of pasture contamination by copper, chromium, arsenic timber preservative on soil biological-activity. *Biology and Fertility of Soils* 18, 200–208.

Yeates, G.W., Bardgett, R.D., Cook, R., Hobbs, P.J., Bowling, P.J. and Potter, J.F. (1997) Faunal and microbial diversity in three Welsh grassland soils under conventional and organic management regimes. *Journal of Applied Ecology* 34, 453–470.

Yuen, P.H. (1966) The nematode fauna of the regenerated woodland and grassland of Broadbalk Wilderness. *Nematologica* 12, 195–214.

Zauner, H., Mayer, W.E., Herrmann, M., Weller, A., Erwig, M. and Sommer, R.J. (2007) Distinct patterns of genetic variation in *Pristionchus pacificus* and *Caenorhabditis elegans*, two partially selfing nematodes with cosmopolitan distribution. *Molecular Ecology* 16, 1267–1280.

Zhou, Z., Sakaue, D., Wu, B. and Hogetsu, T. (2007) Genetic structure of populations of the pinewood nematode *Bursaphelenchus xylophilus*, the pathogen of pine wilt disease, between and within pine forests. *Phytopathology* 97, 304–310.

Zolda, P. (2006) Nematode communities of grazed and ungrazed semi-natural steppe grasslands in Eastern Austria. *Pedobiologia* 50, 11–22.

Zullini, A. and Peneva, V. (2006) Order Mononchida. In: Eyualem, A., Traunspurger, W. and Andrassy, I. (eds) *Freshwater Nematodes: Taxonomy and Ecology*. CAB International, Wallingford, UK, pp. 468–496.

Zullini, A. and Peretti, E. (1986) Lead pollution and moss-inhabiting nematodes of an industrial area. *Water, Air and Soil Pollution* 27, 403–410.

3 Molecular Markers, Indicator Taxa, and Community Indices: the Issue of Bioindication Accuracy

KLEMENS EKSCHMITT[1] AND GERARD KORTHALS[2]

[1]*Justus Liebig University, IFZ – Department of Animal Ecology, Giessen, Germany;* [2]*Wageningen UR, Applied Plant Research, Lelystad, The Netherlands*

Introduction

Amongst soil organisms, nematodes are seen as the most promising candidates for bioindication of soil status (Cortet *et al*., 1999; Achazi, 2002). Using the well-established classifications of nematode feeding types and cp-groups as well as various indices of nematode community structure (Yeates *et al*., 1993; Bongers and Bongers, 1998; Ferris *et al*., 2001), researchers have consistently exploited nematodes to investigate the propagation of disturbance effects and fertilization effects through the soil ecosystem (Freckman and Ettema, 1993; Villenave *et al*., 2001). In addition, it has repeatedly been shown that soil nematodes respond differentially to a range of xenobiotic substances (Bongers *et al*., 2001; De Nardo and Grewal, 2003; Jonker *et al*., 2004; Ekschmitt and Korthals, 2006). During the decade from 1999 to 2008, approximately 200 papers were published in international journals, where nematode indices were used to evaluate and indicate the status of soils and sediments.

Recently, the bioindication of soil status has received particular interest in Europe because indicators were explicitly stated as monitoring tools within the EU soil protection strategy. Table 3.1 summarizes the reports of the technical working groups and the report on risk area identification (Van-Camp *et al*., 2004; Eckelmann *et al*., 2006). Eight major soil threats were identified, and methods were specified by which to assess these threats. The table illustrates that two of the threats, erosion and flooding (threats 1 and 7), cannot be sufficiently assessed through measurement or indication and shall therefore be predicted through modelling. The risk of landslides can best be assessed from past occurrence of landslides and shall therefore be evaluated through registration of such events. Three further threats (threats 3, 5 and 6) can be directly measured with relatively little effort. The decline of soil organic matter can be assessed from the soil content of organic carbon

Table 3.1. Summary of methods and parameters for the assessment of soil threats, as specified in reports on the EU soil protection strategy (Van-Camp *et al.*, 2004; Eckelmann *et al.*, 2006).

	Soil threat	Monitoring (Levels 1, 2, 3)	Risk area identification	Important methods and parameters
1	Erosion	x	R	Modelling
2a	Diffuse pollution	L1 L2 L3		Measurement, sensors
2b	Contamination			Measurement, indicators
3	Loss of organic matter	L1 L2	R	Measurement: Corg, Norg
4	Loss of biodiversity	L3		Measurement: microflora, fauna
5	Compaction	L1	R	Measurement: Bulk density
6	Sealing	x		Remote sensing
7	Flooding and landslides	x	R	Modelling, registration
8	Salinization	x	R	Measurement, indicators

(Corg) and organic nitrogen (Norg). Soil compaction can be assessed from soil bulk density and soil sealing can be monitored through remote sensing. Salinization (threat 8) can be measured directly; in addition good plant indicators are available. Recent approaches exploit indicator variables from satellite imagery to generate prediction maps of salinization risk (Dehaan and Taylor, 2003; Metternicht and Zinck, 2003; Tweed *et al.*, 2007). For example, a low variance of the spectral signal of photosynthetic activity (NDVI) was used to indicate discharge of groundwater to the surface and the associated salinization risk. This table thus illustrates that the term 'indicator' does not automatically imply 'bioindicator'.

In consequence, bioindication with nematodes can primarily contribute to the assessment of two soil threats, namely contamination and biodiversity loss (threats 2 and 4). Under the perspective that bioindication shall be applied within a regulatory framework, the issue of reliability of indication rises to critical importance. This chapter focuses on the bioindication of soil contamination and analyses the levels of accuracy that are achieved with molecular markers, indicator taxa, and community indices. We start by discussing the fundamental difference between 'forward' and 'backward' bioindication. A case study is used to illustrate the difference, and simulated data serve to illustrate its generality. Finally, empirical results are compiled from the literature and discussed.

Forward and Backward Bioindication

Expressed in the most general terms, bioindication is based on a simple reverse conclusion. As, for example, some nematode groups show a negative numerical response to pollution, it is concluded that a decline of susceptible nematode populations is indicative of this pollution. Empirical evidence has made it clear that this reverse conclusion is not as straightforward as it may appear at first glance. One difficulty comes from the fact that the nematode populations under consideration will, most likely, be simultaneously subjected to other impacts than pollution, such as soil conditions, resource availability, disturbance and interactions with other organisms. These other impacts may obscure or even overrule the pollution effects (Schratzberger *et al.*, 2000; Kelaher *et al.*, 2003). A second difficulty arises because, over time, nematode populations may be selected for tolerance towards repeated or permanent stresses, and may therefore change their response to such stresses (Millward and Grant, 2000; Arts *et al.*, 2004; Morgan *et al.*, 2007). While pollution induced community tolerance (PICT) can be used to reveal previous exposure to pollutants, a PICT analysis always requires thorough calibration and comparison against unaffected control sites, which often proves an unsatisfiable requirement in practice (Blanck, 2002). A third problem arises because some of the forces driving nematode populations may relate to events in the past, or to spontaneous intrinsic population dynamics, or may in some other way be obscured from analysis by the investigator. Therefore a relevant proportion of population variation may be left unexplained (Ekschmitt *et al.*, 2003). As a consequence, the reverse conclusion, i.e. the indication of soil impacts from observations on nematodes, is generally more equivocal than desired. This is perfectly congruent with the rule of Aristotelian propositional logic that reverse conclusions do not generally hold.

To avoid possible conceptual confusion, it is proposed here to discriminate between: (i) proving *forward* the effects of environmental or anthropogenic factors on soil populations; and (ii) inferring *backwards* the operation of such factors in the soil from observations on soil populations. The statement that 'a pollutant invokes a response in nematodes' presents an overly simplified concept of reality (Fig. 3.1a). A more realistic concept needs to consider that the bioavailability of the pollutant is modified along the food-chain or through immobilization, that the sensitivity of nematodes towards the pollutant may alter through adaptation, that a nematode's response to a contamination comprises a cascade of several behavioural and physiological reactions, and that nematodes are subject to a multitude of other environmental factors, which induce overlapping responses (Fig. 3.1b).

The more realistic model of Fig. 3.1b pinpoints the difference between forward and backward bioindication. Although *forward* bioindication can be substantially hindered by a multitude of disturbing factors, it seems possible to track the causal chain from a pollutant to a particular response of the nematode. For example, one could analyse the correlation between a known or experimentally controlled contamination and the observed population decline in nematodes. Conversely, *backward* bioindication appears categorically more difficult, because in this case, an unknown contamination must be predicted from the observed population decline, whilst securing that the population decline does not result from one of the many other factors.

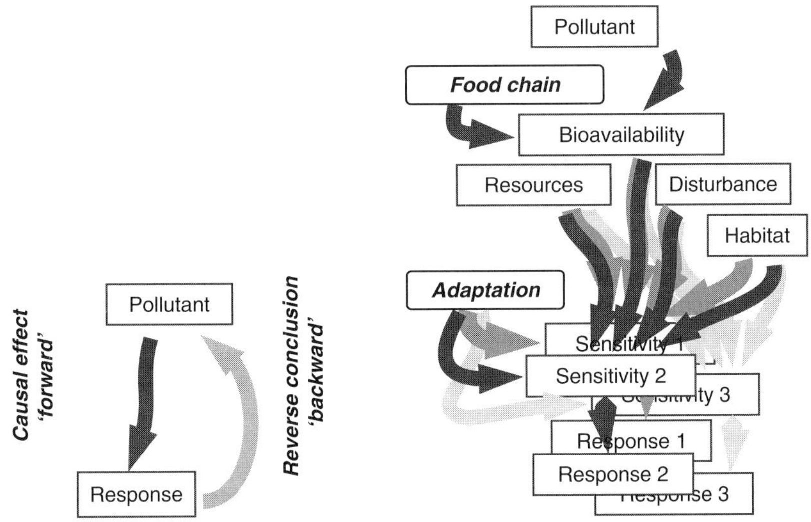

(a) Simple concept of bioindication (b) More realistic concept of bioindication

Fig. 3.1. Bioindication as reverse conclusion. (a) Using a simple concept of bioindication, the reverse conclusion from the organism's response to the pollutant appears to be simple. (b) A more realistic concept of bioindication suggests that due to the multitude of possible alternative causes, the reconstruction of a pollution effect from an organism's response can be difficult.

Case Study: the Dutch Copper Experiment

In this section we present a case study, the Copper Field Experiment, which illustrates the quantitative discrepancy between *forward* and *backward* bioindication. The Copper Field Experiment (Korthals *et al.*, 1996) was established near Wageningen, The Netherlands in 1982. The experiment had a factorial design of four copper levels (0, 250, 500 and 750 kg Cu per ha) and four pH levels (4.0, 4.7, 5.4 and 6.1 pH-KCl) and comprised eight replicate plots (6 × 11 m) of each treatment, arranged in randomized blocks. The dominant soil type was a fimic anthrosol and a continuous crop rotation of maize, potato and oat was maintained on the site.

In 1992, i.e. ten years after the setup of the experiment, 30 soil cores (17 mm diameter, 10 cm depth) were taken from each of the 128 plots, mixed and analysed for pH, copper content and nematodes. In these samples, the experimental copper and pH manipulations were clearly visible in increased concentrations of bioavailable Cu, as indicated by CaCl-extractable Cu (Fig. 3.2a). Nematode abundance decreased from approximately 5000 ind./100 g dm soil in the control plots to less than 1000 ind./100 g dm soil in the most intensive treatments (Fig. 3.2b).

We selected this dataset for its exceptional size, consistency and methodological rigour to analyse issues of bioindication accuracy. In order to isolate the sensitivity of the nematodes themselves, and to eliminate effects transferred

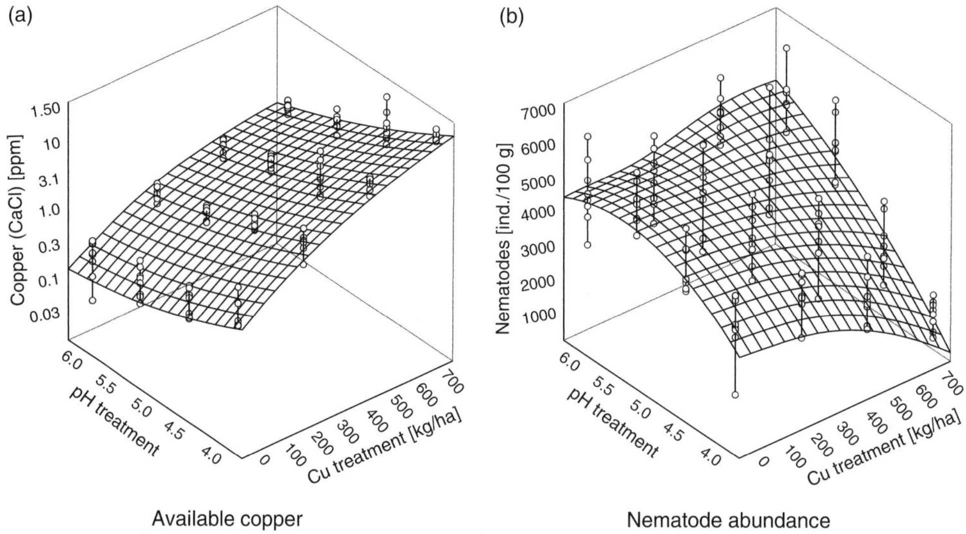

Fig. 3.2. The Cu Field Experiment. (a) Bioavailable (CaCl-extractable) copper 10 years after setup of the experiment. (b) Nematode total abundance 10 years after setup of the experiment. (Data from Korthals *et al.*, 1996.)

via the feeding sources of the different nematode feeding types, we expressed the importance of nematode genera as the relative proportion of each genus within its feeding type (Ekschmitt and Korthals, 2006). Then, we analysed the responses of all genera to the copper concentration by means of Spearman rank correlation. Of the 74 genera observed in total, 35 were represented frequently enough to enable statistical analysis, and of these, 17 genera exhibited correlations with CaCl-extractable copper (Fig. 3.3a). With increasing Cu concentration, 11 genera showed a relative decrease of proportion within their feeding type, and 6 increased in proportion. The correlations were highly significant, error probabilities for the individual genera ranged from 10^{-11} to 10^{-4}, and cumulative error was 0.0003. This leaves very marginal doubt that in each of those single genera, the relative proportion of population size within their feeding type was affected by the copper contamination. Thus, *forward* bioindication successfully proved effects of copper on nematodes.

In a second step, we evaluated the potential of predicting soil Cu contamination from nematode populations. From several technical alternatives such as log-transformations and non-linear regressions, we chose the best multiple regression model for Cu concentration using the 17 genera showing significant responses as predictor variables. Explained variance of the model was fairly high and statistical significance was excellent ($r^2 = 0.80$, $P < 0.0001$). Nevertheless, the proportion of 20% unexplained variance meant that a predicted value of 1 ppm Cu corresponded to a 95% confidence range from 0.33 to 3.02 ppm of observed Cu concentration (Fig. 3.3b). Thus, *backward* bioindication had a disappointing low accuracy of roughly one order of magnitude in Cu concentration, and this was in sharp contrast to the encouraging results previously obtained from *forward* bioindication.

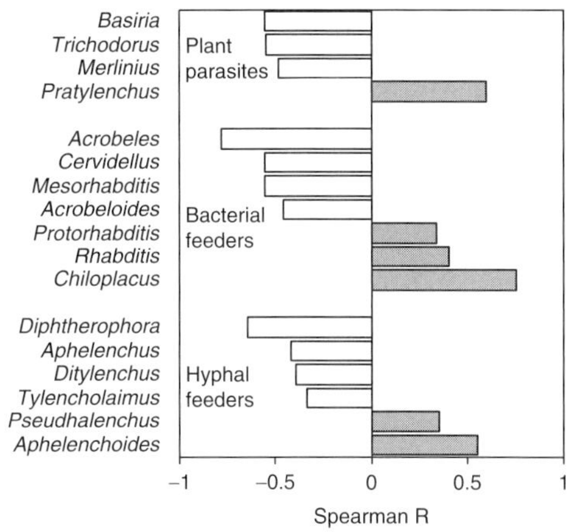

(a) Forward bioindication: significant responses of nematode genera to Cu concentration

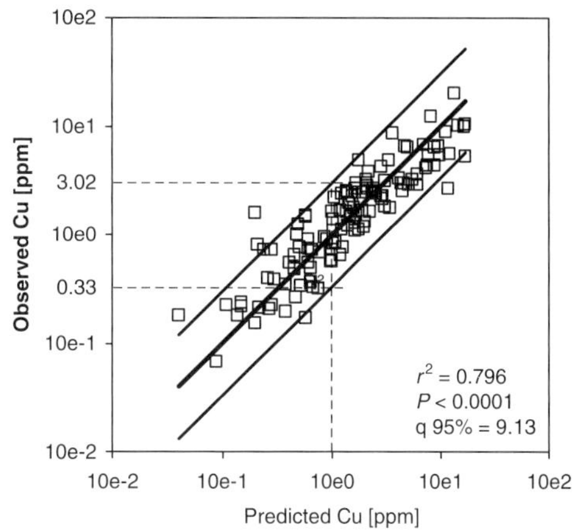

(b) Backward bioindication: reconstruction of Cu concentration from nematode responses

Fig. 3.3. Bioindication of Cu contamination. (a) Responses of 17 nematode genera to the concentration of CaCl-extractable Cu were significant (Spearman, $P < 0.0001$). (b) The bioindication index composed from these genera was linearly and significantly correlated with Cu concentration on a logarithmic scale ($r^2 = 0.80$, $P < 0.0001$). Prediction accuracy was roughly one order of magnitude in Cu concentration (q 95% = 9.1). (Data from Ekschmitt and Korthals, 2006.)

Simulation: General Criteria for Accuracy of Bioindication

We believe that the Dutch copper experiment is not a singular exception but a representative example of the principal discrepancy between *forward* and *backward* bioindication. To further clarify this discrepancy we performed numerical simulations with artificial data, which were free of any suspicious condition of site or season that may potentially have impeded a better *backward* bioindication in the field experiment.

It is a classic notion in the ecological literature that soil animals show an aggregated pattern in space. As a broadly generalized rule of thumb the spatial variance of soil animal species can be predicted according to the geometric distribution, which is a special case of the negative binomial distribution (Ekschmitt *et al.*, 1997). This general rule, together with the observation that data of species groups tend to exhibit lower spatial variance due to stochastic compensation of the distribution patterns of individual species (Ekschmitt, 1998), was used to imitate bioindication through numerical simulation. It was assumed that some kind of impact, scaled from 0 to 1, linearly reduced nematode abundance from 1000 ind./sample to 0 ind./sample, i.e. the impact ranged from no effect to complete extinction of the population. Thirty soil samples were simulated along an impact gradient and the linear regression of impact strength versus nematode abundance was evaluated. Fig. 3.4 shows typical simulation results for three sampling scenarios: (a) impact on a single species, (b) impact on an ensemble of 10 species with identical response; and (c) impact on 10 species obtained from bulk samples composed of 10 subsamples each. *Forward* and *backward* bioindication are illustrated for each scenario.

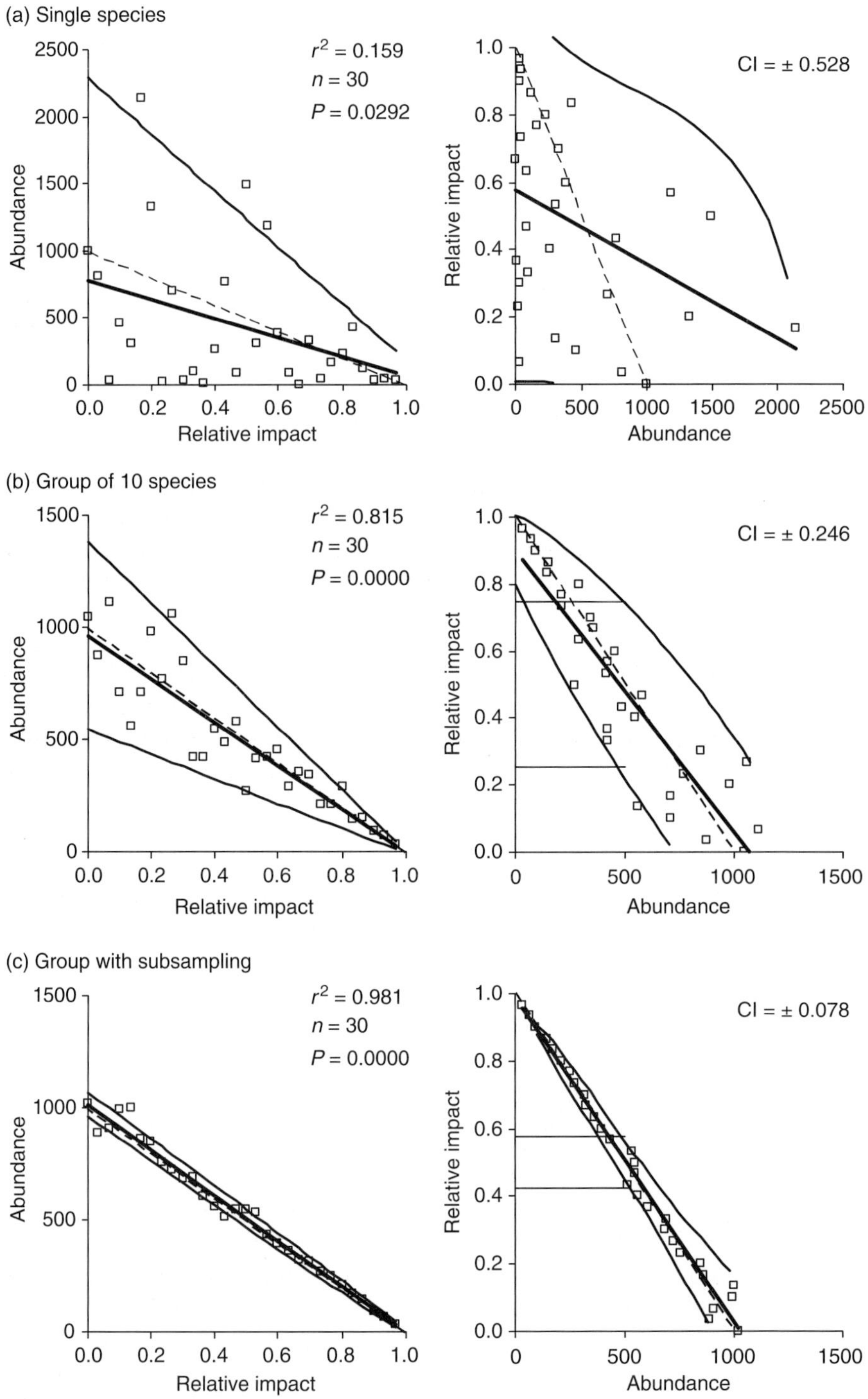

Fig. 3.4. Simulation examples. (a) Single indicator species. (b) Group of 10 indicator species. (c) Group of 10 indicator species registered in bulk samples of 10 subsamples each. Left column: forward bioindication, effect of impact on abundance. Right column: backward bioindication, prediction of impact from abundance.

The simplest scenario with a single species and no subsampling has already sufficed to prove statistically the impact effect (Pearson, $P = 0.03$). The statistical proof was substantially strengthened in scenario (b) where 10 species were evaluated (Pearson, $P < 0.001$). However, in both scenarios (a) and (b) the prediction of impact strength from nematode abundance was far below acceptable levels of accuracy. In scenario (a), the 95% confidence range of prediction was on average ±0.5, which meant that prediction uncertainty was as large as the entire impact gradient ranging from 0 to 1. In scenario (b), the 95% confidence range still amounted to ±0.25, which meant that only two levels of impact strength could be separated along the impact gradient. Only if the sampling effort was radically increased did the uncertainty of *backward* bioindication narrow down to a practically useful dimension. In scenario (c) with ten-fold subsampling, the 95% confidence range of prediction was reduced to ±0.08, which enabled discriminating six to seven grades of impact intensity through bioindication.

The simulation models are built on realistic assumptions on statistical distributions. What they do not contain is the faintest spark of life or biological interaction. The models therefore explain that the discrepancy between *forward* and *backward* bioindication is not primarily based in ecology but in statistics, namely in regression statistics. Statistical significance of a regression, and confidence range of a regression are both related to explained variance, however they exhibit very different characteristics. Fig. 3.5a summarizes the results of 10^5 simulated samples covering a wide range of group size (1–100 species), and sample sizes (1–100 subsamples). The figure depicts the number of samples required to achieve significance ($P < 0.05$) together with

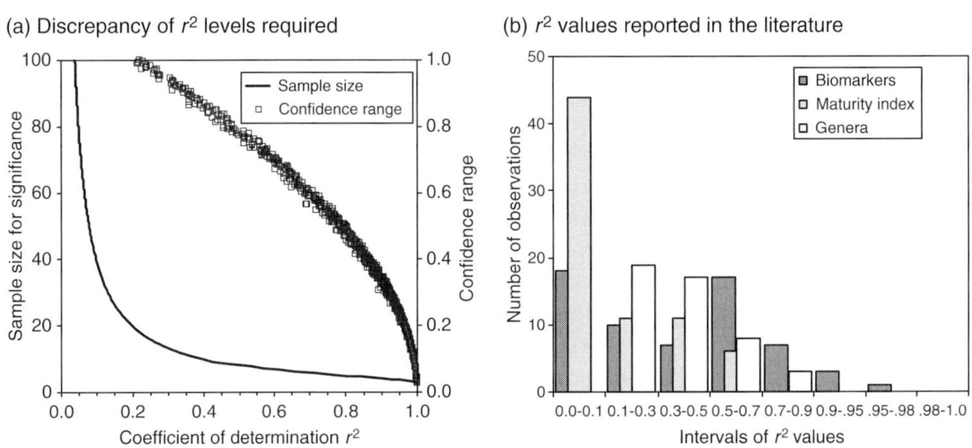

Fig. 3.5. Bioindication quality. (a) Discrepancy of r^2 levels required in forward and backward bioindication. In a survey with 20 samples, an r^2 of 0.20 suffices to prove significant correlation between impact and response. To achieve a 95% confidence range of 0.1 requires an r^2 higher than 0.99, independently of sample size. (b) Frequency of r^2 values reported for molecular biomarkers, for a community index (Maturity Index) and for indicator genera in the literature.

the width of the 95% confidence range (upper limit – lower limit), as they relate to the coefficient of determination (r^2).

The figure illustrates that in *forward* bioindication, in a survey based on 20 samples, an r^2 value as low as 0.20 suffices to prove significant correlation between impact and response of the indicator. However, in *backward* bioindication, in order to achieve a prediction accuracy of ±5% corresponding to a confidence range of 0.1, the impact gradient requires an r^2 value higher than 0.99. It should be mentioned here that these results are of a general nature. They are equally valid for other organisms than nematodes, for other parameters than population size, for positive impacts, for complex impacts composed of several factors, and for non-linear responses of populations. In fact, most researchers are well familiar with the 0.99 rule in that they would never accept calibration of a photometer or gas-chromatograph in their laboratory with a calibration curve inferior than $r^2 = 0.99$. The question that immediately arises from this analysis is: how good a regression can we expect to obtain in environmental field data?

Analysis of Published Data

To evaluate the quality of correlations commonly observed between contaminants and bioindicators in field surveys, we performed a scan of the bioindication literature. Generally, we found that the majority of publications focus on contrasts between polluted and control sites, whilst correlations along continuous pollution gradients are rarely documented. Specifically, we found little documentation of correlations between environmental variables and nematode parameters in terrestrial soils. We therefore included other organisms and aquatic sediments in our analysis. Here, we present examples of biomarkers in eel and starfish, of nematode indicator genera for heavy metals in terrestrial soils, and of the nematode maturity index from aquatic sediments.

The literature on biomarkers in fish was recently reviewed by Van der Oost *et al.* (2003), Au (2004) and Tom and Auslander (2005). Van der Oost *et al.* (1997) published an extensive list of correlations between three groups of organic toxicants measured in freshwater sediments around Amsterdam, The Netherlands, and 21 molecular markers in feral eel sampled at the same sites. r^2 values ranged from 0.00 to 0.96, with an intermediate peak around 0.6 (Fig. 3.5b). Danis *et al.* (2004) report in an extensive analysis of starfish along the Belgian coast that sediment concentrations of six organic pollutants and four heavy metals were correlated with concentrations of pollutants in a variety of starfish tissues with r^2 values ranging up to 0.55.

Ekschmitt and Korthals (2006) reviewed nematode data in soils contaminated with heavy metals. From these data, we recalculated the regressions between the abundances of nematode genera and the concentrations of five heavy metals. r^2 values of significant regressions ranged from 0.18 to 0.82 with a continuous decline in frequency towards higher values (Fig. 3.5b).

While Neilson *et al.* (1996) claimed that in marine nematode assemblages the nematode maturity index (Bongers, 1990) often showed no response to the environment, good regressions were found by Gyedu-Ababio *et al.* (1999), Wu *et al.* (2004), Sanchez-Moreno *et al.* (2006), Bert *et al.* (2007) and Heininger *et al.* (2007). All these are works in aquatic systems contaminated by organic pollutants and/or heavy metals, and r^2 values ranged from 0.00 to 0.65 with an intermediate accumulation around 0.4 (Fig. 3.5b).

The collected data illustrate that a regression of $r^2 = 0.99$, which is required for accurate *backward* bioindication, was hitherto rarely observed in field datasets, independently of the type of bioindicator. According to the curve depicted in Fig. 3.5a, regression quality did not enable separating more than two levels of contamination in the majority of empirical cases.

Discussion

In three independent analyses we found in great correspondence that a substantial discrepancy exists between two possible goals of bioindication: (i) detecting effects of environmental factors on organisms; and (ii) predicting environmental factors from the biological status of organisms. We introduced the terms *forward* bioindication for the former, and *backward* bioindication for the latter goal. In conceptual terms, *backward* bioindication appeared more demanding because of the necessity to isolate the factor to be predicted from potential alternative causes of a similar indicator response. In a case study on soil copper contamination, copper effects on nematodes were detected with high statistical certainty, whilst prediction of soil copper concentrations from nematode data was hampered by a disappointingly low level of accuracy. Simulations with artificial data illustrated that generally, relatively low coefficients of determination ($r^2 > 0.2$) sufficed to statistically prove a correlation between environmental factors and biological indicators. In contrast, very high coefficients of determination ($r^2 > 0.99$) were required to predict environmental factors from biological indicators with reasonable accuracy. We assume that this radical difference in the requirements for *forward* and *backward* bioindication was often underestimated by researchers and has led to unrealistic expectations into *backward* bioindication.

A review of the bioindication literature revealed that in the majority of field investigations, and independently of whether bioindicators were molecular markers, indicator taxa, or community indices, r^2 values resided below 0.8. This is just adequate to separate with 95% certainty two levels of effect, such as low and high contamination. It seems theoretically possible, as indicated by the simulations, to narrow spatial and biological variation, and thereby to increase r^2, through massive averaging by extensive subsampling and by combining many indicators into a composite indicator. For example, one could combine several hundreds or thousands of congeneric biomarkers in a micro-array, and thereby eventually overcome the r^2 hurdle of *backward* bioindication. As long as such a brute force tool is not available, it seems however more promising to focus bioindication on the *forward* mode.

From our analysis we conclude that at the present state of technology, bioindication of soils and sediments should be used less to attempt detection of hitherto unknown contaminations or other soil threats. Bioindication should be applied to analyse the biological consequences of known environmental impacts, to specify threshold levels of pollutants and disturbances, and to evaluate the success of soil remediation measures.

References

Achazi R.K. (2002) Invertebrates in risk assessment. *Journal of Soils and Sediments* 2, 174–178.

Arts, M.J.S.J., Schill, R.O., Knigge, T., Eckwert, H., Kammenga, J.E. and Kohler, H.R. (2004) Stress proteins (hsp70, hsp60) induced in isopods and nematodes by field exposure to metals in a gradient near Avonmouth, UK. *Ecotoxicology* 13, 739–755.

Au, D.W.T. (2004) The application of histo-cytopathological biomarkers in marine pollution monitoring: a review. *Marine Pollution Bulletin* 48, 817–34.

Bert, W., Messiaen, M., Hendrickx, F., Manhout, J., De Bie, T. and Borgonie, G. (2007) Nematode communities of small farmland ponds. *Hydrobiologia* 583, 91–105.

Blanck, H. (2002) A critical review of procedures and approaches used for assessing pollution-induced community tolerance (PICT) in biotic communities. *Human and Ecological Risk Assessment* 8, 1003–1034.

Bongers, T. (1990) The maturity index: An ecological measure of an environmental disturbance based on nematode species composition. *Oecologia* 83, 14–19.

Bongers, T. and Bongers, M. (1998) Functional diversity of nematodes. *Applied Soil Ecology* 10, 239–251.

Bongers, T., Ilieva-Makulec, K. and Ekschmitt, K. (2001) Acute sensitivity of nematode taxa to $CuSO_4$ and relationships with feeding-type and life-history classification. *Environmental Toxicology and Chemistry* 20, 1511–1516.

Cortet, J., Gomot-De Vauflery, A., Poinsot-Balaguer, N., Gomot, L., Texier, C. and Cluzeau, D. (1999) The use of invertebrate soil fauna in monitoring pollutant effects. *European Journal of Soil Biology* 35, 115–134.

Danis, B., Wantier, P., Dutrieux, S., Flammang, R., Dubois, P. and Warnau, M. (2004) Contaminant levels in sediments and asteroids (*Asterias rubens* L., Echinodermata) from the Belgian coast and Scheldt estuary: polychlorinated biphenyls and heavy metals. *Science of the Total Environment* 333, 149–165.

De Nardo, E.A.B. and Grewal, P.S. (2003) Compatibility of *Steinernema feltiae* (Nematoda: Steinernematidae) with pesticides and plant growth regulators used in glasshouse plant production. *Biocontrol Science and Technology* 13, 441–448.

Dehaan, R. and Taylor, G.R. (2003) Image-derived spectral endmembers as indicators of salinisation. *International Journal of Remote Sensing* 24, 775–794.

Eckelmann, W., Baritz, R., Bialousz, S., Bielek, P., Carre, F., Houšková, B., Jones, R.J.A., Kibblewhite, M.G., Kozak, J., Le Bas, C., Tóth, G., Tóth, T., Várallyay, G., Yli Halla, M. and Zupan, M. (2006) Common criteria for risk area identification according to soil threats. European Soil Bureau Research Report No. 20, EUR 22185 EN.

Ekschmitt, K. (1998) Population assessments of soil fauna: general criteria for the planning of sampling schemes. *Applied Soil Ecology* 9, 439–445.

Ekschmitt, K. and Korthals, G.W. (2006) Nematodes as sentinels of heavy metals and organic toxicants in the soil. *Journal of Nematology* 38, 13–19.

Ekschmitt, K., Weidemann, G. and Wolters, V. (1997) Spatial heterogeneity in the density of soil animals. *Recent Research Developments in Soil Biology and Biochemistry* 1, 21–38.

Ekschmitt, K., Stierhof, T., Dauber, J., Kreimes, K. and Wolters, V. (2003) On the quality of soil biodiversity indicators: abiotic and biotic parameters as predictors of soil faunal richness at different spatial scales. *Agriculture, Ecosystems and Environment* 98, 273–283.

Ferris, H., Bongers, T. and de Goede, R.G.M. (2001) A framework for soil food web diagnostics: extension of the nematode faunal analysis concept. *Applied Soil Ecology* 18, 13–29.

Freckman, D.W. and Ettema, C.H. (1993) Assessing nematode communities in agroecosystems of varying human intervention. *Agriculture, Ecosystems and Environment* 45, 239–261.

Gyedu-Ababio, T.K., Fürstenberg, J.P., Baird, D. and Vanreusel, A. (1999) Nematodes as indicators of pollution: a case study from the Swartkops River system, South Africa. *Hydrobiologia* 397, 155–169.

Heininger, P., Höss, S., Claus, E., Pelzer, J. and Traunspurger, W. (2007) Nematode communities in contaminated river sediments. *Environmental Pollution* 146, 64–76.

Jonker, M.J., Piskiewicz, A.M., Castella, N.I.I. and Kammenga, J.E. (2004) Toxicity of binary mixtures of cadmium-copper and carbendazim-copper to the nematode *Caenorhabditis elegans. Environmental Toxicology and Chemistry* 23, 1529–1537.

Kelaher, B.P., Levinton, J.S., Oomen, J., Allen, B.J. and Wong, W.H. (2003) Changes in benthos following the clean-up of a severely metal-polluted cove in the Hudson River estuary: Environmental restoration or ecological disturbance? *Estuaries* 26, 1505–1516.

Korthals, G.W., Alexiev, A.D., Lexmond, T.M., Kammenga, J.E. and Bongers, T. (1996) Long-term effects of copper and pH on the nematode community in an agroecosystem. *Environmental Toxicology and Chemistry* 15, 979–985.

Metternicht, G.I. and Zinck, J.A. (2003) Remote sensing of soil salinity: potentials and constraints. *Remote Sensing of Environment* 85, 1–20.

Millward, R.N. and Grant, A. (2000) Pollution-induced tolerance to copper of nematode communities in the severely contaminated Restronguet Creek and adjacent estuaries, Cornwall, United Kingdom. *Environmental Toxicology and Chemistry* 19, 454–461.

Morgan, A.J., Kille, P. and Sturzenbaum, S.R. (2007) Microevolution and ecotoxicology of metals in invertebrates. *Environmental Science & Technology* 41, 1085–1096.

Neilson, R., Boag, B. and Palmer, L.F. (1996) The effect of environment on marine nematode assemblages as indicated by the maturity index. *Nematologica* 42, 232–242.

Sanchez-Moreno, S., Camargo, J.A. and Navas, A. (2006) Ecotoxicological assessment of the impact of residual heavy metals on soil nematodes in the Guadiamar River Basin (Southern Spain). *Environmental Monitoring and Assessment* 116, 245–262.

Schratzberger, M., Rees, H.L. and Boyd, S.E. (2000) Effects of simulated deposition of dredged material on structure of nematode assemblages – the role of contamination. *Marine Biology* 137, 613–622.

Tom, M. and Auslander, M. (2005) Transcript and protein environmental biomarkers in fish – a review. *Chemosphere* 59, 155–162.

Tweed, S.O., Leblanc, M., Webb, J.A. and Lubczynski, M.W. (2007) Remote sensing and GIS for mapping groundwater recharge and discharge areas in salinity prone catchments, southeastern Australia. *Hydrogeology Journal* 15, 75–96.

Van der Oost, R., Vindimian, E., Van den Brink, P.J., Satumalay, K., Heida, H. and Vermeulen N.P.E. (1997) Biomonitoring aquatic pollution with feral eel (*Anguilla anguilla*). III. Statistical analyses of relationships between contaminant exposure and biomarkers. *Aquatic Toxicology* 39, 45–75.

Van der Oost, R., Beyer, J. and Vermeulen, N.P.E. (2003) Fish bioaccumulation and biomarkers in environmental risk assessment: a review. *Environmental Toxicology and Pharmacology* 13, 57–149.

Van-Camp L., Bujarrabal B., Gentile A.-R., Jones R.J.A., Montanarella L., Olazabal C. and Selvaradjou S.K. (2004) Reports of the Technical Working Groups Established under the Thematic Strategy for Soil Protection. EUR 21319 EN / 1–6.

Villenave, C., Bongers, T., Ekschmitt, K., Djigal, D. and Chotte, J.L. (2001) Changes in nematode communities following cultivation of soils after fallow periods of different length. *Applied Soil Ecology* 17, 43–52.

Wu, J.H., Fu, C.Z., Liang, Y.L. and Chen, J.K. (2004) Distribution of the meiofaunal community in a eutrophic shallow lake of China. *Archiv für Hydrobiologie* 159, 555–575.

Yeates G.W., Bongers T., de Goede R.G.M., Freckman D.W. and Georgieva S.S. (1993) Feeding habits in soil nematode families and genera – an outline for soil ecologists. *Journal of Nematology* 25, 315–331.

4 General Community Indices that can be used for Analysis of Nematode Assemblages

DEBORAH A. NEHER AND BRIAN J. DARBY

Department of Plant and Soil Science, University of Vermont, USA

Introduction

The objective of classical community indices is to condense community data into one or a few variables to simplify analysis, interpretation, or review. To be successful as an indicator, a single index must be able to perform one of two functions: either reflect a past ecological process or predict a future ecological process. The success of community indices to reflect ecological processes or predict patterns depends on the relative completeness of ecological knowledge. A limitation of community indices is that they rely on *pattern* to reflect process, and often several processes can result in similar patterns. Productivity, resilience, and stability are some of the ecological functions relevant to ecosystem management, and some early successful attempts to link diversity with function include Rosenberg (1976) and Schafer (1973). However, the link between ecosystem process and diversity is not always clear even for well-studied communities, so it is not surprising that linkages between ecosystem processes and nematode diversity are also unclear (Ettema, 1998).

Two ecological approaches are necessary for community analysis, both *autoecology*, the study of an individual species and its interaction with the environment, and *synecology*, the study of a community of species interacting together in a predictable manner for several groups of organisms, including nematodes. Nematode communities differ in the degree to which their autecology and synecology are understood and, thus also vary in the potential for classical community indices to reflect ecological processes. Often simple univariate indices are more successfully applied to communities in which the autecology of community members and the synecology of the system results in processes that have distinct and well known patterns. This is not always the case for nematode communities, however. When the ecology of the community or system is poorly understood, more complex community assemblage and multivariate community analysis is required to discern

 Nematodes as Environmental Indicators (eds M.J. Wilson and T. Kakouli-Duarte)

patterns. Application of diversity indices to describe nematode communities is insufficient as a stand-alone indicator of ecological processes because the ecology of nematode communities is simply not known well enough for most habitats.

For ecological community assays, a few routine diversity indices can be reported in ecological studies to benefit meta-analyses in linking past, present, and future studies. However, the current state of knowledge does not permit univariate diversity indices to conclusively reveal ecological processes. Therefore, it is imperative to complement univariate identity-independent approaches with multivariate identity-explicit approaches to improve our understanding of both the autecology of individual community members and synecology of the community as a whole. In this chapter, we first offer recommendations on performing some of the common identity independent ('diversity') indices and, second, suggest methods of incorporating community data into identity-explicit analysis with community assemblage and multivariate techniques.

Univariate Identity Independent Indices

In the broadest sense, diversity can refer to the sum of differences in form and function of life, including multiple scales of organization (ranging from the gene to the biome), space (with alpha diversity reflecting localities, beta diversity reflecting landscapes, and gamma diversity reflecting regions), and diversity of habitat and environmental disturbance types. The following section is concerned mostly with the representation of alpha diversity at the biological organization level of species and above. Although general ecological studies apply the following indices in the context of species, most nematode communities are enumerated at coarser resolutions because species identifications based on morphology are difficult (Neher, 2001). Besides, functional groups are a practical necessity because the effect of individual species on ecosystem processes has yet to be determined (Chapin *et al.*, 1992).

Diversity indices have their roots in post-Second World War information theory with the goal of optimizing code length for digital communication. Theoretically, alpha diversity, richness, and evenness indices are applicable to any taxonomic level, which is thought to convey information, whether it is species, genus, family, or trophic group. The appropriate resolution should be determined by the objectives of the study. From an information-theory perspective, if information is lost at coarser resolutions then the corresponding index would be unlikely to distinguish among samples and statistical populations. From an ecological perspective, however, if ecological information is lost at fine resolution, the corresponding index may also be unlikely to distinguish among samples and treatments. To illustrate this, we have computed Shannon's diversity index at the species, genus, family, and trophic group level (Table 4.1) from data published by Yeates and Cook (1998). Each soil type exhibits a unique pattern of diversity between management practices when viewed at various levels of taxonomic resolution.

Table 4.1. Mean (and standard deviation) of Hill's N1 diversity (N1 = exp[H']) at the species, genus, family, and trophic level for nematodes from organic and conventional grassland management regimes of three Welsh soils: Conway fine silt, Moor Gate coarse loam and Newport sand (n = 10 in all cases). Data from Yeates and Cook (1998).

	Silt[a]		Loam[a]		Sand[a]	
	Conventional	Organic	Conventional	Organic	Conventional	Organic
$N1_{species}$	10.7	13.1	14.0	** 17.4	17.8	15.8
	(3.70)	(3.48)	(3.02)	(2.61)	(3.70)	(2.47)
$N1_{genus}$	9.8	12.2	12.6	** 15.4	16.1	* 13.9
	(3.38)	(3.10)	(2.57)	(2.38)	(3.40)	(1.96)
$N1_{family}$	8.2	8.6	9.7	* 11.1	12.7	** 10.8
	(2.74)	(2.02)	(1.74)	(1.63)	(2.42)	(0.95)
$N1_{trophic}$	2.7	2.8	3.0	3.1	3.1	* 3.5
	(0.57)	(0.41)	(0.41)	(0.30)	(0.47)	(0.43)

[a]non-adjusted t-test between management regimes of similar soil type. * $P < 0.1$, ** $P < 0.05$.

One conceptual challenge with applying diversity, richness, or evenness indices to trophic or functional groups is that a 'nematode community' does not represent an entire soil or aquatic community but rather parts of several communities that may or may not interact with each other directly; the remainder of the communities are comprised of organisms that may be considered outside the scope of the study or simply not enumerable quantitatively from a nematode extraction. For example, soil systems include several bacterivorous water-film faunal groups. Only part of this group is represented by bacterivorous nematodes and the rest of the group is composed of amoebae, flagellates, ciliates, rotifers, and other taxa. There is a paucity of data that compare how changes in the composition of nematode trophic groups parallel those of protozoan communities in the sample.

A second conceptual challenge with diversity indices is the interpretation of evenness in reflecting community structure; the ecological implication of a uniform distribution of species from multiple trophic groups is unknown. For example, ecologists may agree that a community with low evenness (e.g. 19 species of very low relative abundance dominated by one or two enrichment bacterivorous nematodes of high relative abundance) might represent a disturbed community, relative to a community with intermediate evenness. However, a community with evenly distributed abundance (e.g. where predators are equally abundant as microbivores) might also reflect a recent disturbance. Furthermore, the mechanisms for how predators and competitors (e.g. protozoa, tardigrades, and microarthropods) influence the diversity of nematode species are still unclear.

Identity-independent indices and their calculation

A variety of identity-independent indices is available to serve different purposes in different circumstances (Hill, 1973; Peet, 1974; Pielou, 1975). The total number of species collected from a sample can be referred to as *species richness* (if representative of a known number of individuals) or *species density* (if representative of a known number per mass or volume of soil). *Evenness* is the equitable distribution of proportions or relative abundance. Diversity, then, is a combination of both richness and evenness elements. Each diversity index weights richness and evenness uniquely, but all diversity indices generally function so that an increase in either richness or evenness will always increase diversity. In some reports, the term *diversity* continues to refer simply to the total number of species; it is preferable, however, to restrict the use of 'diversity' to incorporate both the number of species and evenness. Formulae for calculating several common indices are summarized in Table 4.2 and accompanied by a customized SAS macro written to compute all indices (Fig. 4.1).

Table 4.2. Selected richness, diversity and evenness indices that can be calculated for nematode communities.

Name	Equation*	Reference
Margalef's richness	$D_{Marg} = \frac{(S-1)}{\ln(N)}$	Margalef (1958)
Shannon's diversity	$H' = -\Sigma\,(p_i \ln p_i)$	Shannon (1948)
Hill's diversity	$N1 = \exp\,[\,-\Sigma\,(p_i \ln p_i)] = \exp\,(H')$	Hill (1973)
Simpson's dominance (infinite community)	$D = \Sigma\, p_i^2$	Simpson (1949)
Simpson's dominance (finite community)	$\lambda = \frac{\sum n_i\,(n_i - 1)}{N(N-1)}$	Simpson (1949)
Hill's reciprocal of D	$N_2 = (\Sigma\, p_i^2)^{-1} = 1/D$	Hill (1973)
Brillouin's diversity	$H = \frac{1}{N}\log\frac{N!}{\prod N_i!}$	Brillouin (1962) Pielou (1975)
Brillouin's maximum diversity	$H_{max} = \frac{1}{N}\ln\frac{N!}{(X!)^{S-r}(Y!)^r}$	Brillouin (1962) Pielou (1975)
Brillouin's minimum diversity	$H_{min} = \frac{1}{N}\ln\frac{N!}{(N-S+1)!}$	Brillouin (1962) Pielou (1975)
Brillouin's evenness	$J = \frac{H}{H_{max}}$ or $J' = \frac{H'}{\ln S}$	Brillouin (1962) Pielou (1975)
Brillouin's relative evenness	$V = \frac{H - H_{min}}{H_{max} - H_{min}}$	Hurlbert (1971) Pielou (1975)
Hill's evenness	$E_{2,1} = \frac{(N_2)}{(N_1)}$	Hill (1973)
Heip's evenness	$E_{Heip} \frac{(e^{H'} - 1)}{(S-1)}$	Heip (1974)

*p_i represents the proportion of the i-th taxa in a sample, or n_i the number, with N individuals and S total species. X (in Brillouin's maximum diversity) is the integer portion of (N/S), Y = X+1, and r = the remainder of X.

```
%let species = a b c d e f g h;
data countdata;
    input sample &species;
cards;
1     12    18    17    12    3     4     18    15
2     10    11    28    8     26    3     7     8
3     10    11    19    22    18    6     3     13
4     12    18    9     13    18    15    12    4
5     19    5     4     26    4     11    17    14
6     15    18    0     11    14    13    14    14
7     14    12    19    12    6     9     13    16
8     16    15    19    18    5     9     2     16
9     20    4     16    7     26    3     3     21
10    14    23    27    5     0     12    14    4
;
proc IML;
use countdata;
read all var {&species} into data;
/* N = column vector of count sums */;
N = data[,+];
/* p = matrix of proportions */;
p = j(nrow(data),ncol(data),0); *Pre-allocate space;
p = data # (1/N); *elementwise division of data by sums;
/* richness = column vector of taxa present */;
richness = (data>0)[,+]; *only data > 0 are used;
MargalefsD = (richness - 1) # (1 / log(N)); /* Margalef's D index as richness corrected for N */;
/* Shannon's H index as the opposite of the sum of all proportions times ln(proportions) */;
nonzeros = loc(p>0); *nonzeros = a row vector of elements of p that are present;
ShannonH = j(nrow(data),ncol(data),0); *Pre-allocate space to speed up computation;
ShannonH[nonzeros] = p[nonzeros] # log(p[nonzeros]); *all absent species remain zero;
ShannonH = -ShannonH[,+]; *opposite to the sum of all columns;
/* Simpson's D dominance index (community and sample) as the sum of all proportions squared */;
SimpsonD = p[,##];
Simpsonfinitelambda = ((data # (data - 1))[,+]) / (N # (N - 1));
/* Hill's diversity (N1 and N2) and evenness (E21) */;
HillsN1 = exp(ShannonH); *exponent of Shannon's H ;
HillsN2 = 1/SimpsonD; *inverse of SimpsonD;
HillsE21 = HillsN2 / HillsN1; *ratio of N2 to N1;
/* Heips alternative evenness */;
BrillouinJprime = ShannonH/log(richness);
HeipE = (HillsN1 - 1) / (richness - 1);
/* Brillouin's indeces */;
if any(N>=100) then largeN = loc(N>=100); *largeN = a row vector of locations where N >= 100;
if any(N<100) then smallN = loc(N<100); *smallN = a row vector of locations where N < 100;
/* SAS fact(n) may not compute factorials for large n (> 100) so it is necessary to run */;
/* an alternate module to compute the log of n! by summing a vector of 1:n */;
/* IML module to compute natural log of factorial of large (>100) n */;
/* ---------------------------------------------------------------- */;
/* use: factorial(n) returns: log(n!) */;
start factorial(n);
factorial = j(nrow(n),1,0);
do k = 1 to nrow(n);
```

Fig. 4.1. Annotated SAS/IML code to illustrate an approach to implementing the univariate indices of Table 4.2 from within SAS.

```
        a = 1:n[k];
        temp = log(a);
        factorial[k] = temp[,+];
    end;
    return (factorial);
    finish;
    BrillouinH = j(nrow(data),ncol(data),0); *Pre-allocate space;
    BrillouinH = log(fact(data))[,+]; *Sum of log(Ni!), which equals the logarithm of the products of Ni!;
    logNfact = factorial(N); *Call IML module factorial(n) for n > 100;
    BrillouinH = (1/N) # (logNfact - BrillouinH); *final BrillouinH calculation
    /* Brillouin's Hmax */;
    intBrillouinHmax = int(N # (1/richness)); *integer portion;
    r = mod(N,richness); *remainder, or modulus;
    BrillouinHmax = j(nrow(data),1,0); *Pre-allocate space;
    do j = 1 to nrow(data); *repeat for each row;
    BrillouinHmax[j] = log((factorial(intBrillouinHmax[j]) ## (richness[j] - r[j])) # (factorial((intBrillouinH
      max[j] + 1)) ## r[j]));
    end;
    BrillouinHmax = (1/N) # (logNfact - BrillouinHmax); *final computation
    /* Brillouin's Hmin */;
    BrillouinHmin = factorial(N - richness + 1);
    BrillouinHmin = (1/N) # (factorial(N) - BrillouinHmin);
    /* Brillouin's evenness (J for samples, Jprime for collections) and relative evenness (Vrel) */;
    BrillouinJ = BrillouinH / BrillouinHmax; * ;
    BrillouinJprime = ShannonH / log(richness); * ;
    BrillouinVrel = (BrillouinH - BrillouinHmin) / (BrillouinHmax - BrillouinHmin); * ;
    print N richness MargalefsD ShannonH SimpsonD Simpsonfinitelambda HillsN1 HillsN2 HillsE21
        BrillouinH BrillouinHmax BrillouinJ BrillouinJprime BrillouinHmin BrillouinVrel;
    CREATE indices var {richness MargalefsD ShannonH SimpsonD Simpsonfinitelambda HillsN1
        HillsN2 HillsE21
        BrillouinH BrillouinHmax BrillouinJ BrillouinJprime BrillouinHmin BrillouinVrel};
    APPEND;
    quit;
    data final;
        merge countdata(keep=sample) indices(keep = richness MargalefsD ShannonH SimpsonD
        Simpsonfinitelambda HillsN1 HillsN2 HillsE21
        BrillouinH BrillouinHmax BrillouinJ BrillouinJprime BrillouinHmin BrillouinVrel);
    run;
    proc print data = final;
    run;
```

Fig. 4.1. Continued

The procedure for enumerating nematodes should be standardized with each experiment or sampling regime to prevent artifacts of sampling effort when reporting richness and diversity indices. Nematodes are enumerated differently than, for example, vascular plant surveys, in that nematodes are not enumerated as they appear *in situ*, but rather are extracted from soil, benthos, or water samples. Nematode density generally varies widely from sample to sample, so the number of nematodes enumerated is a representative subset of the total number extracted, i.e., an unknown number at the time of sampling. Therefore, species richness is the appropriate term to refer to the total number of species found when enumerating a uniform *number of extracted individuals* (e.g. 200 from each sample) from samples of a uniform initial mass or volume. Species density differs by referring to total number of species expressed as a uniform *portion of all extracted individuals* (e.g. 20% of the individuals from each sample). Regardless, for quantitative comparisons, it is preferable to begin with approximately the same initial mass or volume of soil, water, or benthos. As a rough guide, the range of initial mass or volume for all samples of comparison should be within 5% of the mean. Notice that species richness and density are not necessarily linear in relationship. For example, 20 species found among 200 individuals may not extrapolate to 40 species from 400 individuals. For this case, *rarefaction* of original data is necessary to estimate the number of species collected from a hypothetical number of individuals from the same population and species abundance curve (Gotelli and Colwell, 2001). It is essential to explicitly state the conditions in which richness indices were computed. Sometimes it is neither practical nor possible to enumerate a uniform number or portion of individuals, such as nematodes collected from small, isolated habitats such as pitcher plants, epiphytes, or an insect. In such a case, Margalef's index (= [S – 1] / ln[N]) can be used to adjust the number of species (S) for the number of individuals enumerated (N).

Evenness indices appear infrequently in the literature. Heip (1974) proposed an evenness index (= [exp(H') –1] / [S – 1]) which standardizes the Shannon's diversity index (H') by total number of species (S). Alternatively, Brillouin (1962) developed a series of statistics for censused communities that are computationally complex. Brillouin's maximum theoretical diversity is computed with the assumption that all individuals are distributed as uniformly as possible, and minimum theoretical diversity is computed assuming all individuals are distributed as asymmetrically as possible. Two forms of evenness can be computed, the first as diversity relative to maximum diversity and the second ('relative evenness') as diversity relative to maximum diversity but scaled to minimum diversity. The former relative evenness (not scaled to minimum diversity) can be based on two estimates of diversity depending on whether the user wishes to assume a finite or infinite community enumeration. Use Brillouin's sample diversity relative to Brillouin's maximum diversity when assuming a finite community enumeration, or Shannon's population diversity relative to the natural logarithm of richness when assuming infinite community enumeration. The second 'relative' form of evenness (scaled to minimum diversity) uses Brillouin's calculation of diversity from a censused community. Although nematode communities are rarely, if ever, fully censused in nature, the assumption of complete enumeration may be appropriate in some unique applications, e.g. small isolated habitats or virtual individuals in a computationally simulated model community.

Ecologists disagree on the best method to incorporate both richness and evenness, as well as the degree to which dominant and rare species, respectively, should influence the index. Therefore, exercise caution in application and interpretation of diversity indices. Shannon's diversity (= $-\Sigma[\,p_i \ln p_i]$, Shannon, 1948) is a popular diversity index. The exponent of Shannon's index (=exp[H'], also called Hill's N1) can be interpreted as the number of uniformly distributed species that would produce an identical Shannon's index as the non-uniformly distributed community. For example, consider a community with 20 non-uniformly distributed species and a Shannon's index of

2.3. The exponent of 2.3 (Hill's N1) equals 9.97, so, intuitively, approximately 10 uniformly distributed species would be needed to produce a Shannon's index similar to the community of 20 non-uniformly distributed species. Furthermore, Heip's evenness index (above) = ([9.97 – 1] / [20 – 1]) = 0.47, indicating again that roughly half of the observed species would be necessary to produce a similar Shannon's index if they were uniformly distributed. Simpson's index (D ($= \Sigma p_i^2$), Simpson, 1949), is considered a dominance index because it increases as species are distributed more unevenly (an increase in dominance) and can be interpreted intuitively as the probability that two randomly selected individuals from an infinite community will be the same. The reciprocal of Simpson's index (Hill's N2 = 1 / D) is often reported as a diversity index, and like Hill's N1, Hill's N2 can be interpreted as the number of uniformly distributed species that would produce a Simpson's index identical to that of the non-uniform community. Notice that the minimum Simpson's D possible (i.e. least dominance by any taxa) is 1/S and the maximum Hill's N2 possible (greatest equitability) is S, so we could compute an evenness index similar to Heip's approach as N2 / S.

Community Assemblage Models

Ecological succession

Ecological succession refers to a relatively predicable or directional sequence of spatio-temporal patterns of ecological interactions within a community. As species composition changes, it alters the abiotic environment, which in turn selects against the existing community favoring a community composition that performs better under the newly created abiotic environment. The concept originated in plant ecology (Whittaker, 1975), but also applies to invertebrate communities in soil and sediment. Succession usually progresses directionally unless set back by an environmental disturbance such as cultivation, pollution, or nutrient enrichment (Neher, 1999). Therefore, quantitative measures of ecological succession can serve as indicators of disturbance. With improved knowledge of synecology of nematode communities, one could identify the type and intensity of disturbance based on an index of succession.

Bongers (1990) proposed an index of ecological succession for application to nematodes whereas Ruf (1998) applied a similar approach to mesostigmatid mites. An alternative approach is to quantify species assemblage patterns. This can be achieved by repeated sampling methods or a Mantel test (Manly, 1997). These approaches are computationally intensive but practical given the speed of current computer systems. Repeated sampling methods include techniques referred to as bootstrap, resampling, jackknife, randomization, and Monte Carlo (Manly, 1997). A Mantel test computes a correlation coefficient among matrices. Each matrix can represent an assemblage of species in a community through time. Data can be entered as raw or transformed in variables that are continuous, ordinal or binary (Peres-Neto and Jackson, 2001). One can test hypotheses that concern the (dis)similarity of order and composition between two communities or treatments. A third variable, e.g. spatial pattern can be adjusted by using a partial Mantel test. These approaches are rank or distribution-free which allow them to be applied to small and unbalanced data. Data are reshuffled or resampled repeatedly for 10,000 to 100,000 times to compute a *P*-value and confidence intervals. The level of significance possible is affected by the choice of distance measure (Jackson, 1995). Distance can be quantified in Euclidean and non-Euclidean spaces (e.g. genetic distance, Bray Curtis). The methods vary in their sampling approach (i.e. with or without replacement) and the sample size (i.e. replacing the whole or subset of original sample). In addition to ecological succession, these approaches can also be applied to quantifying other ecological phenomena including intrinsic rate of increase (*r*), estimate of genetic distance, ecological divergence or phylogenies, microarrays, and biogeography (Felsenstein, 1985; Hillis and Bull, 1993; Jackson, 1995; Efron *et al.*, 1996; Rossi, 1996, 2003; Diniz-Filho *et al.*, 1998; Kerr and Churchill, 2001).

An alternative approach is the Procrustean superimposition approach (Gower, 1971). It differs from Mantel by scaling raw data or their ordination solutions to find optimal superimposition rather than transformation. This avoids the problem that the space between distances of transformed variables may not be necessarily equivalent to ones taken from the space of the original data. This approach is more powerful and results in lower type I error rates than the Mantel test (Peres-Neto and Jackson, 2001). Commercial computer software is available to compute most of these indices (Table 4.3).

Neutral community assemblage models

In addition to the niche-based models that are the impetus for the successional, seasonal, disturbance, and habitat-based studies that dominate historical nematode community analyses, non-neutral models present a necessary alternative perspective to spatio-temporal dynamics of communities. Neutral models have been in use for some time but were brought to the forefront of ecology as Hubbell (2001) presented a neutral model of community dynamics whereby individuals immigrate from a metacommunity into a local community at random with ecologically equivalent fitness. The spatio-temporal dynamics of neutral communities resemble neutral drift, analogous to random genetic drift. The surprising result of the neutral community model is that the distribution of species abundance closely resembles natural abundance distributions. This finding is controversial because, although neutral models can predict many ecological patterns, Hubbell's implicit assumption of neutrality (i.e. ecological equivalence) challenges nearly 150 years of niche-based ecology that sought to delineate the niche boundaries of what were believed to be non-neutral species. However, many authors suggest that neutral and non-neutral models of community assemblage may not necessary be contradictory, but rather complementary (Chave, 2004). Neutral dynamics may occur when non-neutral interactions play out on a reciprocal

Table 4.3. Software packages containing univariate statistical procedures.

		EstimateS[a]	Primer-E[b]	R[c]	Canoco 4.5	PC-ORD 4	PROTEST[d]	NTSYS-PC[e]
Univariate	Diversity		×		×	×		
	Shannon	×						
	Simpson	×						
	Margalef's							
	Evenness				×	×		
	Brillouin							
Ecol. Succ.	Maturity							
	Simple Mantel		×	×		×		×
	Partial Mantel			×				
	Procrustes						×	×
	Jackknife	×				×		×
	Bootstrap	×						×

[a]EstimateS, Statistical Estimation of Species Richness & Shared Species from Samples (viceroy.eeb.uconn.edu/estimates).
[b]Primer-E, version 5 available, Plymouth Routines in Multivariate Ecological Research (www.primer-e.com).
[c]The R Package, (http://www.bio.umontreal.ca/Casgrain/en/labo/R/v4/index.html), French version also available.
[d]PROTEST software available on http://www.zoo.utoronto.ca/jackson/software/.
[e]NTSYS-PC (Rohlf, 1994), Version 2.2 was initially released Sept. 2005 (http://www.exetersoftware.com/cat/ntsyspc/ntsyspcfaq.html), Exeter Software Inc., 100 North Country Road, Building B, Setauket, NY 11733.

non-uniform fitness landscape, in effect, equalizing fitness. Neutral processes likely occur within nematode communities because many species appear to be functionally redundant. However, nematodes may not be suitable to test neutral theories because most representative neutral models are spatially and temporally explicit (Holyoak and Loreau, 2006) and the destructive nature of nematode enumerations prevent truly repeated samplings of the same volume. It is important to remember that neutral dynamics may occur over the course of an experiment and it may not be advantageous to force niche-based explanations onto what may be neutral dynamics.

Multivariate Techniques

Multivariate analysis offers both descriptive and inferential procedures to analyse multiple variables simultaneously so as to reveal the collective interactions of all variables and the effect each variable has on the others. *Descriptive* procedures help to illustrate the overall structure of a dataset while *inferential* procedures help to test hypotheses of interactions. Therefore, multivariate analysis has two complementary applications, *exploratory hypothesis-generating* and *inferential hypothesis-testing*, that can be combined into a two-phase approach that might begin with an exploratory phase that seeks patterns in nature by asking 'to what can I ascribe the variation in my data?'. The second phase, then, tests the hypotheses that were generated by asking 'can I reject the null hypothesis that species are unrelated to each other or postulated environmental factor(s)?'. In this way, multivariate analysis is useful in evaluating nematode community structure as a biological indicator by keeping the identity of individual taxa explicit throughout the analysis. Below, we discuss two types of multivariate analysis commonly applied to nematode communities, cluster analysis and ordination (see also Trett *et al.*, Chapter 12, this volume). Commercial software packages that compute these procedures are summarized in Table 4.4.

Cluster analysis

Cluster analysis treats each multivariate observation (sample) as a vector and attempts to group vectors that are similar to each other into clusters (see Figure 12.5). Cluster analysis begins with a (dis)similarity matrix, often computed as the Euclidean distance among all pairs of vectors. *Hierarchical clustering* algorithms are either agglomerative or divisive. *Agglomerative clustering* begins with each vector representing a unique cluster and sequentially combining the two nearest clusters into one until an optimal number of clusters have been obtained. Divisive clustering begins with one cluster containing all vectors and sequentially divides the cluster into two until an optimal number of clusters have been obtained. Agglomerative clustering is most common and there are several methods of determining the distance of vector clusters from each other. The single linkage (or nearest neighbour) method determines

Table 4.4. Software packages containing multivariate statistical procedures.

Category	Statistic	SAS[a]	SPSS[b]	Statistica[c]	SYSTAT[d]	Canoco[e]	Primer-E[f]	PC-ORD[g]	SYN-TAX
	Cluster	CLUSTER FASTCLUS	CLUSTER	Join	Join		×	×	×
	Discriminant	DISCRIM STEPDISC	DISCRIMINANT	Discriminant	MGLH		×		
Direct Gradient	Canonical correspondence analysis (CCA)					×		×	×
	Nonmetric multidimensional scaling (NMDS)						×	×	×
	Redundancy Analysis (RDA)					×			×
	Detrended canonical correspondence analysis (DCCA)					×			
	Canonical correlation analysis	CANCORR							
Indirect gradient	Polar (= Bray-Curtis) Ordination (PO)					×		×	
	Principal coordinates analysis (PCoA)					×			
	Principal components analysis (PCA)	PRINCOMP FACTOR	FACTOR	Factor	Factor	×	×	×	×
	Correspondence analysis (CA)	CORRESP				×		×	×
	Detrended correspondence analysis (DCA)					×		×	
	Principal response curves (PRC)					×			

[a]SAS Version 9.1 (http://www.sas.com/).
[b]SPSS Version 15 (http://www.spss.com/).
[c]Statistica Version 8 (http://www.statsoft.com/).
[d]SYSTAT Version 12 (http://www.systat.com/).
[e]Canoco Version 4.5 (http://www.microcomputerpower.com/).
[f]Primer-E: Plymouth Routines in Multivariate Ecological Research (http://www.primer-e.com).
[g]PC-ORD, Version 4, MjM Software Design.

the distance between two clusters as the minimum distance (e.g. Euclidean) between the two most similar vectors of each cluster, while the complete linkage (e.g. farthest neighbour) method determines the distance between two clusters as the maximum distance (e.g. Euclidean) between the two most dissimilar vectors of each cluster. The average linkage method defines the distance between two clusters as the average distance of all elements from each cluster, while the *centroid method* defines the distance between two clusters as the distance between the two mean (or median) vectors of a cluster, called the centroids. Finally, *Ward's method* joins clusters so as to minimize the increase in sum of squares within and between clusters. The result of hierarchical cluster analysis is a dendrogram (tree diagram) that shows each step of the clustering procedure and the distance at which the clusters merge.

Discriminant Analysis is a related approach based on an a priori expectation of group members whereas cluster analysis has no preconceived expectation of group members and therefore conducts a posteriori aggregation. With discriminant analysis, one hypothesizes that there are two or more distinct groups, and then determines whether the observations divide significantly among those two predicted groups (Afifi and Clark, 1997; McGarigal *et al.*, 2000; Gotelli and Ellison, 2004).

Ordination

Ordination techniques are popular in community analysis due to their ability to visualize data in two-dimensional space. There are two main classes of ordination techniques, direct and indirect gradient analysis. *Indirect gradient analysis*, also called unconstrained, seeks to interpret patterns from within a dataset. *Direct gradient analysis*, also called constrained, seeks to extract patterns from known gradients and is therefore *constrained* by the environmental variables supplied. Indirect gradient analysis is divided into distance-based and eigenanalysis-based methods whereas all direct gradient analyses are eigenanalysis-based methods. Examples of distance-based indirect gradient ordination include Polar Ordination (PO), Principal Coordinates Analysis (PCoA) and Nonmetric Multi-Dimensional Scaling (NMDS). In polar ordination, two samples most different from each other based on their species composition serve as endpoints and all other samples are plotted relative to them. In this way, new samples can be added to polar ordination without changing the structure of the ordination diagram. Principal Coordinates Analysis simply maximizes linear distance measures of the ordination in metric space (using a distance matrix), while NMDS is analogous to a non-parametric variant of PCoA by maximizing *rank* distance measures of the ordination in non-metric space. Computer software packages are commercially available to compute any of these methods (Table 4.4).

The concept of eigenanalysis, used in the remaining indirect and direct gradient analyses, is important but somewhat more tedious. Eigenanalysis is a procedure to reduce the dimensionality of data that also begins with a square distance, similarity, correlation or covariance matrix. The result of eigenanalysis, an *eigenvalue* and its corresponding *eigenvector*, describes the data matrix as a multidimensional volume. Consider a cluster of samples with three species (x, y, and z) plotted in three-dimensional space (i.e. along axes x, y, and z) that take the shape of a ball. Eigenanalysis of these data circumscribes a volume around the points and the dominant eigenvalue (one of three) describes the length of the longest dimension and so describes the greatest amount of variance in the data. If the second eigenvalue, which is the length of the second longest dimension, is much shorter than the first, the ellipse around the points in these two dimensions is oblong, and the three-dimensional volume resembles the shape of a rugby football. Just as eigenvalues describe the shape of the volume of data points, eigenvectors describe the orientation of the data points, with the first eigenvector defining the orientation of the first eigenvalue, and so on. In this way, datasets with multiple variables can be visualized in multi-dimensional space derived from latent gradients.

Both direct and indirect eigenanalysis methods are available to model either linear or unimodal (humped or convex) responses to environmental gradients. Within indirect eigenanalysis methods, Principal Components Analysis (PCA) is a special eigenanalysis expression of PCoA using Euclidean distance and models a linear response of variables while Correspondence Analysis (CA) models a unimodal response of variables. Within direct (constrained) methods, Redundancy Analysis (RDA) models a linear response of variables while Canonical Correspondence Analysis (CCA) models a unimodal response of variables.

Data transformation

The multivariate normal distribution is analogous to the univariate normal distribution. As multivariate ordination is an extension of multiple regression, the transformation of variables for multivariate analysis is similar to the problem of transforming variables for regression. Lepš and Šmilauer (2003) advise against strict adherence to traditional Gaussian (i.e. normal) distributions, but rather recommend 'the semantics of the hypothesis you are testing'. To them, the effect of violating multivariate normality on the results of multivariate analysis and ordination is unclear and often considered insignificant. If one wishes to interpret the association between variables on a scale of *one measurement unit*, then it is acceptable to use non-transformed variables. However, many animal populations follow alternative scales, such as a logarithmic or square-root. In these cases, it is appropriate to apply a logarithm or square-root transformation for species data. In the case of log transformations, a natural logarithm (ln or $\log_e$) is used more often than a $\log_{10}$-transformation, although both give similar results. The only computational restrictions to transformations are zeros and negative values. Linear multivariate models (as in PCA and RDA) can employ negative values so that the data can even be centered and standardized. However, unimodal models (as in CA and CCA) cannot employ negative numbers and so logarithm transformations employ a constant such as $\log(x + b)$ where b is some small constant to accommodate zeros. Again, NMDS is a non-parametric analog to multivariate analysis.

Visualization

The result of ordination is a biplot or diagram that illustrates the data, either species scores or sample points or both, plotted in multivariate space typically along two axes, but sometimes three. The goal is to explain up to 80% of the variation. If environmental variables are included to constrain the analysis, such as with direct gradient analysis, it is possible to overlay the environmental variables as vectors through the observations. In this case the length of the vector represents its descriptive importance, the direction indicates the vector's correlation with various sites or species, and the angle between vectors reflects the variable's correlation with other variables. These ordination plots are generally used to represent snapshot datasets taken at one point in time. Principal Response Curves (PRC) is a novel approach to representing repeated measures multivariate data, especially in the context of community responses to stress (Van den Brink and ter Braak, 1998, 1999; Van den Brink *et al.*, 2003). The ordination begins with redundancy analysis (the linear model of direct gradient analysis) with time as a covariate. The resulting diagram is a community coefficient along the y-axis plotted against time along the x-axis with the community coefficient representing the relative change in community composition (as indicated by an accompanying list of species scores) of treatment populations against a control population.

Conclusion

Classical community composition can be analysed using metrics that either disregard or preserve the identity of taxa within the community. Identity-independent methods such as diversity and evenness indices are relatively simple to compute and analyse statistically. However, the user must exercise caution by selecting the form of index most appropriate to the goals of the study, and resisting the temptation to singularly extrapolate to a greater ecological meaning without substantial supplementary evidence. Alternatively, indices that incorporate and/or maintain taxon identity can more convincingly be linked to ecological process and function. Measures of ecological succession and species assemblage are univariate forms that can be analysed using traditional statistical tools such as regression and analysis of variance. Given the advances in computer technology, a variety of multivariate methods are accessible through commercial software packages. Many multivariate approaches capture a one-time 'snapshot' of community composition. However, repeated measures approaches are becoming available to evaluate changes in community composition through time. Practitioners should be aware of the many limitations, assumptions, and caveats of community assemblage and multivariate techniques by consulting with expert statisticians. We recommend Afifi and Clark (1997), Legendre and Legendre (1998), Lepš and Šmilauer (2003), and Gotelli and Ellison (2004) as helpful texts for further study of these techniques.

References

Afifi, A.A. and Clark, V. (1997) *Computer-Aided Multivariate Analysis*. 3rd edn. Chapman and Hall, New York.

Bongers, T. (1990) The maturity index: An ecological measure of environmental disturbance based on nematode species composition. *Oecologia* 83, 14–19.

Brillouin, L. (1962) *Science and Information Theory*, 2nd edn. Academic Press, New York.

Chapin, F.S., Schultz, E.D. and Mooney, H.A. (1992) Biodiversity and ecosystem processes. *Trends in Ecology and Evolution* 7, 107–108.

Chave, J. (2004) Neutral theory and community ecology. *Ecology Letters* 7, 241–253.

Diniz-Filho, J.A.F., Oliveira, L.G., and Silva, M.M. (1998) Explaining the beta diversity of aquatic insects in 'Cerrado' streams from central Brazil using multiple Mantel test. *Revista Brasileira de Biologia* 58, 223–231.

Efron, B., Halloran, E., and Holmes, S. (1996) Bootstrap confidence levels for phylogenetic trees. *Proceedings of the National Academy of Sciences USA* 93, 13429–13434.

Ettema, C.H. (1998) Soil nematode diversity: Species coexistence and ecosystem function. *Journal of Nematology* 30, 159–169.

Felsenstein, J. (1985) Confidence limits on phylogenies: an approach using the bootstrap. *Evolution* 39, 783–791.

Gotelli, N.J. and Colwell, R.K. (2001) Quantifying biodiversity: procedures and pitfalls in the measurement and comparison of species richness. *Ecology Letters* 4, 379–391.

Gotelli, N.J. and Ellison, A.M. (2004). *A Primer of Ecological Statistics*. Sinauer Associates, Sunderland, Massachusetts.

Gower, J.C. (1971) Statistical methods of comparing different multivariate analyses of the same data. In: Hodson, F.R., Kendall, D.G., and Tautu, P. (eds) *Mathematics in the Archaeological and Historical Sciences*. Edinburgh University, Edinburgh, pp. 138–149.

Heip, C. (1974) A new index measuring evenness. *Journal of the Marine Biological Association* 54, 555–557.

Hill, M.O. (1973) Diversity and evenness: a unifying notation and its consequences. *Ecology* 54, 427–432.

Hillis, D.M. and Bull, J.J. (1993) An empirical test of bootstrapping as a method for assessing confidence in phylogenetic analysis. *Systematic Biology* 42, 182–192.

Holyoak, M. and Loreau, M. (2006) Reconciling empirical ecology with neutral community models. *Ecology* 87, 1370–1377.

Hubbell, S.P. (2001) *The Unified Theory of Biodiversity and Biogeography*. Princeton University Press, Princeton, New Jersey.

Hurlbert, S.H. (1971) The nonconcept of species diversity: a critique and alternative parameters. *Ecology* 52, 577–586.

Jackson, D.A. (1995) PROTEST: A PROcrustean randomization TEST of community environment concordance. *Ecoscience* 2, 297–303.

Kerr, M.K. and Churchill, G.A. (2001) Bootstrapping cluster analysis: Assessing the reliability of conclusions from microarray experiments. *Proceedings of the National Academy of Sciences USA* 98, 8961–8965.

Legendre, P. and Legendre, L. (1998) *Numerical Ecology*. 2nd English edition. Elsevier Science, Amsterdam.

Lepš, J. and Šmilauer, P. (2003) *Multivariate Analysis of Ecological Data using CANOCO.* Cambridge University Press, Cambridge.

Manly, B.F.J. (1997) *Randomization, Bootstrap and Monte Carlo Methods in Biology*, 2nd edition. Chapman and Hall, New York.

Margalef, R. (1958) Information theory in ecology. *General Systems* 3, 36–71.

McGarigal, K., Cushman, S. and Stafford, S. (2000) *Multivariate Statistics for Wildlife and Ecology Research.* Springer-Verlag, New York.

Neher, D.A. (1999) Soil community composition and ecosystem processes: Comparing agricultural ecosystems with natural ecosystems. *Agroforestry Systems* 45, 159–185.

Neher, D.A. (2001) Role of nematodes in soil health and their use as indicators. *Journal of Nematology* 33, 161–168.

Peet, R.K. (1974) The measurement of diversity. *Annual Review of Ecology and Systematics* 5, 285–307.

Peres-Neto, P.R. and Jackson, D.A. (2001) How well do multivariate data sets match? The advantages of a Procrustean superimposition approach over the Mantel test. *Oecologia* 129, 169–178.

Pielou, E.C. (1975) *Ecological Diversity.* Wiley, New York.

Rohlf, F.J. (1994) NTSYS-PC: Numerical taxonomy and multivariate analysis system, version 1.80. Exeter Software, Setaukat, NY.

Rosenberg, R. (1976) Benthic faunal dynamics during succession following pollution abatement in a Swedish estuary. *Oikos* 27, 414–427.

Rossi, J.P. (1996) Statistical tool for soil biology. 11. Autocorrelogram and Mantel Test. *European Journal of Soil Biology* 32, 195–203.

Rossi, J.P. (2003) The spatiotemporal pattern of a tropical earthworm species assemblage and its relationship with soil structure. *Pedobiologia* 47, 497–503.

Ruf, A. (1998) A Maturity Index for predatory soil mites (Mesostigmata: Gamasina) as an indicator of environmental impacts of pollution on forest soils. *Applied Soil Ecology* 9, 447–452.

Schafer, C.T. (1973) Distribution of foraminifera near pollution sources in Chaleur Bay. *Water Air Soil Pollution* 2, 219–233.

Shannon, C.E. (1948) A mathematical theory of communication. *Bell System Technical Journal* 27, 379–423, 623–656.

Simpson, E.H. (1949) Measurement of diversity. *Nature* 163, 688.

ter Braak, C.J.F. (1988) CANOCO – an extension of DECORANA to analyze species–environment relationships. *Vegetatio* 75, 159–160.

ter Braak, C.J.F. (1990) Update notes: CANOCO version 3.10. Agricultural Mathematics Group, Wageningen, The Netherlands.

Van den Brink, P.J. and ter Braak, C.J. (1998) Multivariate analysis of stress in experimental ecosystems by principal response curves and similarity analysis. *Aquatic Ecology* 32, 163–178.

Van den Brink, P.J. and ter Braak, C.J.F. (1999) Principal response curves: Analysis of time-dependent multivariate responses of biological community to stress. *Environmental Toxicology and Chemistry* 18, 138–148.

Van den Brink, P.J., Van den Brink, N.W. and ter Braak, C.J.F. (2003) Multivariate analysis of ecotoxicological data using ordination: demonstrations of utility on the basis of various examples. *Australasian Journal of Ecotoxicology* 9, 141–156.

Whittaker, R.H. (1975) *Communities and Ecosystems.* 2nd edn. Macmillian, New York.

Yeates, G.W. and Cook, R. (1998) Nematode fauna of three Welsh soils under conventional and organic grassland farm managements. In: de Goede, R.G.M. de and Bongers, T. (eds) *Nematode Communities of Northern Temperate Grassland Ecosystems.* Focus Verlag, Giessen, pp. 305–313.

5 Indices Developed Specifically for Analysis of Nematode Assemblages

Howard Ferris[1] and Tom Bongers[2]

[1]Department of Nematology, University of California, USA; [2]Laboratory of Nematology, Wageningen University, The Netherlands

History

Back in the early 20th century, about 50 years into the era in which the diversity and abundance of nematodes became generally recognized, Cobb (1914) calculated that if the nematodes resident in one acre (0.4 ha) of soil near San Antonio, Texas, USA, were to proceed head-to-tail to Washington DC, some 3200 km distant, the first nematode would reach Washington before the rear of the procession left San Antonio! Eighty years later, Jairajpuri and Ahmad (1992) estimated that nematodes constitute nearly 90% of the world's multicellular animals while Platt (1994) asserted that four out of every five multicellular animals on the planet are nematodes.

Awareness of the seething hordes of worms was at first constrained by the microscopic nature of most of the phylum Nematoda. The study of human and animal parasites, usually larger than soil and aquatic forms and of immediate interest to their hosts, was first to develop (Dujardin, 1845; Baird, 1853). Besides patches of poor growth or 'tired soil' of undetermined cause, early recognition of the prevalence and impact of plant-parasitic nematodes was of those for which there are characteristic signs and symptoms in buds, seed heads, stems, foliage and roots (see examples in Christie, 1959 and Filipjev and Schuurmans Stekhoven, 1941). Around 1850, marine biologists began to recognize nematodes; there were, for example, studies on the nematodes of Iceland (Leuckart, 1849), the Mediterranean (Eberth, 1863), the English coast (Bastian, 1865), the coast of Brittany (Villot, 1875) and on nematodes collected by various expeditions (Von Linstow, 1876). Freshwater nematodes began to receive attention around 1890 with the papers of Daday (1897) on the Hungarian fauna. The development of information on free-living soil nematodes is well reviewed by Filipjev and Schuurmans Stekhoven (1941), Overgaard Nielsen (1949) and Paramonov (1962). Early work on the free-living nematodes included careful descriptions of *Enoplus*, *Oncholaimus*, *Rhabditis* and *Dorylaimus* (Dujardin, 1845). That work was followed by a period of discovery during what Filipjev (1918) called 'the period from Bastian (1865) to de Man (1884-1893)' in which Bastian's studies on terrestrial, freshwater and marine nematodes were followed by further contributions to the knowledge of soil nematodes (e.g. Bütschli, 1873). Awareness of the diversity and abundance of nematodes was further accelerated with the impressive series of papers by de Man (see Karssen, 2006) and by Cobb (see Spenneman, 2003). Unfortunately, despite his legendary productivity, Cobb was sometimes unaware of the European literature, which resulted in a confusion of terminology, species descriptions and classification schemes (Overgaard Nielsen, 1949; Paramonov, 1962).

Further impetus to the study of nematology was provided by technological developments: advances in microscopy, biochemistry and molecular biology. Awareness of the impact of nematodes on plants increased as a result of the development and wide-scale use of nematicides; crop yields were improved in previously less-productive soils. Consequently, researchers and practitioners with varied backgrounds and interests were drawn into nematology. From the 1970s onwards, research on nematodes exploded following the fortuitous connections and interactions among Margaret Briggs, Ellsworth Dougherty, Warwick Nicholas and Sydney Brenner that resulted in the selection of *Caenorhabditis elegans* as a model organism for deciphering the genetic code (http://plpnemweb.ucdavis.edu/Nemaplex).

Prior to the eruption of research on *C. elegans*, other than the activities of a few taxonomists, most studies on soil nematodes centred on the biology and management of those that cause damage to higher plants. A milestone in the ecology of free-living soil nematodes was the seven-year study in Denmark by Overgaard Nielsen (1949) on nematode faunae of different soils, their physiological ecology and even their ecosystem services. Further notable ecological contributions emerged in the 1970s and 1980s (e.g. Nicholas, 1975). Centres of ecological study on nematodes developed in Sweden (e.g. Sohlenius, 1973), Poland (e.g. Prejs, 1970; Wasilewska, 1970), Italy (e.g. Zullini, 1976), Germany (e.g. Sudhaus, 1981), and Russia (e.g. Tsalolikhin, 1976). In the USA, there was a surge of activity in soil ecology at the National Resource Ecology Laboratory in Colorado Springs, led by Coleman and others (e.g. Yeates and Coleman, 1982), and similar activity at the Institute of Ecology of the University of Georgia, led by Crossley and colleagues (e.g. Stinner and Crossley, 1982). In the same time period, Yeates was developing a very productive programme on the ecology of soil nematodes in New Zealand (e.g. Yeates, 1979). A significant contribution was the publication of the PhD research of Ingham, with its accompanying review of preceding studies, in which the functional significance of bacterivore and fungivore nematodes was established by the demonstration that their excretion of nitrogen in excess of structural and metabolic needs stimulated plant growth (Ingham *et al.*, 1985).

During the latter part of the 20th century, analyses of nematode communities in aquatic environments revealed that the incidence and prevalence of species in the community reflect the nature and quality of the environment (e.g. Micoletzky, 1925; Prejs, 1970; 1977; Zullini, 1976; Callahan *et al.*, 1979). Many subsequent studies have expanded our understanding that various nematode species differ in response to degradation of aquatic environments. The nature and magnitude of changes in community structures of aquatic nematodes are recognized as excellent indicators of water and sediment quality in relation to pollution or enrichment (Beier and Traunspurger, 2001; Barbuta and Zullini, 2005).

So, interest in the role of nematodes in soil and aquatic ecosystems was established. In parallel, work continued in nematode taxonomy and systematics with understanding and interest accelerated by electron microscopy and molecular tools. In his oft-quoted passage reflecting that if all non-nematode matter was swept away, topography and land use patterns would be recognizable from the remaining nematodes, Cobb (1915) speculated that, had we sufficient knowledge, location and species of the various plants and animals would be decipherable by examination of their erstwhile nematode parasites. As documented above, we now have sufficient knowledge to concur with Cobb's speculation and, in fact, to expand on it in relation to environmental quality (Bongers and Ferris, 1999).

The development of nematodes as bioindicators in soil and aquatic systems required determination of appropriate ways to assess and quantify their contributions to ecological processes, and the validation of their utility as indicators of environmental condition. Several unique characteristics of nematodes facilitated those developments (see Yeates *et al.*, Chapter 1, this volume and Trett *et al.*, Chapter 12, this volume). Key among those characteristics is diversity, both taxonomic and functional, of nematodes.

Diversity Indices

Many indices have been developed and applied to assess the biodiversity of ecosystems (see Neher and Darby, Chapter 4, this volume for fuller explanation). In the purest sense, the indices are based on assessment of all organisms at the species level. In practice, they are usually applied at a resolution determined by available taxonomic knowledge and sometimes using data of differing taxonomic resolution, which confounds comparison among studies. They have seldom, if ever, been applied to an ecosystem in absolute terms; rather they are more likely applied to numbers of species of aboveground vertebrates, soil nematodes, etc.

Indices of taxonomic diversity have been described previously (Neher and Darby, Chapter 4, this volume); in summary they include: Species richness (S) (sometimes referred to as Hill's N_0 index (Hill, 1973)), Simpson's diversity index (D) (Simpson, 1951), Shannon's diversity index (H') (Shannon and Weaver, 1949), Hill's N_1 index (Hill, 1973) which is the exponential of the Shannon's index, Hill's N_2 index (Hill, 1973) which is the reciprocal of Simpson's index, and Pielou's J' evenness index (Pielou, 1966). These and other diversity indices provide assessment of organism heterogeneity in a system but no direct indication of organism or system function.

Table 5.1. Diversity index calculations for two nematode assemblages (Samples A and B) of the same number of taxa and the same number of individuals.

	Sample A	Sample B	Sample A Diversity calculations			Sample B Diversity calculations		
			p_i	p_i^2	$log(p_i)$	p_i	p_i^2	$log(p_i)$
Hoplolaimidae	5	15	0.045	0.002	–3.109	0.134	0.018	–2.010
Pratylenchidae	5	15	0.045	0.002	–3.109	0.134	0.018	–2.010
Aphelenchidae	15	5	0.134	0.018	–2.010	0.045	0.002	–3.109
Cephalobidae	15	5	0.134	0.018	–2.010	0.045	0.002	–3.109
Plectidae	2	15	0.018	0.000	–4.025	0.134	0.018	–2.010
Rhabditidae	5	50	0.045	0.002	–3.109	0.446	0.199	–0.806
Dorylaimidae	50	5	0.446	0.199	–0.806	0.045	0.002	–3.109
Discolaimidae	15	2	0.134	0.018	–2.010	0.018	0.000	–4.025
Totals	112	112						
Hill N_0	8	8						
Simpson	0.26	0.26						
Shannon	1.66	1.66						
Hill N_1	5.24	5.24						
Hill N_2	3.85	3.85						
Pielou J′	0.80	0.80						

p_i is the number of individuals of a taxon as a proportion of the total number of individuals.

The indices do not assess or measure the abundances or proportions of organisms that are food or feeders, prey or predators (Table 5.1). Rather, ecosystem function is inferred, for example, in the expectation that if there are many different types of organisms feeding on each other, the system has some internal regulation and therefore overall stability. While that conclusion may bear weight in Sample A with high proportions of omnivores (Dorylaimidae) and predators (Discolaimidae), it would be less valid for Sample B. In Sample B there are greater proportions of opportunists, including bacterial-feeding Rhabditidae, and plant feeders (Hoplolaimidae and Pratylenchidae) which may be responding to enhanced host vigor in the more enriched environment.

Indices of Ecosystem Function

While there are several indices of community and ecosystem structure (Table 5.1), there are fewer indicators of ecosystem function. Certainly, it is a common observation that a preponderance of herbivores in the nematode assemblage of soil is an indicator that management practices have diminished *functional diversity* in the soil food web. The first papers on the effect of disturbances on soil nematode assemblages appeared around 1960 with studies on the effect of liming and fertilization (Bassus, 1960, 1967), on nematodes of regenerating woodland and grassland (Yuen, 1966), in sand dunes, forests and grasslands (Yeates, 1968, 1972, 1974). Wasilewska (1974) noted that omnivorous dorylaimids of forest soil were most sensitive to disturbance. In the same period, Johnson *et al.* (1972, 1973, 1974), also studying nematode assemblages of forests, similarly concluded that dorylaimids are very sensitive to disturbance and should be considered K-strategists.

The potential for macrofauna to indicate water quality has a long history; the use of nematodes as indicators of water pollution was initiated in rivers by Zullini (1976) and in lakes by Prejs (1977). Zullini's focus on nematodes as bioindicators in freshwater systems emerged when he was appointed as an ecological expert in a lawsuit on river pollution and asked if he could prove that pollution affected the river biology (A. Zullini, 2007, personal communication). The early indicator work was followed by several studies on the relationships between water quality and nematode assemblages (e.g. Cantelmo and Rao, 1978; Boucher, 1980; Tietjen, 1980; Lambshead, 1986; Vranken *et al.*, 1988). The use and formalization of nematodes as bioindicators in freshwater systems has continued to expand (e.g. Hodda and Nicholas, 1986; Samoiloff, 1987; Yeow *et al.*, 1999; Bazzanti, 2000; Beier and Traunspurger, 2001, 2003; Barbuta and Zullini, 2005).

In the early 1980s, concerns regarding soil pollution and its impact on the functioning of soils were increasing. In The Netherlands, the Dutch National Institute for Public Health and the Environment (RIVM) started a search for groups of organisms with potential as bioindicators of soil quality similar to those used in biological water quality assessment. Nematodes and bacteria appeared to be the most promising organisms. In 1984, not having experience with nematodes, RIVM officials asked the second author of this chapter to bridge the gap between nematode taxonomists and soil ecologists by composing a user-friendly identification key in 'De Nematoden van Nederland' (Bongers, 1988), and to study the relation between soil type, vegetation type and nematode assemblages (Bongers *et al.*, 1989). One project of the RIVM soil ecology group monitored the biological recovery of contaminated soils after heat sterilization and amendment with organic material (Kappers and van Esbroek, 1988). Building on studies of ecological succession in cow dung (Sudhaus, 1981; Sudhaus *et al.*, 1988) and on the ecological studies of Johnson *et al.* (1972, 1973, 1974), Wasilewska (1970, 1974); Zullini (1976), Zullini and Peretti (1986) in a variety of ecosystems, Bongers *et al.* (1989) arranged nematode taxa into five categories along an r-K scale. That arrangement evolved into the Maturity Index for terrestrial and marine nematodes (Bongers, 1990; Bongers *et al.*, 1991).

Further catalytic activities followed, including evolution and testing of the ideas (De Goede *et al.*, 1993, Ettema and Bongers, 1993; Korthals *et al.*, 1996a, b, c) and cataloging of the feeding habits of soil nematodes (Yeates *et al.*, 1993). A testament to the biological insights underpinning the r-K scale is that, over the years, there have been few adjustments. One example of a change is based on the notion that opportunists can be distinguished as enrichment or general opportunists and that the former are characterized by having a dauerlarva stage. Since Monhysteridae do not have a dauerlarva stage and are tolerant of unfavorable conditions, the family was moved to the 'general opportunist' category (Bongers *et al.*, 1995).

Basics of the c-p series and the maturity index family

For calculation of maturity indices, soil nematodes are categorized into a 1-5 colonizer-persister series; ranging from extreme r- to extreme K-strategists. 'Colonizer' nematodes at the lower end of the c-p scale are considered enrich ment opportunists and therefore indicate resource availability; 'persister' nematodes at the high end of the scale indicate system stability, food web complexity and connectance. Each nematode taxon, usually at family level, is classified into one of the five c-p classes. Genera and species within a taxon have the same c-p value as their family, or genus in the case of some marine taxa. For the terrestrial and freshwater taxa, the following groups can be distinguished:

c-p1

Nematodes with a short generation time and a large proportion of the body occupied by gonads which produce many small eggs. Population growth under food-enriched conditions is explosive. The nematodes are primarily bacterial feeders with high metabolic activity. They are tolerant of pollutants and of products of organic matter decomposition. These enrichment opportunists form dauerlarvae when microbial biomass and activity decreases.

c-p2

Nematodes with a short generation time and relatively high reproduction rates, although lower than those in c-p1, consequently, they are slower to respond to environmental enrichment than c-p1 nematodes. These nematodes do not form dauerlarvae and occur in all environments, including those in which resources are abundant and those in which resources are scarce. They are very tolerant of pollutants and other disturbances. They include bacterial feeders, fungal feeders and a few predators.

c-p3

Nematodes with longer generation time than c-p2 nematodes and greater sensitivity to disturbances. They include bacterial feeders, fungal feeders and some predators.

c-p4

Small dorylaims and the large non-dorylaimids with a low ratio of gonad to body volume. These nematodes are characterized by a long generation time, permeable cuticle and high sensitivity to pollutants. The non-carnivorous nematodes in this group are relatively sessile whereas the carnivores actively seek prey. The group is composed of larger carnivores, smaller omnivores and some bacterial feeders.

c-p5

Large dorylaimid nematodes with a long life span, low reproduction rates, low metabolic activity and slow movement. The gonads are small relative to the body volume and produce a small number of large eggs. They have a permeable cuticle and are very sensitive to pollutants and other disturbances. This group is composed of the larger omnivores and predators.

As recognized early in the development of the c-p series (Bongers, 1990; Bongers *et al.*, 1991), a c-p classification at the genus or species level would be more informative. However, early attempts to assign c-p values at the genus level (Bongers *et al.*, 1989) proved difficult due to lack of information on the biology and sensitivity of the individual genera. Consequently, family level assignments to c-p classes were used in the formal introduction of the MI (Bongers, 1990). The relevance of the family level assignments has been justified on the basis that nematodes with similar morphology and feeding habits, and with similar life history traits, have a high probability of similar sensitivity and responsiveness to environmental change (Bongers and Ferris, 1999). As information emerges on the biology and sensitivity of individual genera and species, greater resolution in c-p assignments will be possible. The most recent descriptions of c-p class assignments for families of terrestrial nematodes is Bongers and Bongers (1998) and of marine nematodes is Bongers *et al.* (1991).

Calculation and use of the Maturity Index family

All the indices are based on the weighted proportion of nematodes in the fauna that meet the index criteria. A generic formula for calculation of indices in the MI family is:

$$XI = \sum_{i=1,f} v_i n_i \Big/ \sum_{i=1,f} n_i$$

where XI is the index of interest, v_i is the colonizer-persister (c-p) value assigned to taxon *i*, and n_i is the number of nematodes in each of the *f* taxa *that meet the criteria of the index.*

MI

The Maturity Index is based on non-plant-feeding taxa and considered a measure of environmental disturbance; low MI values indicate a disturbed and/or enriched environment, high MI values a stable environment (Bongers, 1990). In essence, the MI is an ecological indicator of the state of succession of a system whereby disturbance and its consequent enrichment effects result in a setback of succession to an earlier state (Odum, 1985). In the case of the nematode assemblage, the successional setback is reflected in a lower MI (Bongers *et al.*, 1997).

The dauerlarvae of enrichment opportunists, animal parasites such as mermithids, and entomopathogenic nematodes are excluded from the calculation of MI (Bongers and Bongers, 1998) as their presence does not provide information about the present functioning of the soil food web. An abundance of dauerlarvae indicates a system that has been enriched and has now declined to a less enriched phase. The ratio of dauerlarvae to active stages of rhabditids, as an indicator of resource availability, was introduced and tested by Sohlenius (1969, 1973) and comparisons of that ratio over time may provide insights into the resource dynamics of the system. However, a difficulty with such an approach would be the problem of identifying dauerlarvae of different nematode taxa. For example, dauerlarvae of entomopathogenic nematodes often are found in soil but are not indicators of food web enrichment.

PPI

The Plant Parasite Index, is comparable to the MI but computed only for the plant-feeding nematodes with the rationale that their abundance is determined by the vigor of their host plants which, in turn, is determined by system enrichment. Consequently, under nutrient poor conditions of natural ecosystems, often associated with a high proportion of Tylenchidae (c-p2) in the nematode assemblage, the PPI is lower than under enriched agricultural conditions, the inverse of the response of the MI to enrichment (Bongers, 1990; Bongers *et al.*, 1997). The reports that *Filenchus misellus* feeds on fungi (Brzeski, 1998; Okada *et al.*, 2002, 2005) underscores the need for further study on the feeding habits of the many genera and species in this ubiquitous Tylenchidae.

PPI/MI

The PPI/MI ratio is lower under nutrient poor conditions than under nutrient rich conditions. It is a sensitive indicator of enrichment in agroecosystems (Bongers and Korthals, 1995; Bongers *et al.*, 1997).

MI2-5

This index is identical to the MI but excludes the c-p1 enrichment opportunists. The index was derived during studies of the relationship between MI and copper concentration under agricultural conditions. In those studies, it was apparent that there was a strong relationship between decrease in higher c-p value nematodes and pollution-induced stress while the c-p1 nematodes responded to the presence of decomposing organic material. In some cases, the pollutant may become a resource for a component of the microbial community which, in turn, acts as a resource for the c-p1 nematodes. The MI2-5 was first discussed at the Crop Protection Symposium in Ghent (Bongers and Korthals, 1993).

ΣMI

This was proposed by Yeates (1994) and is equivalent to the Total MI of Wasilewska (1994). The index is the MI for all nematodes in the system, including plant feeders, based on the assertion that the complete assemblage provides integral information with regard to disturbance and environmental condition. If a soil ecosystem receives nutrient input, opportunistic bacterial- and fungal-feeding nematodes respond rapidly to the corresponding increase in their resources. Plant parasites do not respond in the short term but may increase later as a result of higher plant vigor. Since many are c-p3 or higher, the expected decrease of MI in response to enrichment is offset by inclusion of plant parasites in ΣMI. Further, many plant feeders, such as the c-p3 Pratylenchidae, are tolerant of pollutant stress (Korthals *et al.*, 1996a, b) which, in ΣMI, offsets the impact of pollution registered by the MI or MI2-5 (Bongers and Bongers, 1998; Bongers, 1999).

ΣMI2-5

This index computes the MI for all nematodes in the c-p2–5 range (Neher and Campbell, 1996). The index recognizes that the higher c-p value plant-feeding species also provide information of environmental stress but bears some of the burden of the ΣMI in situations of nutrient enrichment.

In all cases, the indices of ecosystem function in the Maturity Index family show differences between the two samples of identical species richness and abundance (Table 5.2) that were not apparent in the diversity and evenness indices calculated for the same data (Table 5.1).

Sometimes the Maturity Index has been expressed as

$$MI = \sum_{i=1,f} v_i p_i$$

which has led to some unfortunate miscalculations in manuscripts submitted for publication. The errors commonly arise when the proportions of all taxa present are calculated in a spreadsheet, as for the calculation of ΣMI, and then the same proportions, excluding those that are not relevant, are used to calculate incorrectly the other indices in the family. To obtain the correct index values, it is necessary to recalculate the proportions to be weighted with respect to the total number of nematodes in the sample which meet the specific criteria of each index.

Conceptual evolution underlying functional indices based on nematode faunae

In their equilateral c-p triangles, (graphical representations of faunal composition), De Goede *et al.* (1993) and Ettema and Bongers (1993) accommodated two enrichment (%c-p1 and %c-p2) axes, and an ecosystem complexity (%c-p3–5) axis, based on unweighted proportions of the nematode fauna in each grouping (Fig. 5.1). The right-angled triangle representation of Bongers *et al.* (1995) depicted the proportional representation of c-p1 and c-p3–5 nematodes, further emphasizing the indicator distinction between the two groups. The graphical representations advanced the recognition that the c-p classes are indicators of ecosystem structure and function that are not necessarily aligned on a common trajectory. However, since each axis of the triangles indicates a proportion of the whole nematode fauna, an increase along one axis is accompanied by a decrease along another. The notion that enrichment

Table 5.2. Calculation of the MI, MI2-5, ΣMI, ΣMI2-5, PPI and PPI/MI for two nematode assemblages (Samples A and B) of the same number of taxa and the same number of individuals (cf. Table 5.1).

					A	B	A	B	A	B	A	B	A	B
					MI	MI	MI2-5	MI2-5	ΣMI	ΣMI	ΣMI2-5	ΣMI2-5	PPI	PPI
Nematode taxon	Sample A	Sample B	c-p	Feeding	c-p wtd propn	c-p wtd propn	c-p wtd propn	c-p wtd propn	c-p wtd propn	c-p wtd propn	c-p wtd propn	c-p wtd propn	c-p wtd propn	c-p wtd propn
Hoplolaimidae	5	15	3	H					0.1	0.4	0.1	0.8	1.5	1.5
Pratylenchidae	5	15	3	H					0.1	0.4	0.1	0.8	1.5	1.5
Aphelenchidae	15	5	2	F	0.3	0.1	0.3	0.3	0.3	0.1	0.3	0.2		
Cephalobidae	15	2	2	B	0.3	0.1	0.3	0.1	0.3	0.0	0.3	0.1		
Plectidae	2	15	2	B	0.0	0.4	0.0	1.0	0.0	0.3	0.0	0.5		
Rhabditidae	2	50	1	B	0.0	0.6			0.0	0.5				
Dorylaimidae	50	5	4	O	2.0	0.3	2.1	0.7	1.8	0.2	1.9	0.3		
Discolaimidae	15	2	5	P	0.8	0.1	0.8	0.3	0.7	0.1	0.7	0.2		
Relevant total	109	109			99	79	97	29	109	109	107	59	10	30
MI	3.4	1.6												
MI2-5	3.5	2.6												
ΣMI	3.4	2.0												
ΣMI2-5	3.4	2.8												
PPI	3.0	3.0												
PPI/MI	0.9	1.9												

H=plant feeders, F=fungal feeders, B=bacterial feeders, O=omnivores, P=predators. In the columns in which each index is calculated, the total number of nematodes meeting the criteria of the index is first determined (relevant total) and then the proportion of that total in taxa meeting the criteria is weighted by the c-p values of those taxa. The index value is the sum of the weighted proportions.

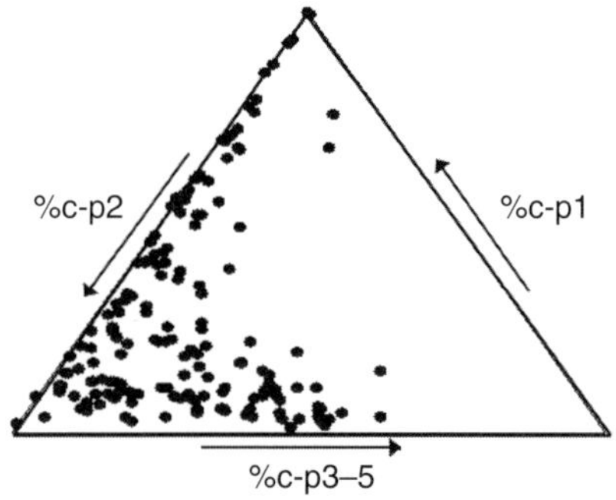

Fig. 5.1. C-p triangles, based on unweighted proportional representation of c-p1, c-p2, and c-p3–5 groupings of the nematode fauna, were a first step in distinguishing between basal fauna, enrichment indicators and structure indicators (modified from De Goede *et al.*, 1993).

should be independent of complexity led to development of separate trajectories of enrichment and structure to assess the magnitude of disparate services (Ferris *et al.*, 2001, 2004).

The nematode fauna provides information on two major characteristics of the soil environment and its resident communities. One characteristic is the flow of resources into the food web system as indicated by enrichment opportunist species; the other is the trophic connectance (*sensu* Cohen, 1989) of the system as indicated by prevalence and abundance of higher trophic level organisms. Ferris *et al.* (2001, 2004) considered general opportunist c-p2 nematodes to be representative of organisms that persist in most soil food webs, always present, and the survivors of the most adverse conditions. Two axes can be conceived as emerging from this basal state of the nematode fauna, one defined as an enrichment index, indicated by the weighted abundance of the proportion of all c-p1 and c-p2 nematodes that are c-p1 bacterivores and c-p2 fungivores, and the other as a structure index, derived from the proportional contribution of the weighted c-p3–5 nematodes to the c-p2–5 grouping. Also calculated in this system is a basal index, the relative proportion of the basal (c-p2) component of the fauna to all nematodes present (Berkelmans *et al.*, 2003). Further resolution to the enrichment component is provided by assessing the relative flow of resources into the food web through fungally- and bacterially-mediated decomposition channels (Ruess and Ferris, 2004).

Indicators of ecosystem function: enrichment, structure, basal and channel indices

The evolution of concepts, research and model validation associated with development of the Maturity Index family led to a functional guild classification of nematodes as a basis for studying and comparing ecosystem processes (Bongers and Bongers, 1998; Bongers and Ferris, 1999). The functional redundancy represented in the diversity of nematode faunae creates a high probability that the absence of a guild is a reliable indicator of disturbance and that the presence of a guild is a reliable indicator of lack of perturbation or of recovery from perturbation. In the case of organic enrichment of soil, opportunistic guilds (*r*-strategists) respond reliably (Sánchez-Moreno *et al.*, 2006). Considering soil nematode taxa as representatives of functional guilds generates an indicator profile that is not constrained by population distribution patterns and micro-environment effects (Ferris and Bongers, 2006).

The Enrichment Index and the Structure Index, both based on the indicator importance of functional guilds of nematodes, are descriptors of food web condition. Functional guilds are defined as a matrix of nematode feeding habits with the biological, ecological and life history characteristics embodied in the c-p classification. Thus, the Ba3 functional guild comprises c-p3 bacterivores such as those in the Teratocephalidae or Prismatolaimidae. Nematodes of all feeding habits classified as c-p2 are considered basal (b) to both enrichment and structure trajectories. Bacterial-feeding c-p1 and fungivores in c-p2 are indicators of enrichment (e) while nematodes of all feeding habits in c-p3–5 are indicators of structure (s). Functional guild indicators are weighted according to growth and metabolic rates (resource utilization) on the enrichment axis (Figs 5.2 and 5.3), and according to estimates of the degree of connectance, as determined by numbers of nematodes in higher c-p classes, in food webs of increasing complexity (Figs 5.2 and 5.4). Greater detail on derivation of the structure and enrichment weightings (W_i) is provided in Ferris *et al.*, 2001.

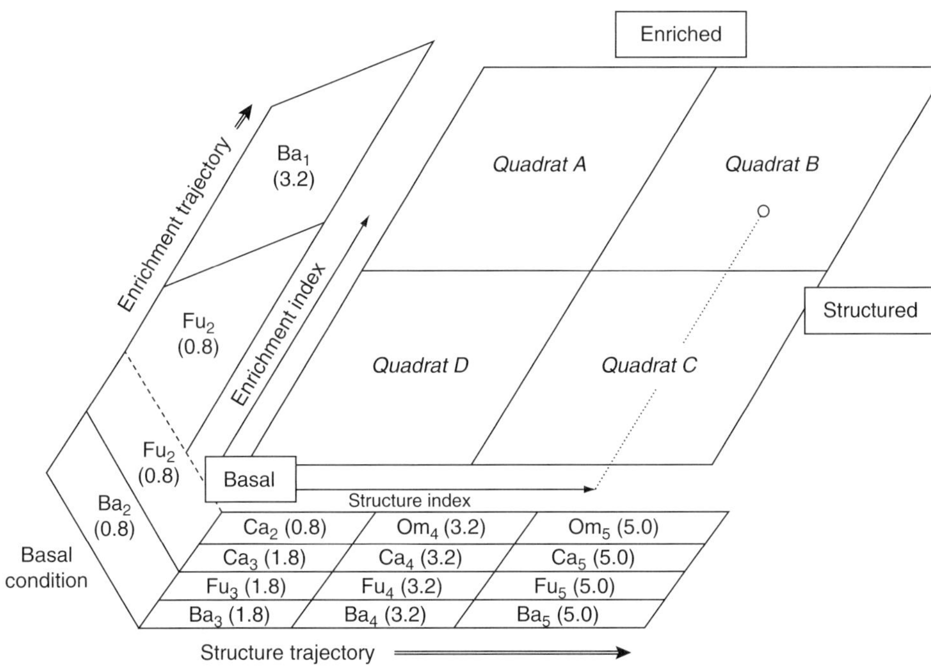

Fig. 5.2. A graphic representation of the nematode faunal profile indicates whether the soil community is enriched but unstructured (Quadrat A), enriched and structured (Quadrat B), resource-limited and structured (Quadrat C), or resource-depleted with minimal structure (Quadrat D). Functional guilds of soil nematodes are characterized by feeding habit (trophic group) and by life history characteristics, after Bongers and Bongers, 1998. Indicator guilds of soil food web condition (basal, structured, enriched) are designated and weightings of the guilds along the structure and enrichment trajectories are provided, for determination of the Enrichment Index and Structure Index of the food web. (Modified from Ferris *et al.*, 2001.)

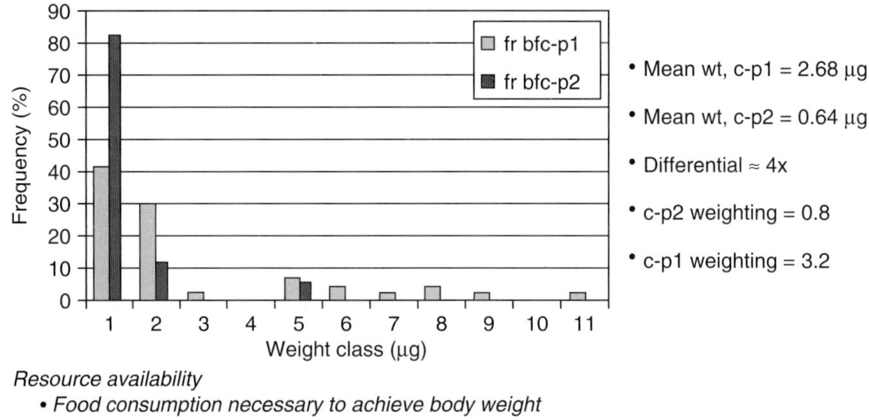

Fig. 5.3. Weighting system for enrichment and basal indicator soil nematodes as determined by mean weight of adults based on their frequency representation in different weight classes (adapted from Ferris *et al.*, 1996a,b).

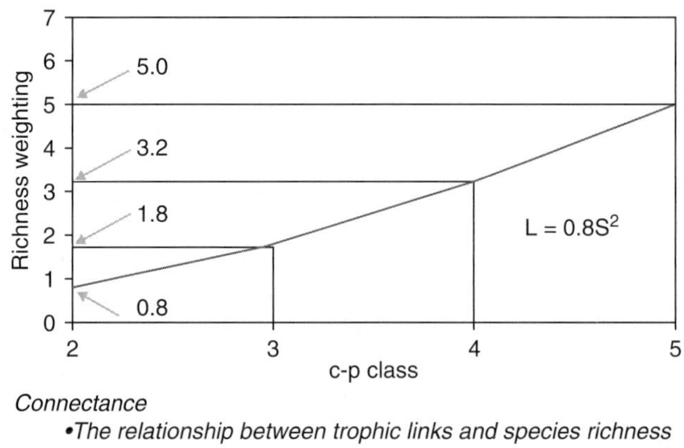

Fig. 5.4. Weighting system for structure and basal indicator soil nematodes as determined by taxonomic richness in food webs of different complexity (adapted from Ferris *et al.*, 2001).

The nematode fauna is comprised of basal, enrichment and structural components (b,e,s):

$b=(Ba2+Fu2)*W_2$, where $W_2.= 0.8$,
$e=(Ba1*W_1)+(Fu2*W_2)$, where $W_1 = 3.2$ and $W_2.= 0.8$
$s=(Ba_n*W_n+Ca_n*W_n+Fu_n*W_n+Om_n*W_n)$
where n=3-5, $W_3.= 1.8$, $W_4.= 3.2$, $W_5.= 5.0$.

The Enrichment (EI), Structure (SI), Basal (BI), and Channel (CI) indices are calculated from the weighted faunal components (Ferris *et al.*, 2001; Berkelmans *et al.*, 2003):

$EI =100*e/(e+b)$
$SI = 100*s/(s+b)$

$BI = 100*b/(e+s+b)$
$CI = 100\ Fu2*W_2/(Ba1*W_1+Fu2*W_2)$.

The EI, SI and BI represent an evolution of the concepts embodied in the c-p triangles of De Goede *et al.* (1993) and provide higher resolution to the enrichment, disturbance and contamination effects on the ecosystem (Table 5.3). Clearly, Sample A represents an environment with an abundance of omnivore and predator nematodes, suggesting greater connectance in the soil food web and the probable top-down regulation of opportunistic species (Sánchez-Moreno *et al.*, 2006; Sánchez-Moreno and Ferris, 2007). Sample B represents a disturbed and enriched condition in which the disturbance has had detrimental effects on higher trophic levels. Faunal analyses based on these indices provide insights into food web enrichment and structure and allow derivation of testable hypotheses based on the relative enrichment and structure of the system (Table 5.4).

The CI provides a means to partition flow of resources through fungal and bacterial decomposition channels. Indices of fungal and bacterial activity based on the relative abundance of fungal- and bacterial-feeding nematodes have been proposed several times following the calculation of their relative proportions in grasslands, woodlands and cultivated fields by Twinn (1974). The indices have included a ratio of F/B (Sohlenius and Boström, 1984) which has been defined most recently as NCR = B/(B+F) where NCR is the Nematode Channel Ratio, and B and F represent the abundance of bacterial- and fungal-feeding nematodes, respectively (Yeates, 2003). The CI differs in including weighting parameters for the size and metabolic rates of the nematode indicators.

When resources become available to soil organisms through external input, disturbance, organism mortality, turnover, or environment changes there is an enrichment pulse of opportunistic guilds. The pulse is followed by heterotrophic succession whereby the predominance of organisms changes through time depending on trophic roles, life course dynamics, and prevailing environmental conditions (Sudhaus, 1981; Ferris *et al.*, 1996b; Ferris and Matute, 2003). Substrates rich in labile carbon but deficient in nitrogen may favor the fungal rather than the bacterial decomposition channel (Ruess and Ferris, 2004).

Similar to the MI (Bongers and Bongers, 1998), the EI and CI are calculated excluding dauerlarvae to provide an index of the present state of the system. Rather than proliferate indices calculated with and without dauerlarvae, we consider that the ratio of dauerlarvae to active forms, as proposed by Sohlenius (1969, 1973), provides a clear metric of resource availability to functional guilds of bacterivores and fungivores. When the proportion of dauerlarvae is low, the resource supply is probably stable; when it is high, the system is probably in a state of resource-driven succession from bacterial to fungal domination of decomposition channels. However, considering the short lifespan of many enrichment opportunist nematodes (Ferris *et al.*, 1996a), frequent sampling will be necessary for using such calculations to model resource flow rates through the lower levels of the soil food web.

Further understanding of enrichment is determined by relative flow through and activity in fungal, bacterial and herbivore channels using total biomass of

Table 5.3. Calculation of the BI, EI and SI for two nematode assemblages (Samples A and B) of the same number of taxa and the same number of individuals (cf. Tables 5.1 and 5.2).

								A			B		
Nematode taxon	Sample A	Sample B	c-p	Feeding	b wt	e wt	s wt	b	e	s	b	e	s
Hoplolaimidae	5	15	3	H				0.00	0.00	0.00	0.00	0.00	0.00
Pratylenchidae	5	15	3	H				0.00	0.00	0.00	0.00	0.00	0.00
Aphelenchidae	15	5	2	F	0.80	0.80		12.00	12.00	0.00	4.00	4.00	0.00
Cephalobidae	15	2	2	B	0.80			12.00	0.00	0.00	1.60	0.00	0.00
Plectidae	2	15	2	B	0.80			1.60	0.00	0.00	12.00	0.00	0.00
Rhabditidae	2	50	1	B		3.20		0.00	6.40	0.00	0.00	160.00	0.00
Dorylaimidae	50	5	4	O			3.20	0.00	0.00	160.00	0.00	0.00	16.00
Discolaimidae	15	2	5	P			5.00	0.00	0.00	75.00	0.00	0.00	10.00
Total	109	109											
Index Components								25.60	18.40	235.00	17.60	164.00	26.00
BI	9.18	8.48											
EI	41.82	90.31											
SI	90.18	59.63											

H=plant feeders, F=fungal feeders, B=bacterial feeders, O=omnivores, P=predators. In the columns in which each index is calculated, the total number of nematodes in each taxon meeting the criteria of the index is weighted in terms of the basal, enrichment and structure characteristics of that taxon. The index values are calculated from the index components, that is the sum of the b, e and s values (see text).

Table 5.4. Inferred condition of the soil food web and its environment based on weighted nematode faunal analysis. Quadrats refer to faunal ordination in the faunal profile (Fig. 5.2) (from Ferris *et al.*, 2001).

General diagnosis	Quadrat A	Quadrat B	Quadrat C	Quadrat D
Disturbance	High	Low to mod.	Undisturbed	Stressed
Enrichment	N-enriched	N-enriched	Moderate	Depleted
Decomposition channels	Bacterial	Balanced	Fungal	Fungal
C-to-N ratio	Low	Low	Mod. to high	High
Food web condition	Disturbed	Maturing	Structured	Degraded

bacterivore (B), fungivore (F) and herbivore (H) nematodes. That provides the basis for developing the enrichment profile of the food web (Fig. 5.5). Changes through time in the abundance and type of organisms in the soil community may be considered structural succession; changes in food web function, not necessarily concurrent with community composition, are considered functional succession. The mass of available C diminishes with each trophic interchange, effectively dictating the abundance and biomass of organisms at each trophic level. Sustained organic enrichment may halt the succession and maintain a

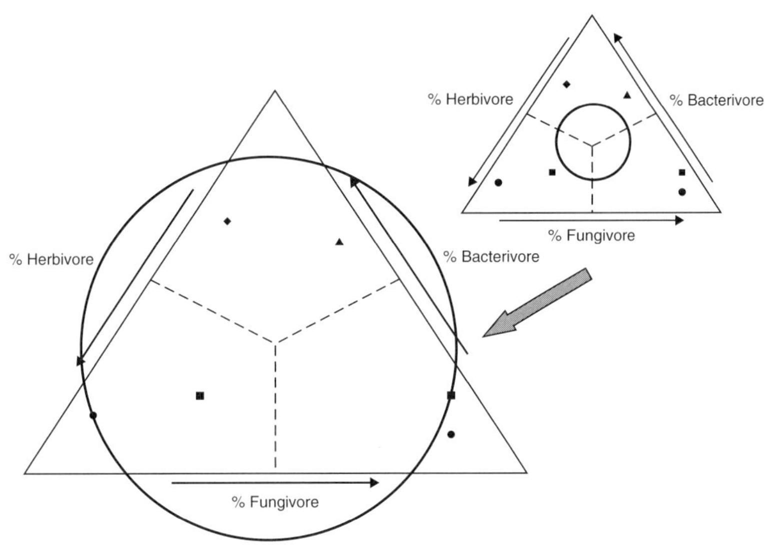

Fig. 5.5. Intake biomass of the soil food web partitioned into relative flows through herbivore-, fungivore- and bacterivore-mediated channels as indicated by the nematode fauna. The size of the triangle indicates the magnitude of resource flow. The circles indicate the biomass of generalist and specialist predators supported by the lower trophic levels.

consistent structure among functional guilds of soil organisms (Ferris and Bongers, 2006). Some of the organism responses to enrichment are ephemeral; others, including responses of certain guilds of nematodes, are more persistent and can be measured reliably (Ferris *et al.*, 1996b; Bongers and Ferris, 1999).

Abundance and biomass

The indices developed from nematode faunal analysis are all based on proportions of the faunae in various functional guilds. They provide an indication of the relative proportions of services and functions, but not of their magnitude. The biomass or abundance of organisms in various functional guilds must be important in determining the magnitude of services. Resource inflow into the soil food web can be represented as a subdivided triangle with the subdivisions indicating the proportion of inflow through separate channels. If the size of the triangle is based on the biomass of nematodes functioning in the inflow channels (Fig. 5.5), we are provided with a clearer understanding of the resources available to soil food web organisms and of the likely magnitude of services provided.

The constraints of resource inflow on higher trophic biomass are apparent when the biomass of higher trophic levels is represented as a circle superimposed on the intake triangle (Fig. 5.5). Also evident is the likelihood that the higher trophic level of the food web will provide the service of regulating populations of opportunistic organisms in the inflow channels. A low predator biomass relative to the intake (prey) biomass, as represented by a low MI, may indicate an environmental contamination or disturbance constraint on nematodes of higher c-p classes. A high predator biomass relative to the intake biomass, as represented by a high MI, indicates the possibility of top-down regulation of opportunistic species and, while there are sufficient intake resources to sustain the predator biomass, a relatively stable system.

Conclusions

The evolution of functional indices based on nematode faunal analysis provides insights into functioning and services of ecosystems. It has been greatly advanced by inference and observation of nematode feeding habits in relation to stomal architecture and by knowledge of the life history traits of nematode functional guilds. Undoubtedly, refinement and finetuning of the system is warranted and will occur as further information is developed on feeding habits and life history traits and the assignment of taxa to functional guilds. There are other examples of the use of the community structure of various organism groups for environmental monitoring. The advantage of those based on nematode functional guilds derives from the abundance and ubiquity of nematodes, the relationships between form and function, the differences among families in sensitivity to environmental disturbance, and the ease with which nematodes can be separated from substrate and categorized into taxonomic groups or functional guilds.

References

Baird, W. (1853) *Catalogue of the species of Entozoa or intestinal worms contained in the collection of the British Museum*, Woodfall and Kinder, London, pp. 132.

Barbuta, M. and Zullini, A. (2005) The nematode community of two Italian rivers (Taro and Ticino). *Nematology* 34 (2), 667–675.

Bassus, W. (1960) Die Nematodenfauna des Fichtenrohhumus unter der Einfluss der Kalkdüngung. *Nematologica* 5, 86–91.

Bassus, W. (1967) Der Einfluss von Meliorations- und Düngungsmassnahmen auf die Nematodenfauna verschiedener Waldböden. *Pedobiologia* 7, 280–295.

Bastian, H. Ch. (1865) Monograph on the Anguillulidae or free Nematoïds, marine, land and freshwater; with descriptions of 100 new species. *Transactions Linnaean Society London* 25, 73–184.

Bazzanti, M. (2000) Macrobenthic nematodes as biological indicators in a Mediterranean lowland river in Central Italy: a case study. *Archiv für Hydrobiologie* 148, 59–70.

Beier, S. and Traunspurger, W. (2001) The meiofauna community of two small German streams as indicator of pollution. *Journal of Aquatic Ecosystem Stress and Recovery* 8, 387–405.

Beier, S. and Traunspurger, W. (2003) Seasonal distribution of free-living nematodes in the Kraehenbach, a fine-grained submountain carbonate stream in southwest Germany. *Nematology* 5, 113–136.

Berkelmans, R., Ferris, H., Tenuta M., and Van Bruggen, A.H.C. (2003) Effect of long-term crop management on nematode trophic levels other than plant feeders disappear after one year of disruptive soil management. *Applied Soil Ecology* 23, 223–235.

Bongers, T. (1988) *De Nematoden van Nederland*. Pirola, Schoorl.

Bongers, T. (1990) The maturity index: an ecological measure of environmental disturbance based on nematode species composition. *Oecologia* 83, 14–19.

Bongers, T. (1999) The Maturity Index, the evolution of nematode life history traits, adaptive radiation and cp-scaling. *Plant and Soil* 212, 13–22.

Bongers, T. and Bongers, M. (1998) Functional diversity of nematodes. *Applied Soil Ecology* 10, 239–251.

Bongers, T. and Ferris, H. (1999) Nematode community structure as a biomonitor in environmental monitoring. *Trends in Ecology and Evolution* 14, 224–228.

Bongers, T. and Korthals, G. (1993) The Maturity Index, an instrument to monitor changes in the nematode community structure. *Summaries of the 45th International Symposium on Crop Protection, May 4, 1993, Ghent, Belgium*. 80.

Bongers, T. and Korthals, G. (1995) The behavior of MI and PPI under enriched conditions. *Nematologica* 41 (3), 286.

Bongers, T., Goede, R.G.M. de, Kappers, F.I. and Manger, R. (1989) Ecologische typologie van de Nederlandse bodem op basis van de vrijlevende nematodenfauna. *RIVM-rapport* 718602002.

Bongers, T., Alkemade, R. and Yeates, G.W. (1991) Interpretation of disturbance-induced maturity decrease in marine nematode assemblages by means of the Maturity Index. *Marine Ecology Progress Series* 76, 135–142.

Bongers, T., Goede, R.G.M. de, Korthals, G. and Yeates, G.W. (1995) Proposed changes of c-p classification for nematodes. *Russian Journal of Nematology* 3, 61–62.

Bongers, T.,Van der Meulen, H. and Korthals, G. (1997) Inverse relationship between the nematode maturity index and plant parasite index under enriched nutrient conditions *Applied Soil Ecology* 6, 195–199.

Boucher, G. (1980) Impact of Amoco Cadiz oil spill on intertidal and sublittoral meiofauna. *Marine Pollution Bulletin* 11, 95–101.

Brzeski, M. (1998) Nematodes of Tylenchina in Poland and temperate Europe. Museum and Institute of Zoology, Polish Academy of Sciences, Warsaw, pp. 396.

Bütschli, O. (1873) Beiträge zur Kenntnis der Freilebenden Nematoden. *Nova Acta Academiae Naturae Curiosorum* 36 (5), 1–124.

Callahan, C.A., Ferris, V.R. and Ferris, J.M. (1979) The ordination of aquatic nematode communities as affected by stream water quality. In: Cairns, J., Patil, G.P. and Waters, W.E. (eds) *Environmental Biomonitoring, Assessment, Prediction and Management*. International Cooperative Publishing House, Baltimore, Maryland, pp. 101–106.

Cantelmo, F.R. and Rao, K.R. (1978) Effect of pentachlorophenol (PCP) on meiobenthic communities in an experimental system. *Marine Biology* 46, 17–22.

Christie, J.R. (1959) *Plant Nematodes, their Bionomics and Control*. Agricultural Experiment Stations, University of Florida, Gainesville, Florida.

Cobb, N.A. (1914) North American free-living fresh-water nematodes. Reprinted from *Transactions of the American Microscopical Society* 33 in Cobb, N.A. *Contributions to a Science of Nematology*.

Cobb, N.A. (1915) Nematodes and their relationships. *USDA Yearbook of Agriculture* 1914, 457–490.

Cohen, J.E., (1989) Food webs and community structure. In: Roughgarden, J., May, R.M. and Levin, S.E. (eds), *Perspectives in Ecological Theory*. Princeton University Press, Princeton, New Jersey, pp. 181–202.

Daday, E. (1897) Die freilebenden Süsswasser-Nematoden Ungarns. *Zoologisches Jahrbuch. (Systematik)* 10, 91–134.

Dujardin, F. (1845) *Histoire naturelle des Helminthes ou vers intestinaux*. Libraire encyclopédique de Roret (Suites a Buffon). Paris.

Eberth, C.I. (1863) *Untersuchungen über Nematoden*. W. Engelman, Leipzig.

Ettema, C. H. and Bongers, T. (1993) Characterization of nematode colonization and succession in disturbed soil using the Maturity Index. *Biology and Fertility of Soils* 16, 79–85.

Ferris, H. and Bongers, T. (2006) Nematode indicators of organic enrichment. *Journal of Nematology* 38, 3–12.

Ferris, H., and Matute, M.M. (2003) Structural and functional succession in the nematode fauna of a soil food web. *Applied Soil Ecology*. 23, 93–110.

Ferris, H., Eyre, M., Venette, R.C., and Lau, S.S. (1996a) Population energetics of bacterial-feeding nematodes: Stage-specific development and fecundity rates. *Soil Biology and Biochemistry* 28, 271–280.

Ferris, H., Venette, R.C. and Lau, S.S. (1996b) Dynamics of nematode communities in tomatoes grown in conventional and organic farming systems and their impact on soil fertility. *Applied Soil Ecology* 3, 161–175.

Ferris, H., Bongers, T. and Goede, R.G.M. de (2001) A framework for soil food web diagnostics: extension of the nematode faunal analysis concept. *Applied Soil Ecology* 18, 13–29.

Ferris, H., Bongers, T. and Goede, R. de (2004) Nematode faunal analyses to assess food web enrichment and connectance, pp. 503–510. In Cook R.C., and Hunt D.J., (eds) *Proceedings of the Fourth International Congress of Nematology*. Nematology Monographs and Perspectives 2. Brill, The Netherlands, 866p.

Filipjev, I.N. (1918) Free-living marine nematodes of the Sevastopol area. *Transactions of the Zoological Laboratory and Sevastopol Biological Station, Russian Academy of Sciences*, Series II, No. 4. (translated from Russian, 1968).

Filipjev, I.N. and Schuurmans Stekhoven, J.H. (1941) *A Manual of Agricultural Helminthology*. E.J. Brill, Leiden.

Goede, R.G.M. de, Bongers, T. and Ettema, C. (1993) Graphical presentation and interpretation of nematode community structure: C-P triangles. *Mededelingen Faculteit Landbouwwetenschappen Universiteit Gent*, 58/2b, 743–750.

Hill, M.O. (1973) Diversity and evenness: A unifying notation and its consequences. *Ecology* 54, 427–432.

Hodda, M. and Nicholas, W.M. (1986) Nematode diversity and industrial pollution in the Hunter river estuary, NSW, Australia. *Marine Pollution Bulletin* 17, 251–255.

Ingham, R.E., Trofymow, J.A., Ingham, E.R. and Coleman D.C. (1985) Interactions of bacteria, fungi, and their nematode grazers: Effects on nutrient cycling and plant growth. *Ecological Monographs* 55, 119–140.

Jairajpuri, M.S. and Ahmad, W. (1992) *Dorylaimida, Free-living, Predaceous and Plant-Parasitic Nematodes*. Oxford and IBH Publishing, New Delhi, Bombay, Calcutta.

Johnson, S.R., Ferris, V.R. and Ferris, J.M. (1972) Nematode community structure of forest woodlots: I. Relationships based on similarity coefficients of nematode species. *Journal of Nematology* 4, 175–183.

Johnson, S.R., Ferris, J.M. and Ferris, V.R. (1973) Nematode community structure of forest woodlots: II. Ordination of nematode communities. *Journal of Nematology* 5, 95–107.

Johnson, S.R., Ferris, J.M. and Ferris, V.R. (1974) Nematode community structure of forest woodlots: III. Ordinations of taxonomic groups and biomass. *Journal of Nematology* 6, 118–126.

Kappers, F.I. and van Esbroek, M.L.P. (1988) Ecological recovery of decontaminated soil. *Proceedings of the Second International TNO/BFTM/Conference on Contaminated Soil*. Hamburg, 11–15 April 1988.

Karssen, G. (2006) *Life and work of Dr. Johannes Govertus de Man (1850–1930)*. Brill, Leiden.

Korthals, G.W., Ende, A. Van der, Megen, H. Van, Lexmond, T.M., Kammenga, J.E. and Bongers, T. (1996a) Short-term effects of cadmium, copper, nickel and zinc on soil nematodes from different feeding and life-history strategy groups. *Applied Soil Ecology* 4, 107–117.

Korthals, G.W., Alexiev, A.D., Lexmond, T.M., Kammenga, J.E. and Bongers, T. (1996b) Long-term effects of copper and pH on the nematode community in an agroecosystem. *Environmental Toxicology and Chemistry* 15 (6), 979–985.

Korthals, G.W., Goede, R.G.M. de, Kammenga, J.E. and Bongers, T (1996c) The Maturity Index as an instrument for risk assessment of soil pollution. In: Straalen, N.M. Van, and Krivolutsky, D.A. (eds) *Bioindicator systems for Soil Pollutions,* pp. 85–93.

Lambshead, P.J.D. (1986) Sub-catastrophic sewage and industrial waste contamination as revealed by marine nematode faunal analysis. *Marine Ecology Progress Series* 29, 247–260.

Leuckart, R. (1849) Zur Kenntnis der Fauna von Island. Erster Beitrag (Würmer). *Archiv Naturgeschichte* 15, 148–206.

Linstow, O. von (1876) Helminthologische Beobachtungen. *Archiv Naturgeschichte* 42, 1–18.

Micoletzky, H. (1925) Die freilebenden Süsswasser- und Moornematoden Dänemarks. *Mémoires de l'Académie Royale des Sciences et des Lettres de Danemark, Copenhague, Section de Sciences* (8) 10 (2), 55–310.

Neher, D.A. and Campbell, C.L. (1996) Sampling for regional monitoring of nematode communities in agricultural soils. *Journal of Nematology* 28, 196–208.

Nicholas, W.L. (1975) *The Biology of Free-living Nematodes*. Clarendon Press, Oxford.

Odum, E.P. (1985) Trends expected in stressed ecosystems. *BioScience* 35, 419–422.

Okada, H., Tsukiboshi, T. and Kadota, I. (2002) Mycetophagy in *Filenchus misellus* (Andrassy, 1958) Lownsbery and Lownsbery, 1985 (Nematoda: Tylenchidae), with notes on its morphology. *Nematology* 4, 795–801.

Okada, H., Harada, H. and Kadota, I. (2005) Fungal-feeding habits of six nematode isolates in the genus Filenchus. *Soil Biology and Biochemistry* 37, 1113–1120.

Overgaard Nielsen, C. (1949) Studies on the soil microfauna II. The soil inhabiting nematodes. *Natura Jutlandica* 2, 1–131.

Paramonov, A.A. (1962) *Plant-parasitic Nematodes Vol. 1*. Laboratory of Helminthology, Academy of Sciences of the USSR (translated from Russian, 1968).

Pielou, E.C. (1966) The measurement of diversity in different types of biological collections. *Journal of Theoretical Biology* 13, 131–144.

Platt, H. M. (1994) Foreword. In: *The Phylogenetic Systematics of Free-living Nematodes*, Lorenzen, S. (ed.), pp. i–ii, The Ray Society, London.

Prejs, K. (1970) Some problems of the ecology of benthic nematodes (Nematoda) of Mikolajskie Lake. *Ekologia Polska* 25 (9), 225–242.

Prejs, K. (1977) The littoral and profundal benthic nematodes of lakes with different trophy. *Ekologia Polska* 25 (1), 21–30.

Ruess, L. and Ferris, H. (2004) Decomposition pathways and successional changes, pp. 547–566. In Cook, R.C. and Hunt, D.J. (eds) *Proceedings of the Fourth International Congress of Nematology. Nematology Monographs and Perspectives* 2. Brill, Netherlands, 866p.

Samoiloff, M.R. (1987) Nematodes as Indicators of Toxic Environmental Contaminants. In: Veech, J.A. and Dickson, D.W. (eds) *Vistas on Nematology*. E.O. Painter, DeLeon Springs, Florida, pp. 433–438.

Sánchez-Moreno, S. and Ferris, H. (2007) Suppressive service of the soil food web: Effects of environmental management. *Agriculture, Ecosystem and Environment* 119, 75–87.

Sánchez-Moreno, S., Minoshima, H., Ferris, H. and Jackson, L.E. (2006) Linking soil properties and nematode community composition: effects of soil management on soil food webs. *Nematology* 8, 703–715.

Shannon, C.E. and Weaver, W. (1949) *The Mathematical Theory of Communication*. University of Illinois Press, Urbana.

Simpson, E.H. (1951) The interpretation of interaction in contingency tables. *Journal of the Royal Statistical Society, Series B* 13, 238–241.

Sohlenius, B. (1969) The monoxenic cultivation of some rhabditid nematodes. *Oikos* 20, 287–293.

Sohlenius, B. (1973) Structure and dynamics of populations of Rhabditis (Nematodes: Rhabditidae) from forest soil. *Pedobiologia* 13, 368–375.

Sohlenius, B. and Boström, S. (1984) Colonization, population development and metabolic activity of nematodes in buried barley straw. *Pedobiologia* 27, 67–78.

Spenneman, D.H.R. (2003) *Nathan Augustus Cobb Plant Pathologist. A Bibliography of his Work*. Retrospect, Letao Publishing, Albury, Australia.

Stinner, B.R. and Crossley, D.A., Jr. (1982) Nematodes in no-tillage agroecosystems. In: Freckman, D.W. (ed.) *Nematodes in Soil Ecosystems*. University of Texas Press, Austin, pp. 14–28.

Sudhaus, W. (1981) Über die Sukzession von Nematoden in Kuhfladen. *Pedobiologia* 21, 271–297.

Sudhaus, W., Rehfeld, K., Schlüter, D. and Schweiger, J. (1988) Beziehungen zwischen Nematoden, Coleopteren und Dipteren in der Sukzession beim Abbau von Kuhfladen. *Pedobiologia* 31, 305–322.

Tietjen, J.H. (1980) Population structure and species composition of the free-living nematodes inhabiting sands of the New York Bight Apex. *Estuarine and Coastal Marine Science* 10, 61–73.

Tsalolikhin, S.J. (1976) Free-living nematodes as indicators of polluted freshwaters. In: Skarlato, O.A. (ed.) *Methods of Biological Analysis of Freshwaters*. Akademia Nauk, Leningrad, pp. 118–122.

Twinn, D.C. (1974) Nematodes. In: Dickinson, C.H. and Pugh, G.J.F. (eds) *Biology of Plant Litter Decomposition* II. Academic Press, London, pp. 421–465.

Villot, A. (1875) Recherches sur les Helminthes libres ou parasites des côtes de la Bretagne. *Archives de Zoologie Expérimental et Générale* (1) 4, 451–482.

Vranken, G., Herman, P.M.J. and Heip, C. (1988) Studies of the life-history and energetics of marine and brackish-water nematodes. I. Demography of *Monhystera disjuncta* at different temperature and feeding conditions. *Oecologia* 77, 296–301.

Wasilewska, L. (1970) Nematodes of the sand dunes in the Kampinos Forest. I. Species structure. *Ekologia Polska* 18, 429–443.

Wasilewska, L. (1974) Rola wskaznikowa wszystkozernej grupy nicieni glebowych. *Wiadomosci Ekologiczne* 20, 385–390.

Wasilewska, L. (1994) The effect of age of meadows on succession and diversity in soil nematode communities. *Pedobiologia* 38, 1–11.

Yeates, G.W. (1968) An analysis of annual variation of the nematode fauna in dune sand, at Himantangi beach, New Zealand. *Pedobiologia* 8, 173–207.

Yeates, G.W. (1972) Nematoda of a Danish beech forest. I. Methods and general analysis. *Oikos* 23, 178–189.

Yeates, G.W. (1974) Studies on a climosequence of soils in Tussock grasslands. 2. Nematodes. *New Zealand Journal of Zoology* 1, 171–177.

Yeates, G.W. (1979) Soil nematodes in terrestrial ecosystems. *Journal of Nematology* 11, 213–229.

Yeates, G.W. (1994) Modification and qualification of the Nematode Maturity Index. *Pedobiologia* 38, 97–101.

Yeates, G.W. (2003) Nematodes as soil indicators: functional and biodiversity aspects. *Biology and Fertility of Soils* 37, 199–210.

Yeates, G.W. and Coleman, D.C. (1982) Role of nematodes in decomposition. In: Freckman, D.W. (ed.) *Nematodes in Soil Ecosystems*. University of Texas Press, Austin, pp. 55–81.

Yeates, G.W., Bongers, T., Goede, R.G.M. de, Freckman, D.W. and Georgieva, S.S. (1993) Feeding habits in soil nematode families and genera – an outline for soil ecologists. *Journal of Nematology* 25, 315–331.

Yeow, H.S., Finney-Crawley, J.R., Lawlor, J. and King, I.W. (1999) Free-living nematode assemblages downstream from a pulp and paper mill in Humber Arm, Newfoundland. *Russian Journal of Nematology* 7 (2), 95–103.

Yuen, P.H. (1966) The nematode fauna of the regenerated woodland and grassland of broadbalk wilderness. *Nematologica* 12, 195–214.

Zullini, A. (1976) Nematodes as indicators of river pollution. *Nematologia Mediterranea* 4, 13–22.

Zullini, A. and Peretti, E. (1986) Lead pollution and moss-inhabiting nematodes of an industrial area. *Water, Air and Soil Pollution* 27, 403–410.

6 Case Studies Using Nematode Assemblage Analysis in Aquatic Habitats

ROBERTO DANOVARO,[1] CRISTINA GAMBI,[1] SEBASTIAN HÖSS,[2] SIMONE MIRTO,[3] WALTER TRAUNSPURGER[4] AND ALDO ZULLINI[5]

[1]*Dipartimento di Scienze del Mare, Università Politecnica delle Marche, Ancona, Italy;* [2]*Ecossa, Starnberg, Germany;* [3]*Istituto Ambiente Marino Costiero (IAMC), CNR, Messina, Italy;* [4]*Department of Animal Ecology, University of Bielefeld, Germany;* [5]*Dipartimento di Biotecnologie & Bioscienze, Università di Milano-Bicocca, Milan, Italy*

Introduction

Approximately 71% of the Earth's surface area is covered by water and of this, 97% is salt water contained in oceans and seas and the remaining 3% is freshwater contained in ponds, lakes, streams and rivers. Thus, the marine environment represents the largest habitat on earth. However, although superficial freshwater accounts for a very small portion of the hydrological cycle, the fact that it forms a bottleneck of the entire water cycle means that freshwater plays a major role in ecological quality assessments (European Community, 2000). Furthermore, despite its small area, freshwater supports about 10% of all aquatic species, including more than 25,000 species of insects.

Both habitats are threatened by the activity of mankind, with industrial discharges, sewage disposal and oil spills all representing major threats. Nematodes are abundant and diverse in both marine and freshwater habitats (see Hodda *et al.*, Chapter 2, this volume) and have the potential to be used as bioindicators. Here we review published literature relating to the use of nematode assemblage analysis in aquatic habitats. Further examples can also be found from the commercial studies described by Trett *et al.*, Chapter 12, this volume.

Marine Habitats

Marine nematode response to hydrocarbon impact

In marine systems, the major sources of hydrocarbon contamination are spills of crude oil, discharges of refined fuels and offshore production activities

(Kennish, 1992). Impacts of these pollutants on benthic communities have been studied extensively (e.g. Coull and Chandler, 1992; Peterson *et al.*, 1996; Beyrem and Aïssa, 2000; Mahmoudi *et al.*, 2003; Suderman and Thistle, 2003). The first consequence of the release of oil at sea is the creation of an emulsion, coupled with oil adsorption on suspended matter with subsequent deposition on the seabed. Depending on the quantity of oil released, the bathymetric depth and the distance from the shore, crude oil may accumulate in the sediment, reaching high concentrations. The main effects on the sediments may be summarized in oxygen reduction and changes in sediment properties (porosity and interstitial sedimentary space), affecting benthic fluxes at the water–sediment interface.

Hydrocarbon contamination may negatively influence meiobenthic species (Coull and Chandler, 1992; Peterson *et al.*, 1996; Beyrem and Aïssa, 2000; Mahmoudi *et al.*, 2003) or positively (Stacey and Marcotte, 1987; Feder *et al.*, 1990). Nematodes appear in many cases to be relatively insensitive to heavy oil conditions (Boucher, 1980; Elmgren *et al.*, 1980; Bonsdorf, 1981; Frìcke *et al.*, 1981; Gee *et al.*, 1992). None the less, other authors reported significant decreases in nematode abundance with increased oil content in the sediment (Wormald, 1976; Giere, 1979; Danovaro *et al.*, 1995, 1999), but also a high resilience (i.e. rapid recovery). After the *Agip Abruzzo* oil spill in the Mediterranean Sea, nematode abundance was significantly reduced with increasing oil content in the sediment, and analysis of nematode community structure clearly detected the presence of disturbed faunal assemblages (Danovaro *et al.*, 1995). Some genera, such as *Chromaspirina*, *Hypodontolaimus*, *Onchalaimellus*, *Paracanthonchus*, *Setosabatieria* and *Xyala* disappeared immediately after the oil spill. All these genera recovered rapidly after the oil spill and appeared to be opportunists. Other genera, such as *Daptonema* and *Viscosia*, appeared to be less sensitive or even tolerant to hydrocarbon stress. Also after the accident of the oil tanker *La Coruña* (Atlantic coast, north of Spain) some nematode species became extremely abundant (e.g. *Enoplolaimus litoralis*), and many dominant nematode genera in polluted sediments ingested tar particles covered by a bacterial film that they used as a food source (Giere, 1979). Danovaro *et al.* (1995), did not report clear influences of the oil spill on the trophic diversity (in terms of the Index of Trophic Diversity, ITD) which suggests the oil pollution had a non-selective impact on the different trophic groups. In some cases the nematode response to hydrocarbon spill is not immediate. After the *Amoco Cadiz* oil spill, Renaud-Mornant *et al.* (1981) reported no significant impact of hydrocarbons on nematode mortality 10 days after the accident. However, 9-12 months later, a significant decline of nematode diversity was observed. The community structure changed considerably as the result of the replacement of species associated with sandy sediments by those typical of muddy sediments. Non-selective deposit feeding nematodes (1B) declined after the oil contamination, indicating an apparent alteration of the functional characteristics of the nematode assemblage (Frithsen *et al.*, 1985). This group is likely to be directly affected by oil toxicity as they ingest tar particles and oil emulsion together with the sediment during feeding. By contrast predator/scavengers (group 2B) were found to increase significantly after the oil spill. Epigrowth feeders (group 2A) also increased their relative significance as a result of the increased microphytobenthic biomass in the weeks following the oil spill. Mahmoudi *et al.* (2005) demonstrated that responses of nematode species to the diesel contamination varied: *Chaetonema* sp. seemed to be an intolerant species; *Pomponema* sp. and *Oncholaimus campylocercoides* were categorized as 'diesel-sensitive'. Moreover, *Hypodontolaimus colesi, Daptonema trabeculosum* and *Daptonema fallax* appeared to be 'opportunistic' and *Marylynnia stekhoveni* can be considered as a 'diesel-resistant' species.

Marine nematode response to heavy metal contamination

Industrial discharges still represent one of the most important environmental threats for coastal habitats, with serious consequences for the future. Due to the potential consequences for human health, the requirement to detect the level and distribution of heavy metal pollution has grown in the last few years. Apart from expensive and time-consuming chemical methods, several rapid and cheap proxy methods based on the use of marine organisms have been developed. The characteristics and life cycles of marine nematodes suggest that they would be sensitive to this kind of impact (Platt *et al.*, 1984; Hodda and Nicholas, 1986; Sundelin and Elmegren, 1991; Zhinan *et al.*, 1993; Guo *et al.*, 2001).

Overall nematode abundance is not significantly altered by heavy metal pollution (Heip *et al.*, 1984). Nevertheless, investigations conducted on nematodes at the species level revealed that nematode diversity is sensitive to pollution and indeed decreases significantly after exposure (reviewed in Heip *et al.*, 1985; Somerfield *et al.*, 1994; Millward and Grant, 1995). This is the consequence of a different response of ecologically similar species, belonging to the same genus or family, to heavy metal pollution. The effects of heavy metal pollution on nematodes were the result of the processes of uptake and loss of heavy metals that occurred in these organisms. Howell (1982) investigating the response of two nematode species to heavy metal pollution, showed that *Enoplus brevis* and *Enoplus communis*, exhibited distinct pathways in metal accumulation through the cuticle. In his study, Howell observed higher copper uptake in *E. communis* (accumulation factor of 10.6) whereas in *E. brevis* the copper uptake was relatively low (accumulation factor of 5.0). Howell (1983) also showed that the rapid uptake and loss of heavy metals (i.e. both copper and zinc) occurred through the cuticle. This suggests that the primary event is the surface adsorption whereas the role of the 'other tissues' is difficult to assess. Surface adsorption occurs in the hypodermis which is known to be highly metabolically active and as well as secreting, the cuticle is thought to be the site of many physiological processes. As such it represents a highly specific target for toxicants including heavy metals, which are known to interfere with a wide variety of enzymes and cellular systems. Derycke *et al.* (2007), investigating the effects of Cd concentrations on population growth and genetic diversity of *Pellioditis marina* (Nematoda) from the

Westerschelde estuary (The Netherlands), reported a substantial increase in adult mortality as well as a decrease in motility of *P. marina* at a Cd concentration of 10 mg l^{-1}. The authors showed that sublethal Cd concentrations reduced population development of *P. marina* and that low salinity conditions induced responses at the genetic level. Derycke *et al.* (2007) suggested that low salinities and high Cd concentrations may explain the lower genetic diversity of the nematode assemblages inhabiting the Westerschelde estuary. Millward and Grant (1995), performing toxicity tests on the whole nematode community from Restronguet Creek (a severely contaminated estuary in Cornwall, UK), showed that nematodes were more resistant to copper than those from an adjacent and less contaminated estuary. The authors suggested that this pattern was the result of different processes that occurred in the nematode assemblages of this area such as: (i) an increase in the abundance of Cu-resistant nematode species; (ii) the evolution of enhanced Cu tolerance in some nematode species; and (iii) the probable exclusion of more sensitive nematode species. These findings are in agreement with extensive ecological studies (Somerfield *et al.*, 1994) that suggested that the Restronguet Creek nematode community is distinct from that in the Percuil River due to the different levels of Cu contamination. This study demonstrated that community tolerance to pollution may be used as a tool to evaluate the biological impact of a chronic pollutant on marine benthos.

Millward and Grant (1995) investigated nematode assemblages at ten estuarine sites that ranged from uncontaminated to grossly contaminated by heavy metals. The relative tolerance of these assemblages to Cu was quantified using acute toxicity tests. There were large differences between sites in tolerance to Cu, and this tolerance was strongly correlated with severity of contamination. Comparison with studies of nematode community composition in these same estuaries indicated that nematodes were as sensitive as the best available ecological monitoring methods.

The observed spatial patterns of nematode assemblages and diversity are not always related to any kind of pollution. A detailed knowledge of the environmental conditions of the sampling area is required to better identify the causes of the observed nematode distributions. Gyedu-Ababio *et al.* (1999) reported that Mn, Ti, Fe, Cr and Sn (associated with higher organic carbon contents) played a very important role in the structuring of the nematode community in the Swartkops estuarine system (South Africa). The sites with the highest concentrations of heavy metals were characterized by the lowest values of diversity (i.e. Shannon index). Gyedu-Ababio *et al.* (1999) found that Mn, Ti, Zn and Fe concentrations affect the abundance, diversity and community structure of the nematodes in the investigated area. Among all genera reported in this study, *Axonolaimus*, *Sabatieria*, *Monhystera* and *Theristus* were identified as indicators of stress conditions. The authors concluded that nematodes, for their sensitivity and fast response to the heavy metal pollution, can easily be used in pollution monitoring in marine and estuarine environments.

Guo *et al.* (2001) suggested that the effects of heavy metal pollution on nematode assemblages are strictly connected with the natural concentrations of heavy metals in the sediment. These authors investigated the effects of arsenic on nematode assemblages in the Boai Sea (China), and revealed that nematode diversity is mainly influenced by the grain size composition and the distribution of food sources (i.e. phaeopigments) rather than the presence of As in the sediment. This is due the high concentrations of As in the investigated sediments, controlled by the natural weathering and lithology of the area (Zhang, 1996). Arsenic concentrations in the Bohai Sea are relatively stable and intimately associated with the discharge from the Huanghe river and are not considered harmful for aquatic organisms (Bryan and Langston, 1992).

Raffaelli and Mason (1981) were the first to propose using the nematode to copepod ratio as a tool for biomonitoring the health status of coastal systems, but this technique has been strongly criticized (Coull *et al.*, 1981; Warwick, 1981). Lambshead (1984) also criticized this approach, pointing out the difficulties in separating the effects of pollution from the effects of the other environmental variables on the nematode to copepod ratio. Lee *et al.* (2001) applied this ratio for the first time in relation to metal pollution along the beaches in the Chanaral area (Northern Chile). The authors showed that the ratio was not a good predictor of heavy metal pollution due to the generally low abundance of harpacticoid copepods from the investigated beaches.

Marine nematode response to sewage discharge and organic enrichment

In coastal marine environments, sewage discharge and organic enrichment are common sources of human impacts and their effects have been studied on several ecological systems (macrofauna from soft substrates, fish assemblages, seagrass, sessile and mobile fauna from hard substrates and epiphytes). Several authors reported that sewage pollution can change structural and functional attributes of biodiversity, but effects can vary depending on the response of the investigated variables and the types of data analysis adopted (Smith *et al.*, 1999; Guidetti *et al.*, 2003; Balestri *et al.*, 2004; Piazzi *et al.*, 2004; Terlizzi *et al.*, 2005a,b). Meiofauna (in particular, nematodes) are well suited for environmental impact assessment studies because the characteristics of their life cycles (cosmopolitan distribution, small size, high turnover and lack of larval dispersion; Higgins and Thiel, 1988) make them sensitive to environmental disturbance (from organic enrichment to mining and placement of dredged material; Vincx and Heip, 1991; Danovaro *et al.*, 1995; Ahnert and Schriever, 2001; La Rosa *et al.*, 2001; Mirto and Danovaro, 2004; Fleeger *et al.*, 2006). Several studies have highlighted the sensitivity of nematodes to various kinds of human activities (Sandulli and De Nicola-Giudici, 1990, 1991; Austen *et al.*, 1994; Mazzola *et al.*, 1999; Mirto *et al.*, 2000, 2002; Fraschetti *et al.*, 2006; Schratzberger *et al.*, 2006, 2007). The lack of larval dispersion and the close dependence on local habitat features make nematodes (and total meiofauna) potentially highly sensitive to environmental changes, including modification of habitat complexity at very small spatial scales (Blome *et al.*, 1999; Sandulli and Pinckney, 1999; Steyaert *et al.*, 2003).

The general outcome from the above studies is that this kind of anthropogenic impact can alter meiofaunal abundance, diversity and biomass and nematode species composition. However, these changes are not always predictable or unequivocal. In spite of the increasing impact of sewage discharge pollution in coastal marine environments, information on meiofaunal response to sewage discharge is still rather limited (Sandulli and De Nicola-Giudici, 1990, 1991). These studies, conducted on soft substrates, indicated that the effect of sewage discharge could determine substantial changes in the structure of meiofaunal assemblages, increasing nematodes and decreasing harpacticoid copepods. These changes were explained by the increased organic content of sewage contaminated sediments, which represent a potential food source.

Fraschetti *et al.* (2006) recently examined the possible impact of a sewage outfall on nematode diversity in a rocky subtidal area of Southern Italy. Nematode assemblages are known to live on hard substrates in association with periphytic and epiphytic algae and attached epibiota, but their abundance, diversity and colonizing abilities on hard substrate have been poorly documented (Gibbons 1988a,b; Danovaro and Fraschetti, 2002; Atilla *et al.*, 2003; Mirto and Danovaro, 2004). Fraschetti *et al.* (2006) adopted an asymmetrical sampling design to compare patterns of nematode diversity at one location exposed to sewage discharge to two reference locations. The results of the asymmetrical analyses of variance (PERMANOVA) showed that the number of nematodes differed with lower abundance in the impacted sites whereas no differences between impacted versus control sites were detected for the number of nematode genera and taxonomic composition. The non-metric multidimensional scaling (nMDS) shows quite a clear separation between sites near the sewage and those of controls. The Similarity Percentage (SIMPER) analyses showed that the concurrent effect of decreasing nematode abundance and increasing hydrozoan abundance (Campanularidae) in the impacted location caused major differences between impacted and control sites. The decrease of nematode abundance was unexpected because available literature reports that often nematodes increase in abundance due to their ability to exploit the food released from sewage discharge (Vidakovic 1983; Arthington *et al.* 1986; Bongers *et al.*, 1991). Fraschetti *et al.* (2006) also calculated the Maturity Index (MI; Bongers, 1990 and see also Ferris and Bongers, Chapter 5, this volume) of the nematode assemblages but this did not vary significantly between impacted and control sites (3.5 and 3.2–3.3 for impact and both controls, respectively) and was rather high compared with data reported in other urban pollution studies (e.g. Bongers and Ferris, 1999) suggesting that nematode assemblages were mainly composed of 'persisters'. This pattern was also confirmed by the fact that in both locations, nematode assemblages were dominated by the same trophic guild (>60% epistrate feeders, 2A) that is generally associated with high Maturity Index values.

Marine nematode response to fish farm impact

Intensive fish farming results in the release of large amounts of dissolved and particulate matter to the surrounding environment (Holmer and Kristensen, 1992). The most evident consequences of fish farming on the benthic environment are the increase in total organic carbon accumulation in the sediment that might easily induce eutrophication (Beveridge, 1996; Alongi *et al*., 2003) and a decrease in oxygen availability for the benthos beneath fish cages (Holmer *et al*., 2005; Kalantzi and Karakassis, 2006). These changes, in turn, have significant impact on the abundance and biodiversity of micro-, meio- and macrobenthic organisms (Karakassis *et al*., 2000; La Rosa *et al*., 2001, 2004; Mirto *et al*., 2002; Kalantzi and Karakassis, 2006). Two studies have examined the response of nematode communities to fish farm biodeposition, revealing that nematode abundance is sensitive to fish farm pollution with lower values below cages, when compared to the control site (Duplisea and Hargrave, 1996; Mirto *et al*., 2002). Studies conducted so far on fish biodeposition effects on nematode size have provided conflicting results. Porter *et al*. (1996) and Tsujino (1998) reported the presence of large size nematodes in organic impacted sediments. Lorenzen *et al*. (1987) and Prein (1988) reported that the large Oncholaimidae *Pontonema vulgare* (Platt and Warwick, 1983) accumulated in organically polluted fjords. Also Mirto *et al*. (2002) indicated that the effects of biodeposition might be evident in terms of nematode body size: nematodes had significantly higher body weights in organic enriched sediments beneath the fish cage, than in the non-impacted site. In contrast, Duplisea and Hargrave (1996), studying nematodes in sediment below salmon cages, did not find differences in individual nematode biomass between the cage and control sites. Due to contrasting results, nematode body size is still not a universally accepted parameter for detecting organic pollution; for instance, Monhysterids are small and tolerant whereas Enoploids are large and sensitive (Heip *et al*., 1982). None the less, Mirto *et al*. (2002) observed an initial increase of the individual nematode biomass immediately after the installation of a new fish farm, but when the fish were harvested and consequently biodeposition strongly reduced, average body weight of nematodes in the sediments beneath the cage immediately became indistinguishable from control values. The same study reported that immediately after fish farm deployment, despite the increased individual size, the strong reduction of nematode abundance beneath the cage determined a decrease of the total nematode biomass (Mirto *et al*., 2002).

The analysis of nematodes to genus level is highly efficient for describing changes occurring in sediments beneath the cages due to organic enrichment. Nematode assemblages in farm sediments differ from control nematode assemblages, displaying dominance of the following genera: *Pierrickia*, *Dorylaimopsis*, *Sabatieria*, *Oncholaimellus*, *Oxystomina*, *Ptycholaimellus*, *Comesomoides*, *Daptonema*, *Setosabatieria* and *Polysigma* (Mirto *et al*., 2002). Analysis of the Maturity Index of the nematode assemblages from fish farm sediments provide the further evidence of fish farm effects, dropping when the highest biodeposition occurred. Moreover the MI analysis appears sensitive enough to detect the resilience of nematode assemblages (Mirto *et al*., 2002). In the same paper *K*-dominance curves were used to illustrate temporal changes of the impact on nematode assemblages. While no clear differences were observed within 15 days of cage deployment, after 45-75 days the difference between impacted and control assemblages was clearly evident. A first sign of recovery (not actually 'resilience' as the nematode community was different from pre-pollution conditions) was noticed 105 days after cage deployment. Similar results were obtained from the analysis of species richness and diversity, which both declined in impacted sediments. As for *K*-dominance curves, nematode response to biodeposition impact was evident 45 days after cage deployment and differences between cage and control were evident until 105 days after cage deployment. Only after eight months, both diversity (H', Shannon) and evenness (Pielou's J) clearly increased. These results are in contrast with other investigations on organic pollution, which showed much longer recovery periods for hydrocarbon pollution (more than 2.5 years, Bodin and Boucher, 1983; 2 years, Elmgren *et al.*, 1983; more than 1 year, Wormald, 1976).

The clear impact on nematodes beneath the cage, described above, was not equally evident from the analysis of the Index of Trophic Diversity (ITD; a measure of functional diversity). The analysis of the trophic groups revealed that ITD in the farm sediments increased, due to the increase of the relative importance of the non-selective deposit-feeder nematodes (1B) (Mirto *et al.*, 2002). Analysis of nematode genera in fish farm and control sediments revealed that genera such as *Latronema* and *Elzalia* disappeared almost completely in farm sediments. In contrast, other nematode genera were tolerant to biodeposition and opportunistically profited from the new conditions (i.e. *Dorylaimopsis*, *Sabatieria* and *Oxystomina*), increasing in dominance in enriched conditions (Mirto *et al.*, 2002). Among these genera, *Sabatieria* can be considered a genus indicator of organic enrichment as it becomes dominant in suboxic sediments (Vanreusel, 1990; Vincx *et al.*, 1990, Lampadariou *et al.*, 1997). Results from the few studies on nematode communities affected by fish farm biodeposition indicate that nematode assemblage structure and genus composition are sensitive tools for describing the environmental impact due to fish farming, particularly using the *K*-dominance curves and the Maturity Index on putative impacted stations versus control sites.

Nematode response to anthropogenic physical disturbance

The effects of physical disturbance on soft sediments are expected to cause partial or complete defaunation of the most severely disturbed patches. This might happen through direct mortality due to physical damage, as well as the removal/displacement of the species to nearby unfavourable habitats, which can increase their vulnerability to predation (Ramsay and Kaiser, 1998). In addition there are numerous effects of disturbance on sediment properties such as stability and bed roughness (Hall, 1994). A large amount of sediment is dredged each year from ports, harbours and waterways to maintain and improve the navigation systems for commercial and recreational purposes (Bolam *et al.*, 2003a). At present, the majority of this material (which is generally fine-grained and relatively uncontaminated) is disposed of at sea (Sabat *et al.*, 2002; Waldock *et al.*, 2003).

While a large number of studies investigating benthic infaunal recovery following a range of disturbance types have indicated that invertebrate recolonization of intertidal habitats can be relatively rapid (Harvey *et al.*, 1998; Bolam and Fernandes, 2002; Bolam *et al.*, 2002, 2003b; Lewis *et al.*, 2003), there have been very few studies to explicitly investigate the invertebrate recovery of fine grained disposals (Bolam and Whomersley, 2003, 2005; Schratzberger *et al.*, 2006). Furthermore, most biological investigations into the recolonisation of newly created soft-bottom habitats have traditionally targeted the larger macrofauna that can readily be counted and identified (Levin *et al.*, 1996; Lee *et al.*, 1998; Craft *et al.*, 1999), whereas the smaller-sized meiofauna has been largely neglected (Schratzberger *et al.*, 2006). Somerfield *et al.* (1995) reported that meiofaunal taxa are also sensitive to the disposal of dredged material and may therefore provide an alternative method of assessment to more conventional approaches (Boyd *et al.*, 2000). Boyd *et al.* (2000) found evident changes in the nematode community structure in response to dredged material disposal, in terms of lower values of diversity (H', Shannon) and equitability (Pielou's evenness). By contrast, differences in total nematode abundance are less evident, due to an increase of *Sabatieria pulchra* grp. and *Daptonema tenuispiculum* within the disposal sites, which mask any reduction in the abundance of many other species. Moreover, sediments from within disposal sites differ from those outside due to both the elimination or reduction of a range of species and as a result of a significant increase in the abundances of the non-selective deposit feeders (as *S. pulchra* grp. and *D. tenuispiculum*). Somerfield *et al.*, (1995) also found *D. tenuispiculum* and *Sabatieria punctata* numerically abundant at the same Liverpool Bay dredged material disposal site. The authors also suggested that other members of the *S. pulchra* group might have indicator value because they are found in undisturbed conditions and they often persist as dominants of the impoverished meiofaunal communities. These findings support this assertion and clearly demonstrate the usefulness of such characteristic species occurrences in aiding the assessment of disturbance effects (Boyd *et al.*, 2000). The persistence of these species over time at the disposal sites despite a dramatic change in particle size also implies a high resilience and tolerance to a range of sedimentary conditions. Tietjen (1980) noted the proliferation of *S. pulchra*, a species normally associated with silty sediments, in polluted sands. This phenomenon of adaptive 'generalist' nematodes such as *Sabatieria spp.* and *D. tenuispiculum* rapidly exploiting disturbed sediments is well documented (Heip *et al.*, 1984; Lambshead, 1986; Somerfield *et al.*, 1995).

Sandy beaches are examples of simple ecosystems, defined as physically stressful environments (McLachlan, 1983; Rodil and Lastra, 2004), principally driven by the physical forces of waves, tides and sediment movements (Short, 1999). The major stresses on sandy beaches include the overexploitation of natural resources, pollution, industrialization and erosion (Dronkers and de Vries 1999) while tourism and recreational activities have been largely neglected (Gormsen, 1997). The impact of tourism does not only include the human trampling on the beach, but also all the activities needed to organize and maintain it (beach management operations). The increasing usage of sandy beaches as recreational places has forced regional authorities of many tourist countries to remove all natural wrack and litter of fabricated origin

(Ryan and Swanepoel, 1996). Consequently, a variety of cleaning techniques (front-end loaders, suction devices) have been developed in tourist coastal regions all over the world (Engelhard and Withers, 1997). Mechanical beach cleaning not only removes beach litter, but also alters the sediment, its micro-topography and its inhabitants, therefore creating a uniform habitat with a short durational stability (Gheskiere *et al.*, 2006). Physical disturbance by cleaning activities is already known to cause a decrease or disappearance of macrofauna, but the effects on the interstitial meiofauna are unclear (Gheskiere *et al.*, 2005a). Gheskiere *et al.* (2005a,b), generally found clear changes in the nematode assemblage structure between tourist and non-tourist beaches. Most of the nematode species absent on tourist upper beaches belong to orders of Dorylaimida, Ironidae and Rhabditida. On the other hand orders like Monhysterida and Enoplida, containing species-rich, well-represented genera such as *Enoplolaimus* (with *E. attenuatus*, *E. enoploidiformis*, *E. littoralis*, *E. villosus*, *E. balgenis*) and *Theristus* (with *T. heterospiculum*, *T. heterospiculoides*, *T. inermis*, *T. aculeatus*, *T. pictus*) are encountered in both tourist and non-tourist upper beaches. Changes in sandy sediment nematode assemblages, subjected to continuous and spasmodic perturbations in contrast to unperturbed situations, were also detected by Schratzberger and Warwick (1999) during microcosm experiments. An experimental study conducted by Gheskiere *et al.* (2006), revealed that the only measurable impacts that could be attributed to the beach cleaning operations can provoke an immediate decrease in faunal abundance and change of assemblage structure, as a result of reduced numbers of individuals from dominant nematode species (*Theristus otoplanobius*, *Trissonchulus benepapilosus*, *Chromadorina germanica*). The authors suggested that the susceptibility of species to beach cleaning/grooming is largely determined by their body size and turnover, with large, slowly-reproducing species being more susceptible than smaller, faster-reproducing ones. In this respect, it is not unexpected that some of the larger nematode species like *T. benepapilosus* (body length: 2.5–3.2 mm) are significantly affected by the cleaning as they are probably crushed by the mixer. Moreover, it is likely that turnover rates are potentially more important in affecting the longer-term maintenance of populations following repeated beach cleaning (Gheskiere *et al.*, 2006). Nematodes in dynamic environments such as the beaches, generally exhibit morphological adaptations (e.g. body ornamentations which provide an anchorage) to high turbulence and shifting sediments (Gheskiere *et al.*, 2005a). These morphological adaptations together with physiological adaptations and population growth rates can be expected to contribute significantly to the high resilience of the studied nematode assemblages to mechanical beach cleaning (Gheskiere *et al.*, 2006).

Freshwater Habitats

Historically, analyses of macrobenthic invertebrates was an important component in determinations of the ecological status of freshwater systems. These analyses included various indices, such as the Belgian Benthic Index

(BBI; de Pauw and Vanhooren, 1983), the Biological Monitoring Working Party (BMWP; Armitage *et al.*, 1983), and the Saprobien Index (Friedrich, 1990). Several studies have shown that nematodes are also useful indicators of anthropogenic pollution of freshwater systems (reviewed in Höss *et al.*, 2006), although there is much less experience with nematodes as biological indicators in freshwater than in marine and terrestrial systems (Höss and Traunspurger, 2003). Nematodes might be particularly useful indicators in medium to highly polluted habitats, where only few macroinvertebrate taxa are present. Moreover, due to the short generation time of some nematode taxa, it is possible to monitor short-term effects in microcosm studies (Höss *et al.*, 2004; Bergtold *et al.*, 2007; see also case studies below). In this section we outline several case studies dealing with the influence of pollution on freshwater nematode communities. Cause–effect relationships are discussed by presenting studies that use three different approaches of nematode community assessment.

Freshwater field study: independent samples across different river basins

In a recent study, Heininger *et al.* (2007) investigated nematode communities of three large German rivers (Elbe, Rhine, Oder) at sites with varying quantities and qualities of contamination. Eight sites at three river catchments were investigated in terms of genera composition, feeding types, and life-history strategies. In this approach, differences in pollution did not occur along a spatial gradient within a certain river stretch (e.g. as determined by distance from a specific pollution source), but instead varied between sites that differed in their hydromorphological structures and between different river catchments. The quantity and quality of pollution at the sites were determined by chemical sediment analysis of priority substances.

Heininger *et al.* (2007) clearly showed that nematode generic composition was related to the degree of pollution at the sampled locations (Fig. 6.1a), with heavy metals and arsenic being the most important contaminants. While at sites with low levels of pollution either the two bacterial-feeding genera *Monhystera* and *Daptonema* or the suction feeder *Dorylaimus* predominated, in highly polluted sites either the bacterial feeder *Eumonhystera* or omnivorous and predatory genera, such as *Tobrilus* and *Mononchus*, prevailed. However, it was difficult to define typical nematode communities for clean versus polluted sediments, as the investigated sites also differed in habitat structure, which strongly influences benthic communities. Heininger *et al.* (2007) also found that the hydromorphological structures of the sites partly explained their different nematode community structures (Fig. 6.1b).

The classification of nematode species based on feeding type or life-history strategy (according to the Maturity Index, MI; Bongers, 1990) showed ambiguous results. On the one hand, there were sites with low levels of pollution and a high maturity index (Dömitz, Elbe catchment; MI = 3.01) and sites with high levels of pollution and a low MI (Alte Elbe, Elbe, MI = 2.26), as would be expected by the MI theory. On the other hand, highly polluted

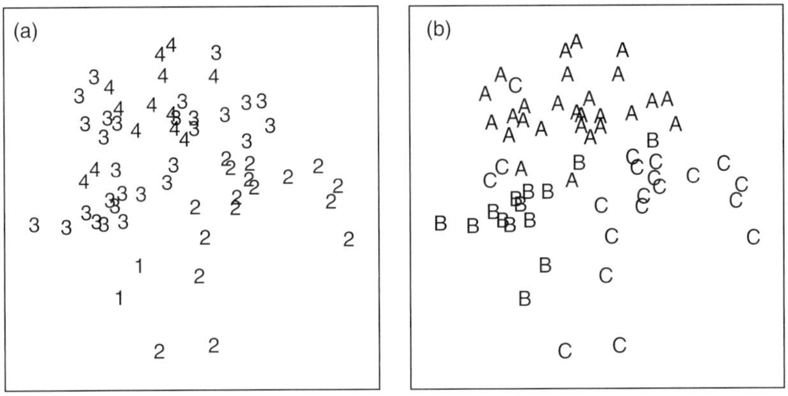

Fig. 6.1. Multidimensional scaling (MDS) ordination of square-root-transformed relative abundances of nematode genera. Sediment samples were obtained from the Elbe, Rhine, and Oder catchments. Data points of the various samples were replaced by information on (a) anthropogenic pollution (pollution classes 1–4; with 1 = lowest pollution level and 4 = highest pollution level) and (b) the site structure (A = groyne field, B = natural, lentic, C = harbour, lock). Reprinted from Heininger *et al.* (2007) with permission from Allen Press Publishing Services.

sites (Fahlberg List, Elbe; Hafen Meissen, Elbe; Ratzdorf, Oder) showed high relative abundances of omnivorous and predatory nematodes, which were classified as persisters (*Tobrilus*: c-p3; *Mononchus* c-p4), but this finding was not in line with the MI theory. Thus, either the c-p classification based on nematode families and developed for terrestrial nematodes, is not or is only partially valid for freshwater nematodes, or indirect food-web effects (e.g. the elimination of macrobenthic competitors) might have benefited predatory species in spite of their higher sensitivity to pollution.

Barbuto and Zullini (2005) compared various Italian rivers in terms of their nematode community structure. These authors applied the Taxonomic Distinctness Index (Δ^+), which is defined by the average taxonomic path length between any two randomly chosen species, traced through a Linnean phylogenetic classification of the full set of species involved (Clarke and Warwick, 1998). Based on literature data, the Δ^+ was calculated for ten samplings from seven different Italian rivers (Fig. 6.2). All the rivers fell into the 95% confidence funnel, except the highly polluted Lambro in which very polluted stretches diverged considerably from the normal situation. Habitats outside the confidence funnel are subject to environmental stresses, usually pollution, but other factors can also affect this index. It is noteworthy that the best river environments (Oglio, Adige, Seveso), i.e. those that were unpolluted, were at or above the average expected value.

Freshwater field study: pollution gradient approach

Investigating pollution gradients within a single river system is one step towards a better understanding of cause–effect relationships. Depending on

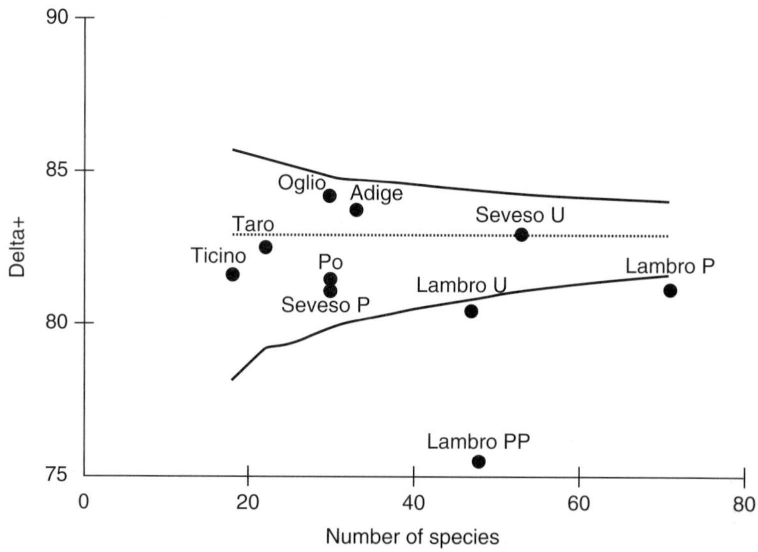

Fig. 6.2. Distinctness indices Δ⁺ from seven Italian rivers. The data were plotted against the number of species as points on the simulated 95% confidence funnel based on random selections from the total Italian species list. U = unpolluted; P = polluted; PP = very polluted. Reprinted from Barbuto and Zullini (2005) with permission from Brill.

the length of the investigated river stretch, hydromorphological factors are relatively constant, while a pollution input (e.g. wastewater or industrial effluents) will increase chemical pollution from a point source. Back in 1976, Zullini published a study on the river Seveso, a small Italian river less than 50 km long. This river has a steep pollution gradient, since it crosses one of the most populated and industrialized areas of northern Italy. In its watershed, there are 400,000 inhabitants and many types of industry. As part of the study, five stations were sampled at monthly intervals during a one-year period. The first station (site 1, with a bed of pebbles and sand) was located about 1.5 km from the river source and its water was only slightly polluted. The fifth station (site 5, with a bed of sand and silt-clay), in the northern outskirts of Milan, consisted of very polluted water, deep grey in colour and with a highly offensive odour.

In Zullini's study, the quantity and quality of pollution were not directly measured as contaminant concentrations in the sediment, but rather by three water-chemical measurements that act as indicators of pollution: (i) dissolved oxygen concentration, which indicates the amount of oxygen left in the water (and available to resident organisms) after its depletion due to organic pollution; (ii) chemical oxygen demand (COD), which reflects the amount of dissolved organic matter that originated from sewage, industrial, and agricultural waste; and (iii) electrical conductivity (μS/cm), which is proportional to the salt content of the river and indicates the amount of inorganic, mainly industrial pollution. Based on these measurements, it was concluded that pollution increased strongly along the Seveso River. Dissolved oxygen dropped from 95% (average at site 1) to only 35% (average at site 5) while COD and conductivity increased from 12 to 235 mg O_2/l and 255 to 881 μS/cm, respectively.

A total of 56 nematode species were detected, with different species predominating at the different sites. Species belonging to the Rhabditida played an important role in discriminating among the various sites. Zullini (1976) considered Rhabditida (referred to as Secernentea) as indicators of mainly organic pollution, since these nematodes are bacterial feeders and typical inhabitants of rotten matter and dung. Diplogasteromorpha peaked in the medium-heavily polluted stretch of the river, whereas Rhabditomorpha peaked in the terminal, most polluted stretch. If the percentage of Rhabditida (excluding *Fictor fictor* as an indicator of clean habitats; Zullini, 1976) is chosen as an index of water pollution, then increasing values of pollution were evident along the river course, from 12% at site 1 to almost 100% at site 3 (Fig. 6.3). These values reflected the increasing degree of pollution as determined according to the above-mentioned chemical parameters. In addition, the MI correlated well with the pollution levels at the different sites, dropping from 2.82 at site 1 to 1.21 at site 3 (Fig. 6.3). The two

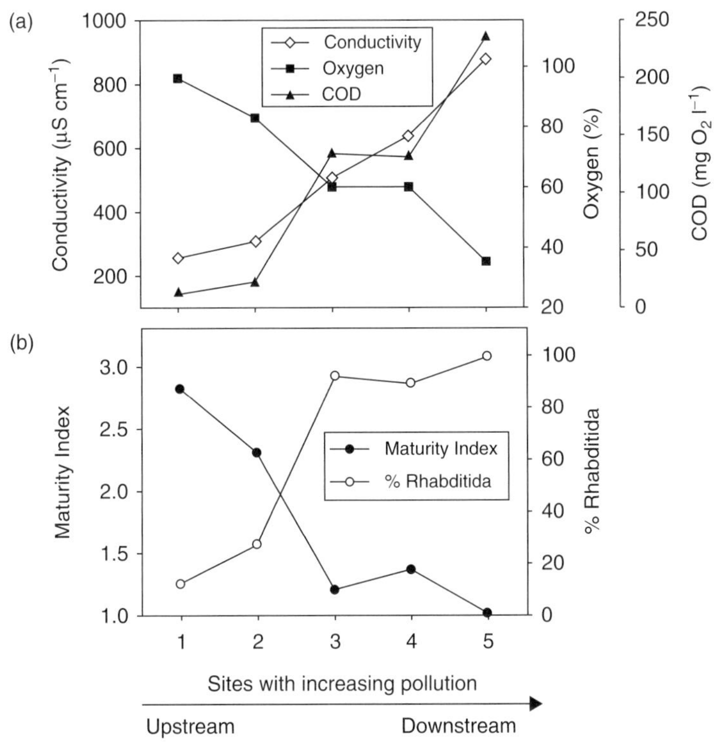

Fig. 6.3. (a) Electrical conductivity (μS/cm), oxygen saturation (%), and chemical oxygen demand (COD, in mg O_2/l) and (b) Maturity Index and % Rhabditida at increasingly polluted sites from the Seveso River (northern Italy). The data were taken from Zullini (1976).

indices, percent Rhabditida and MI, were negatively correlated, as Rhabditid species are usually categorized as colonizers, with a c-p value of 1. Consistent with the above statements, *Fictor fictor* should be assigned a higher c-p score (c-p3 or 4) rather than considered as a typical colonizer, with a c-p value of 1. If a higher c-p score had been applied to *F. fictor*, the difference between site 1 and all other sites would have been greater, as this species only occurred at site 1.

Freshwater experimental approach: microcosm studies

Microcosm studies allow investigation of the effects of single toxicants on a given nematode community under experimental conditions. Although even in such studies, disturbing factors cannot be excluded, community responses can be more accurately assigned to a toxicant effect than is the case in field studies. In the following, two experiments in which freshwater microcosms were spiked with different contaminants, a heavy metal (cadmium; Cd; Bergtold *et al.*, 2007) and an organic chemical (4-nonylphenol; NP; Höss *et al.*, 2004) are discussed in the context of observations from field studies.

In the first study, NP was applied to seven microcosms over a period of six weeks so that maximal sediment concentrations of 0.30–3.37 mg/kg dry weight were achieved (Höss *et al.*, 2004). Nematode community structures in those treatments were compared to those of four controls over a period of 15 weeks. Species composition was analysed using principle response curves (PRC; Van den Brink and Ter Braak, 1999), a multivariate analysis which showed NP-induced changes in species composition over a period of seven weeks, from the end of the application until the end of the experiment (Fig. 6.4). In the highest-dose treatment, the relative abundances of *Monhystera paludicola*, *Tobrilus gracilis*, and *Prodesmodora* sp. decreased whereas those of *Chromadorina bioculata*, *Eumonhystera dispar*, and *Eumonhystera simplex* increased compared to controls. These changes in species composition were also reflected in the MI, with lower values in the highest dose treatment during the last three weeks of the study.

In the second microcosm study (Bergtold *et al.*, 2007), the effect of cadmium on a lentic freshwater nematode community was investigated over a period of 31 weeks (218 days), during which nematodes were exposed to a large range of nominal Cd concentrations (10–1000 mg/kg dry weight). Multivariate analysis (PRC) showed a very clear dose-dependent effect of Cd on the species composition in the microcosms (Fig. 6.5a). In the highest concentration, the initially dominant bacterial-feeding genera, *Daptonema* and *Eumonhystera* (both c-p2), were completely eliminated by Cd, while predatory and omnivorous genera, such as *Dorylaimus*, *Mesodorylaimus*, and *Mononchus* (all c-p4), became dominant. This resulted in a distinct increase of the MI, from an initial 2.21 to 3.92 at the end of the study (Fig. 6.5b). In the medium-level treatment (100 mg/kg dry weight), the relative abundance of *Daptonema* was also reduced; however, this was compensated for by an increased relative abundance of *Eumonhystera*. At the end of the study, the MI

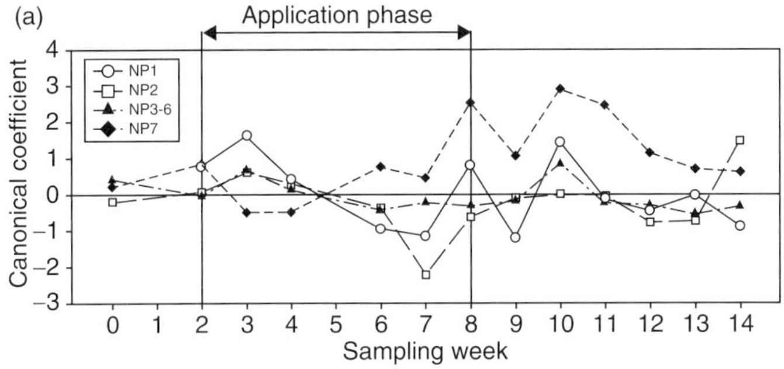

Species	Species score
Chromadorina bioculata	0.88
Eumonhystera dispar	0.63
Rhabdolaimus sp.	0.16
Achromodora sp.	0.14
Prodesmodora sp.	−0.16
Aphanolaimus sp.	−0.18
Tobrilus gracilis	−0.38
Monhystera paludicola	−0.53

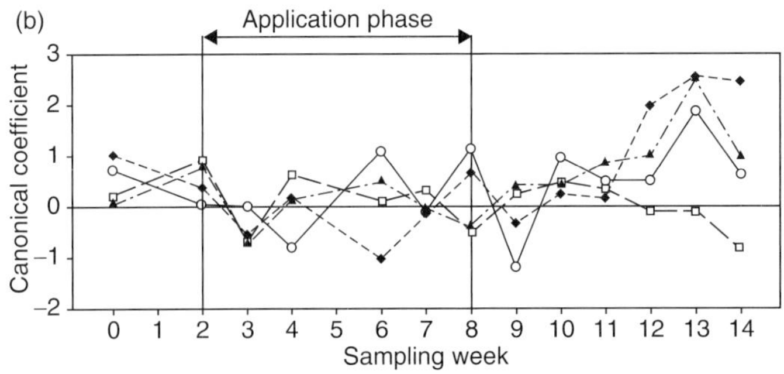

Species	Species score
Eumonhystera simplex	0.51
Rhabdolaimus sp.	−0.12
Tobrilus gracilis	−0.12
Monhystera paludicola	−0.12
Chromadorina bioculata	−0.14
Aphanolaimus sp.	−0.26
Eumonhystera dispar	−0.27
Eumonhystera similis	−0.33
Prodesmodora sp.	−0.77

Fig. 6.4. Principle response curves (PRC) and species scores calculated from the species composition of nematode communities in microcosms dosed with nonylphenol (NP1, NP2, NP3-6 and NP7). (a) First explanatory variable (22.1%; eigenvalue: 0.06). (b) Second explanatory variable (15.4%; eigenvalue: 0.04). For controls and NP 3-6: $n = 4$; for NP1, NP2 and NP7 $n = 1$. Deviation of a PRC from the control line is directly related to a change in species composition compared to controls. The species score provides information on the extent that a species contributes to this difference (positive score = relative abundance shows the same trend; negative score = relative abundance shows opposite trend). Reprinted from Höss *et al.* (2004) with permission from Allen Press Publishing Services; copyright SETAC, Pensacola, Florida, USA.

in the medium-level treatment (2.15 ± 0.11) was found to be lower than in the control (2.74 ± 0.41), reflecting an increase in *Tripyla* species (predators; c-p3) seen in the controls.

Freshwater habitats – discussion and conclusions

As demonstrated by the above-reported case studies, it is difficult to predict a universally valid community response to anthropogenic pollution. Both the quantity and the quality of pollution are crucial to the occurrence of certain changes in freshwater nematode community structure. Organic enrichment due to eutrophication is known to have a major influence on benthic communities (Hellawell, 1978) and it is difficult to distinguish the effects of eutrophication from those of chemical pollution. Bacterivorous nematodes

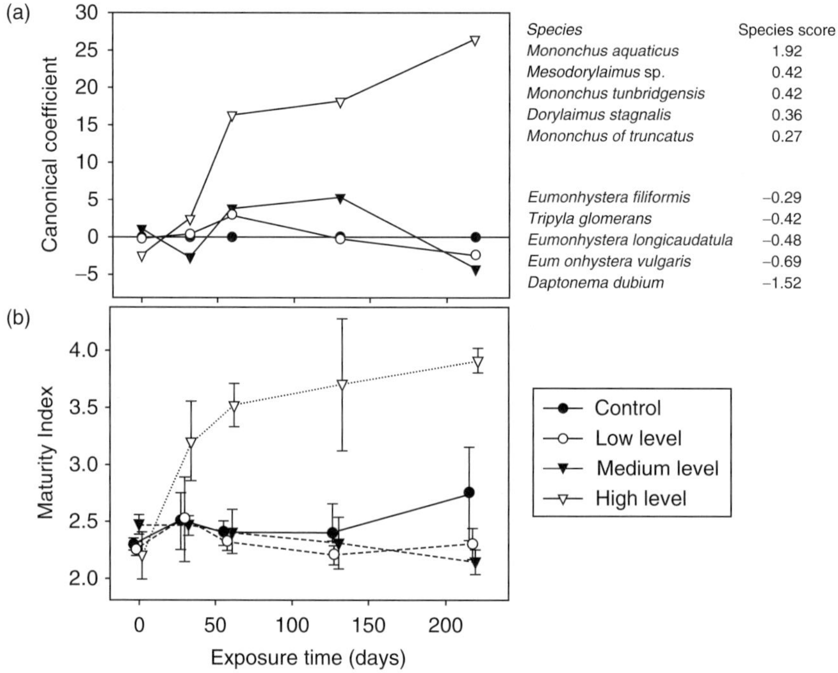

Fig. 6.5. (a) Principle response curves (PRC) with species scores and (b) Maturity Index calculated from the generic composition of nematode communities in microcosms dosed with Cd (low level = 10 mg/kg dry weight; medium level: 100 mg/kg dry weight; high level: 1000 mg/kg dry weight). First explanatory variable (51%; eigenvalue: 0.20), $n = 4$. Deviation of a PRC from the control line is directly related to a change in species composition compared to the controls. The species score provides information on the extent that a species contributes to this difference (positive score = relative abundance shows the same trend; negative score = relative abundance shows opposite trend), data from Bergtold *et al.*, 2007.

belonging to Rhabditida (Secernentea), often originating from terrestrial habitats, may be good indicators of organic enrichment (pollution). These nematodes require less dissolved oxygen, have a high reproductive turnover, exhibit resistant stages and can withstand very high levels of pollution, both organic (such as rotten material and faeces, in which they thrive) and inorganic. Rhabditomorpha abound even in extremely polluted and poorly oxygenated habitats, where they usually become more numerous than Diplogasteromorpha. *Diploscapter coronatus* and *Poikilolaimus oxycerca* are two of the most saprobic species. They are therefore classified as typical colonizers, with the lowest c-p score (=1). The case study from the Seveso River is a typical example of an increase in Rhabditida along a pollution gradient, indicating a strong disturbance of the nematode community. The results from the Seveso case study are supported by data from a study in the Lambro river (Italy), also showing considerably higher proportions of Rhabditita (99%)

and a lower MI (1.03) in the polluted, polysaprobic sites compared to the less polluted, oligotrophic sites (10% and 2.43, respectively; Zullini, 1988). However, as pollution levels in these rivers were estimated by indirect measures (conductivity, oxygen, COD) rather than chemical sediment analysis, it is not clear whether the disturbance was due to general organic enrichment or the toxic effects of contaminants. To distinguish 'chemical pollution stress' from 'eutrophication stress', Korthals *et al.* (1996) suggested the assessment of soil pollution, thereby omitting c-p1 taxa from MI calculations (see Ferris and Bongers, Chapter 5, this volume). The resultant index, MI2-5, was applied in a freshwater study in which the influence of sewage effluents on nematode communities in small streams was investigated (Beier and Traunspurger, 2001). The authors of the study showed that the MI was lower in polluted than in unpolluted sites, whereas the MI2-5 was not affected, indicating eutrophication rather than a pollution effect.

Multivariate analysis of nematode taxon composition is probably the most sensitive tool to assess differences in the community structures of different sites (Warwick and Clarke, 1991). The first freshwater case studies reported in this chapter showed that the composition of nematode taxa sampled from sites distributed across various river catchments was correlated with the degree of contamination at those sites, as determined by multidimensional scaling ordination (MDS). Although these data could not be readily interpreted, the results indicated the dominant nematode taxa, which mainly occurred at reference and polluted sites. In contrast to the second freshwater case study, from the Seveso River, the relation between community structure and pollution was not in accordance with the MI theory, suggesting that, since the index was developed for soils, it might be applicable to freshwater only with restrictions or after revision. Some nematode taxa may be incorrectly categorized within the c-p scale, which would cause a misinterpretation of the MI. For example, the genus *Tobrilus* seems crucial for uncertainties in the allocation of the family Tobrilidae to c-p3. *T. diversipapillatus* is an indicator of sewage effluent pollution and is highly resistant to metals and residual chlorine (Arthington *et al.*, 1986; Beier and Traunspurger, 2001). Moreover, *T. gracilis* was found to be frequent in polluted stretches of the Styrian (Austria) river Mur (Eder and Kirchengast, 1982). Other *Tobrilus* species, such as *T. medius*, however, were found to be sensitive to pollution (Walter Traunspurger, unpublished data). Moreover, the genus *Monhystera* might not be a typical representative of c-p2, since it does not appear to be a typical stress-tolerant 'general opportunist' as other genera in this group (*sensu* Ettema and Bongers, 1993). *Monhystera* is a bacterivorous nematode of aquatic origin that is not strictly a bacterial feeder; instead, it often eats algae or other eukaryote cells living in habitats either moderately or not polluted. In a sediment remediation study, Den Besten and Van den Brink (2005) found higher abundances of *Monhystera* and lower abundances of *Tobrilus* in remediated than in non-remediated sites. Furthermore, in the above-mentioned case study with 4-NP-spiked microcosms, the abundance of *Monhystera* decreased in treatments with high NP concentrations, while *Eumonhystera* (same family: Monhysteridae; c-p2) profited from NP contamination (Höss *et al.*, 2004). The predator *Mononchus* (c-p4), which, unexpectedly, mainly occurred at polluted sites of the rivers Elbe and Oder (Heininger *et al.*, 2007), also showed high cadmium resistance in a microcosm study (see third case study; Bergtold *et al.*, 2007). Until now, these types of uncertainties have impeded proper use of the MI for the study of freshwater nematode communities.

Although the above-reported examples suggest that it is possible to distinguish between nematode communities from polluted and unpolluted habitats by using multivariate analysis (MDS; PRC) or subtle biodiversity indices (taxonomic distinctness index), there is still a need for simple, ecologically meaningful indices, like the MI, for the ecological assessment of freshwater systems. Refinement of the MI, which takes into account theoretical and experimental ecological data (Bongers, 1990), might be achieved by adding ecotoxicological information (Bongers *et al.*, 2001; see also SPEAR Index; Liess and Von der Ohe, 2005) and empirical data from field studies. This would allow the successful use of nematodes as bioindicators, which, in turn, would be a valuable contribution to the aim of the EU-WFD to improve the ecological status of freshwater bodies.

References

Ahnert, A. and Schriever, G. (2001) Response of abyssal Copepoda Harpacticoida (Crustacea) and other meiobenthos to an artificial disturbance and its bearing on future mining for polymetallic nodules. *Deep Sea Research II* 48, 3779–3794.

Alongi, D.M., Chong, V.C., Dixon, P., Sasekumar, A. and Tirend, F. (2003) The influence of fish cage aquaculture on pelagic carbon flow and water chemistry in tidally dominated mangrove estuaries of peninsular Malaysia. *Marine Environmental Research*, 55, 313–333.

Armitage, P.D., Moss, D., Wright, J.F. and Furse, M.T. (1983) The performance of a new biological water quality score based on macroinvertebrates over a wide range of unpolluted running-water sites. *Water Research* 17, 333–347.

Arthington, A.H., Yeates, G.W. and Comick, D.L. (1986) Nematodes, including a new record of *Tobrilus diversipapillatus* in Australia, as potential indicators of sewage effluent pollution. *Australian Journal of Marine and Freshwater Research* 37, 159–166.

Atilla, N., Wetzel, M.A. and Fleeger, J.W. (2003) Abundance and colonization potential of artificial hard substrate-associated meiofauna. *Journal of Experimental Marine Biology and Ecology* 287, 273–287.

Austen, M.C., McEvoy, A.J. and Warwick, R.M. (1994) The specificity of meiobenthic community responses to different pollutants: results from microcosm experiments. *Marine Pollution Bulletin* 28, 557–563.

Balestri, E., Benedetti-Cecchi, L. and Lardicci, C. (2004) Variability in patterns of growth and morphology of *Posidonia oceanica* exposed to urban and industrial wastes: contrasts with two reference locations. *Journal of Experimental Marine Biology and Ecology* 308, 1–21.

Barbuto, M. and Zullini, A. (2005) The nematode community of two Italian rivers (Taro and Ticino). *Nematology* 7, 667–675.

Beier, S. and Traunspurger, W. (2001) The meiofauna community of two small German streams as indicators of pollution. *Journal of Aquatic Ecosystem Health* 8, 387–405.

Bergtold, M., Brinke, M., Ristau, K., Höss, S., Traunspurger, W., Claus, E. and Heininger, P. (2007) The effect of cadmium on a freshwater nematode community: A microcosm study. 1st *International Symposium on Nematodes As Environmental Bioindicators*, 12–13 June 2007, Heriot Watt University, Edinburgh, UK.

Beveridge, M.C.M. (1996) *Cage Aquaculture*, 2nd edn, Fishing News Books, Oxford, 346 pp.

Beyrem, H. and Aïssa, P. (2000) Les nématodes libres, organismes sentinelles de l'évolution des concentrations d'hydrocarbures dans la baie de Bizerte (Tunisie). *Cahiers de Biologie Marine* 41, 329–342.

Blome, D., Schleier, U. and Bernem, K.H. (1999) Analysis of the small-scale spatial patterns of free-living marine nematodes from tidal flats in the East Frisian Wadden Sea. *Marine Biology* 133, 717–726.

Bodin, P. and Boucher, D. (1983) Evolution a moyen terme du meiobenthos et des pigments chlorophylliens sur quelques plage pollutée par les hydrocarbures avec utilization d'un dispersent. *Acta Oecologica* 3, 263–280.

Bolam, S.G. and Fernandes, T.F. (2002) Dense aggregations of tube-building polychaetes: response to small-scale disturbances. *Journal of Experimental Marine Biology and Ecology* 269, 197–222.

Bolam, S.G. and Whomersley, P. (2003) Invertebrate recolonization of fine-grained beneficial use schemes: an example from the southeast coast of England. *Journal of Coastal Conservation* 9, 159–169.

Bolam, S.G. and Whomersley, P. (2005) Development of macrofaunal communities on dredged material used for mudflat enhancement: a comparison of three beneficial use schemes after one year. *Marine Pollution Bulletin* 50, 40–47.

Bolam, S.G., Fernandes, T.F. and Huxham, M. (2002) Diversity, biomass and ecosystem processes in the marine benthos. *Ecological Monograph* 72 (4), 599–615.

Bolam, S.G., Rees, H.L., Murray, L.A. and Waldock, R. (2003a) Intertidal placement of fine-grained dredged material. *Proceedings of the 28th International Conference of Coastal Engineers*, Cardiff, July 2002. American Society of Coastal Engineers, World Scientific Inc., pp. 3606–3615.

Bolam, S.G., Schratzberger, M. and Whomersley, P. (2003b) Macrofaunal recolonization in intertidal mudflats: the effect of organic content and particle size. *Journal of Experimental Marine Biology and Ecology* 306, 157–180.

Bongers, T. (1990) The maturity index: an ecological measure of environmental disturbance based on nematode species composition. *Oecologia* 83, 14–19.

Bongers, T. and Ferris, H. (1999) Nematode community structure as a bioindicator in environmental monitoring. *Trends in Ecology and Evolution* 14 (6), 224–228.

Bongers, T., Alkemade, R. and Yeates, G.W. (1991) Interpretation of disturbance induced maturity decrease in marine nematode assemblages by means of the maturity index. *Marine Ecology Progress Series* 76, 135–142.

Bongers, T., Ilieva-Makulec, K. and Ekschmitt, K. (2001) Acute sensitivity of nematode taxa to $CuSO_4$ and relationships with feeding-type and life-history classification. *Environmental Toxicology and Chemistry* 20, 1511–1516.

Bonsdorf, E. (1981) The Antonio Gramsci oil spill. Impact on the littoral and benthic ecosystem. *Marine Pollution Bulletin* 301–305.

Boucher, G. (1980) Impact of Amoco Cadiz oil spill on intertidal and sublittoral meiofauna. *Marine Pollution Bulletin* 11, 95–101.

Boyd, S.E., Rees, H.L. and Richardson, C.A. (2000) Nematodes as sensitive indicators of change at dredged material disposal sites. *Estuarine, Coastal and Shelf Science* 51, 805–819.

Bryan, G.W. and Langston, W.J. (1992) Bioavailability, accumulation and effects of heavy metals in sediments with special reference to United Kingdom estuaries: a review. *Environmental Pollution* 76, 89–131.

Clarke, K.R. and Warwick, R.M. (1998) A taxonomic distinctness index and its statistical properties. *Journal of Applied Ecology* 35, 523–531.

Coull, B.C., Hicks, G.R.F. and Wells, J.B.Y. (1981) Nematode/copepod ratios for monitoring pollution: a rebuttal. *Marine Pollution Bulletin* 12, 378–381.

Coull, B.C. and Chandler, G.T. (1992) Pollution and meiofauna: field, laboratory and mesocosm studies. *Oceanography and Marine Biology Annual Review* 30, 191–271.

Craft, C., Reader, J., Sacco, J.N. and Broome, S.W. (1999) Twenty-five years of ecosystem development of *Spartina alterniflora* (Loisel) marshes. *Ecological Application* 9, 1405–1419.

Danovaro, R., Fabiano, M. and Vincx, M. (1995) Meiofauna response to the Agip Abruzzo oil spill in subtidal sediments of the Ligurian Sea. *Marine Pollution Bulletin* 30, 133–145.

Danovaro, R., Della Croce, N., Marrale, D., Martorano, D., Parodi, P., Pusceddu, A. and Fabiano, M. (1999) Biological indicators of oil induced disturbance in coastal sediments of the Ligurian Sea. In: Lokman, S., Shazili, N.A.M., Nasir, M.S. and Borowtizka, M.A. (eds), *Assessment & Monitoring of Marine Science*, University Putra Malaysia Terengganu, Kuala Terengganu, Malaysia, pp. 75–85.

Danovaro, R. and Fraschetti, S. (2002) Meiofaunal vertical zonation on hardbottoms: comparison with soft-bottom meiofauna. *Marine Ecology Progress Series* 230, 159–169.

de Pauw, N. and Vanhooren, G. (1983) Method for biological quality assessment of water courses in Belgium. *Hydrobiologia* 100, 153–168.

Den Besten, P.J. and Van den Brink, P.J. (2005) Bioassay responses and effects on benthos after pilot remediations in the delta of the rivers Rhine and Meuse. *Environmental Pollution* 136, 197–208.

Derycke, S., Hendrickx, F., Backeljau, T., D'Hondt, S., Camphijn, L., Vincx, M. and Moens, T. (2007) Effects of sublethal abiotic stressors on population growth and genetic diversity of *Pellioditis marina* (Nematoda) from the Westerschelde estuary. *Aquatic Toxicology* 82, 110–119.

Dronkers, J. and de Vries, I. (1999) Integrated coastal management: the challenge of transdisciplinarity. *Journal of Coastal Conservation*, 5, 97–102.

Duplisea, D.E. and Hargrave, B.T. (1996) Response of meiobenthic size-structure, biomass and respiration to sediment organic enrichment. *Hydrobiologia* 339, 161–170.

Eder, R. and Kirchengast, M. (1982) The nematode-fauna (Nemathelminthes, Nematoda) of a polluted part of the river Mur (Styria, Austria). *Nematologia Mediterranea* 10, 127–134.

Elmgren, R., Hanson. U. and Sundelin, B. (1980) In The Tsesis Oil Spill. In Kineman, J.J., Elmgren R. and Hanson, S. (eds) *Report to the First Year Scientific Study (26 October 1977 to December 1978)* US Dept. of Commerce, Office of Marine Pollutional Assessment, Colorado, pp. 97–126.

Elmgren, R., Hansson, S., Larsson, U., Sundelin, B. and Boehm, P.D. (1983) The Tsesis oil spill: Acute and long-term impact on the benthos. *Marine Biology* 73 (1), 51–65.

Engelhard, T. K. and Withers, K. (1997) *Biological effects of mechanical beach raking in the upper intertidal zone on Padre Island National Seashore, Texas*. Technical report TAMU-CC-9706-CCS. Padre Island National Seashore, Corpus Christi, Texas.

Ettema, C.H. and Bongers, T. (1993) Characterization of nematode colonization and succession in disturbed soil using the Maturity Index. *Biology and Fertility of Soils* 16, 79–85.

European Community (2000) Directive 2000/60/EC of the European Parliament and the council of 23 October 2000 establishing a framework for Community action in the field of water policy. *Official Journal of the European Community* L 327, 1–73.

Feder, H.M., Naidu, A.S. and Paul, A.J. (1990) Trace metal and biotic changes following a simulated oil spill on a mudflat in Port Valdez, Alaska. *Marine Pollution Bulletin* 21, 131–137.

Fleeger, J.W., Tita, G., Carman, K.R., Millward, R.N., Moser, E.B., Portier, R.J. and Gambrell, R.P. (2006) Does bioturbation by a benthic fish modify the effects of sediment contamination on saltmarsh benthic microalgae and meiofauna? *Journal of Experimental Marine Biology and Ecology* 330, 180–194.

Fraschetti, S., Gambi, C., Giangrande, A., Musco, L., Terlizzi, A. and Danovaro, R. (2006) Structural and functional response of meiofauna rocky assemblages to sewage pollution. *Marine Pollution Bulletin* 52, 540–548.

Fricke, A.H., Henning, H.F.H.O. and Orren, M.J. (1981) Relationship between pollution and spammolittoral meiofauna density of two South African beaches. *Marine Environmental Research* 5, 57–77.

Friedrich, G. (1990) Eine Revision des Saprobiensystems. *Zeitschrift für Wasser und Abwasser Forschung* 23, 141–152.

Frithsen, J.B., Elmgren, R. and Rudnick, D.T. (1985) Responses of benthic meiofauna to long-term, low-level additions of No. 2 fuel oil. *Marine Ecology Progress Series. 23*, 1–14.

Gee, J.M., Austen, M., De Smet, G., Ferraro. T., McEvoy, A., Moore, S., Van Gausbeki, D., Vincx, M. and Warwick, R.M. (1992) Soft sediment meiofauna community responses to environmental pollution gradients in the German Bight and at a drilling site off the Dutch coast. *Marine Ecology Progress Series* 91, 289–302.

Gheskiere, T., Vincx, M., Weslawski, J.M., Scapini, F. and Degraer, S. (2005a) Meiofauna as descriptor of tourism-induced changes at sandy beaches. *Marine Environmental Research* 60, 245–264.

Gheskiere, T., Vincx, M., Urban-Malinga, B., Rossano, C., Scapini, F. and Degraer, S. (2005b) Nematodes from wave-dominated sandy beaches: diversity, zonation patterns and testing of the isocommunities concept. *Estuarine, Coastal and Shelf Science* 62, 365–375.

Gheskiere, T., Vincx, M., Pison, G. and Degraer, S. (2006) Are strandline meiofaunal assemblages affected by a once-only mechanical beach cleaning? Experimental findings. *Marine Environmental Research* 61, 245–264.

Gibbons, M.J. (1988a) The impact of sediment accumulations, relative habitat complexity and elevation on rocky shore meiofauna. *Journal of Experimental Marine Biology and Ecology* 122, 225–241.

Gibbons, M.J. (1988b) The impact of wave exposure on the meiofauna of *Gelidium pristoides* (Turner) Kuetzing (Gelidiales: Rhodophyta). *Estuarine, Coastal and Shelf Science* 27, 581–593.

Giere, O. (1979) The impact of oil pollution on intertidal meiofauna. Field studies after the La Coruna spill, May, 1976. *Cahiers de Biologie Marine* 20, 231–251.

Gormsen, E. (1997) The impact of tourism on coastal areas. *GeoJournal* 42, 39–54.

Guidetti, P., Terlizzi, A., Fraschetti, S. and Boero, F. (2003) Changes in Mediterranean rocky-reef fish assemblages exposed to sewage pollution. *Marine Ecology Progress Series* 253, 269–278.

Guo, Y., Soerfield, P.J., Warwick, R.M. and Zhang, Z. (2001) Large-scale patterns in the community structure and biodiversity of freeliving nematodes in the Bohai sea, China. *Journal of Marine Biology Association of United Kingdom* 81, 3867/1–9.

Gyedu-Ababio, T.K., Furstenberg, J.P., Baird, D. and Vanreusel, A. (1999) Nematodes as indicators of pollution: a case study from the Swartkops river system, South Africa. *Hydrobiologia* 397, 155–169.

Hall, S.J. (1994) Physical disturbance and marine benthic communities: life in unconsolidated sediments. *Oceanography Marine Biology Annual Review* 32, 179– 239.

Harvey, M., Gauthier, D. and Munro, J. (1998) Temporal changes in the composition and abundance of the macro-benthic invertebrate communities at dredged material disposal sites in the Anse à Beaufils, Baie des Chaleurs, Eastern Canada. *Marine Pollution Bulletin* 36, 41–55.

Heininger, P., Höss, S., Claus, C., Pelzer, J. and Traunspurger, W. (2007) Nematode communities in contaminated river sediments. *Environmental Pollution* 146, 64–76.

Heip, C., Vincx, M., Smol, N. and Vranken, G. (1982) The systematic ecology of free-living marine nematodes. *Helminthological Abstracts Series B, Plant Nematology* 51, 1–31.

Heip, C., Herman, R. and Vincx, M. (1984) Variability and productivity of meiobenthos in the Southern Bight of the North Sea. *Rapport et procés-verbaux des réunions. Conseil international pour l'Exploration de la Mer* 183, 51–56.

Heip, C., Vincx, M. and Vranken, G. (1985) The ecology of marine nematodes. *Oceanography and Marine Biology. An Annual Review* 23, 399–489.

Hellawell, J. (1978) *Biological Surveillance of Rivers, a Biological Monitoring Handbook.* Water Research Center, Medmenham and Stevenage, UK.

Higgins, R.P. and Thiel, H. (1988) *Introduction to the Study of Meiofauna*. Smithsonian Institution Press, Washington, DC.

Hodda, M. and Nicholas, W.L. (1986) Nematode diversity and industrial pollution in the Hunter River estuary, NSW, Australia. *Marine Pollution Bulletin* 17, 251–255.

Holmer, M. and Kristensen, E. (1992) Impact of fish cage farming on metabolism and sulfate reduction of underlying sediments. *Marine Ecology Progress Series* 80, 191–201.

Holmer, M., Wildish, D. and Hargrave, B. (2005) Organic enrichment from marine finfish aquaculture and effects on sediment processes. In Hargrave, B.T. (ed.) *The Handbook of Environmental Chemistry Volume 5: Water Pollution Environmental Effects Of Marine Finfish Aquaculture*, Springer Verlag, pp.181–206.

Höss, S. and Traunspurger, W. (2003) Nematodes. In: Breure, A.M., Markert, B. and Zechmeister, H. (eds) *Bioindicators/Biomonitors – Principles, Assessments, Concepts*. Elsevier, Amsterdam, pp. 529–554.

Höss, S., Traunspurger, W., Severin, G.F., Jüttner, I., Pfister, G. and Schramm, K.W. (2004) Influence of 4-nonylphenol on the structure of nematode communities in freshwater microcosms. *Environmental Toxicology and Chemistry* 23, 1268–1275.

Höss, S., Traunspurger, W. and Zullini, A. (2006) Freshwater nematodes in environmental science. In: Eyualem, A., Traunspurger, W. and Andrassy, I. (eds) *Freshwater Nematodes – Ecology and Taxonomy*. CABI Publishing, Wallingford, pp. 144–162.

Howell, R. (1982) Levels of heavy metal pollutants in two species of marine nematodes. *Marine Pollution Bulletin* 13, 396–398.

Howell, R. (1983) Heavy metals in marine nematodes: uptake, tissue distribution and loss of copper and zinc. *Marine Pollution Bulletin* 14, 263–268.

Kalantzi, I. and Karakassis, I. (2006) Benthic impacts of fish farming: meta-analysis of community and geochemical data. *Marine Pollution Bulletin* 52, 484–493.

Karakassis, I., Tsapakis, M., Hatziyanni, E., Papadopoulou, K.N. and Plaiti, W. (2000) Impact of cage farming on the seabed in three Mediterranean coastal regions. *ICES Journal of Marine Science* 57, 1462–1471.

Kennish, M.J. (1992) Polynuclear aromatic hydrocarbons. In: *Ecology of Estuaries: Anthropogenic Effects*, CRC Press, Boca Raton, Florida, pp. 133–181.

Korthals, G.W., De Goede, R.G.M., Kammenga, J.E., Bongers, T., Van Straalen, N.M. and Krivolutsky, D.A. (1996) The Maturity Index as an instrument for risk assessment of soil pollution. *Bioindicator Systems for Soil Pollution*. Kluwer Academic Publishers, Dordrecht, The Netherlands, pp. 85–94.

La Rosa, T., Mirto, S., Mazzola, A. and Danovaro, R. (2001) Differential responses of benthic microbes and meiofauna to fish-farm disturbance in coastal sediments. *Environmental Pollution* 112, 427–434.

La Rosa, T., Mirto, S., Mazzola, A. and Maugeri, T.L. (2004) Benthic microbial indicators of fish farm impact in a coastal region of the Tyrrhenian Sea. *Aquaculture* 230, 153–167.

Lambshead, P.J.D., (1984) The nematode/copepod ratio: some anomalous results from Firth of Clyde. *Marine Pollution Bulletin* 15, 256–259.

Lambshead, P.J.D. (1986) Sub-catastrophic sewage and industrial waste contamination as revealed by marine nematode populations. *Marine Ecology Progress Series* 29, 247–260.

Lampadariou, N., Austen, M.C., Robertson, N. and Vlachonis, G. (1997) Analysis of meiobenthic community structure in relation to pollution and disturbance in Iraklion Harbour, Greece. *Vie et Milieu* 47 (1), 9–24.

Lee, J.G., Nishijima, W., Mukai, T., Takimoto, K., Seiki, T., Hiraoka, K. and Okada, M. (1998) Factors to determine the functions and structures in natural and constructed tidal flats. *Water Research* 32, 2601–2606.

Lee, M.R., Correa, J.A. and Castilla, J.C. (2001) An assessment of the potential use of the nematode to copepod ratio in the monitoring of metal pollution. The Chanaral case. *Marine Pollution Bulletin* 42, 696–701.

Levin, L.A., Talley, D. and Thayer, G. (1996) Succession of macrobenthos in a created salt marsh. *Marine Ecology Progress Series* 141, 67–82.

Lewis, L.J., Davenport, J. and Kelly, T.C. (2003) A study of the impact of a pipeline constuction-construction on estuarine benthic invertebrate communities. Part 2. Recolonization by benthic invertebrates after 1 year and response of estuarine birds. *Estuarine Coastal Shelf Science* 57, 201–208.

Liess, M. and von der Ohe, P.C. (2005) Analyzing effects of pesticides on invertebrates in streams. *Environmental Toxicology and Chemistry* 24, 954–965.

Lorenzen, S., Prein, M. and Valentin, C. (1987) Mass aggregations of free living marine nematode *Pontonema vulgare* (Oncholaimidae) in organic polluted fjords. *Marine Ecology Progress Series* 37, 27–34.

Mahmoudi, E., Beyrem, H. and Aïssa, P. (2003) Les peuplements de nématodes libres, indicateurs du degré d'anthropisation des milieux lagunaires: Cas de la lagune de Bou Ghrara (Tunisie). *Vie et Milieu* 53, 47–59.

Mahmoudi, E., Essid, N., Beyrem, H., Hedfi, A., Boufahja, F., Vitiello, P. and Aïssa, P. (2005) Effects of hydrocarbon contamination on a free living marine nematode community: Results from microcosm experiments. *Marine Pollution Bulletin* 50, 1197–1204.

Marine Pollution Monitoring Management Group (MPMMG) (1996) Monitoring and assessment of the marine benthos at UK dredged material disposal sites. Scottish Fisheries Information Pamphlet 21, 35 pp.

Maurer, D., Church, T.M., Lord, C. and Weithe, C. (1985) Marine benthos in relation to pore water chemistry and sediment geochemistry of simulated dredged material. *Internationale Revue der gesamten Hydrobiologie und Hydrographie* 70, 369–377.

Mazzola, A., Mirto, S. and Danovaro, R. (1999) Initial fish-farm impact on meiofaunal assemblages in coastal sediments of the western Mediterranean. *Marine Pollution Bulletin* 38, 1126–1133.

McLachlan, A. (1983) Sandy beach ecology. A review. In: McLachlan, A. and Erasmus, T. (eds) *Sandy Beaches as Ecosystems*. Junk, The Hague, pp. 321–380.

Millward, R.N. and Grant, A. (1995) Pollution-induced tolerance to copper of nematode communities in the severely contaminated Restronguet creek and adjacent estuaries, Cornwall, United Kingdom. *Environmental Toxicology and Chemistry Article* 454–461.

Mirto, S. and Danovaro, R. (2004) Meiofaunal colonisation on artificial substrates: a tool for biomonitoring the environmental quality on coastal marine systems. *Marine Pollution Bulletin* 48, 919–926.

Mirto, S., La Rosa, T., Gambi, C., Danovaro, R. and Mazzola, A. (2000) Microbial and meiofaunal response to intensive mussel-farm biodeposition in coastal sediments of the Western Mediterranean. *Marine Pollution Bulletin* 40, 244–252.

Mirto, S., La Rosa, T., Gambi, C., Danovaro, R. and Mazzola, A. (2002) Nematode community response to fish-farm impact in the western Mediterranean. *Environmental Pollution* 116, 203–214.

Peterson, C.H., Kennicutt, M.C., Green, R.H., Montagna, P., Harper, D.E., Powell, E.N. and Roscigno, P.F. (1996) Ecological consequences of environmental perturbations associated with offshore hydrocarbon production: a perspective on long-term exposures in the Gulf of Mexico. *Canadian Journal of Fisheries and Aquatic Sciences* 53, 2637–2654.

Piazzi, L., Balata, D., Cinelli, F. and Benedetti-Cecchi, L. (2004) Patterns of spatial variability in epiphytes of *Posidonia oceanica*. Differences between a disturbed and two reference locations. *Aquatic Botany* 79, 345–356.

Platt, H.M. and Warwick, R.M. (1983) *A Synopsis of the Freeliving Marine Nematodes. Part 1. British Enoplids*. Cambridge University Press, Cambridge, 307 pp.

Platt, H.M., Shaw, K.M. and Lambshead, P.J.D. (1984) Nematode species abundance patterns and their use in the detection of environmental perturbations. *Hydrobiologia*, 118, 59–66.

Porter, C.B., Krost, P., Gordin, H. and Angel, D.L. (1996) Preliminary assessment of grey mullet (*Mugil cephalus*) as a forager of organically enriched sediments below marine fish farms. *Israeli Journal of Aquaculture-Bamidgeh* 48 (1), 47–55.

Prein, M. (1988) Evidence for a scavenging lifestyle in the free-living nematode *Pontonema vulgare* (Enoplida, Oncholaimidae). *Kieler Meeresforsch* 6, 389–394.

Raffaelli, D.G. and Meson, C.F. (1981) Pollution monitoring with meiofauna using the ratio of nematodes to copepods. *Marine Pollution Bulletin* 12, 158–163.
Ramsay, K. and Kaiser, M.J. (1998) Demersal fishing disturbance increases predation risk for whelks (*Buccinum undatum* L). *Netherlands Journal of Sea Research* 39, 299–304.
Renaud-Mornant, J., Gourbault, N., de Panafiew, J.B. and Heleouet, M.N. (1981) Effets de la pollution par hydrocarbures sur la meiofauna de la baie de Morlaix. In *Amoco Cadiz, Consequences d'une Pollution Accidentale par les Hydrocarbures. Actes coli. Intern. C.O.B.*, Brest (France), 19–22 November 1979, pp. 551–561.
Rodil, I.F. and Lastra, M. (2004) Environmental factors affecting benthic macrofauna along a gradient of intermediate sandy beaches in northern Spain. *Estuarine, Coastal and Shelf Science* 61, 37–44.
Ryan, P.G. and Swanepoel, D. (1996) Cleaning beaches: sweeping the rubbish under the carpet. *South African Journal of Science* 92, 275–276.
Sabat, L.J., Vogt, C. and Holliday, B. (2002) Promoting dredged material as a resource. *Sea Technology* 43, 47–52.
Sandulli, R. and De Nicola-Giudici, M. (1990) Pollution effects on the structure of meiofaunal communities in the Bay of Naples. *Marine Pollution Bulletin* 21, 144–153.
Sandulli, R. and De Nicola-Giudici, M. (1991) Responses of meiobenthic communities along a gradient of sewage pollution. *Marine Pollution Bulletin* 22, 463–467.
Sandulli, R. and Pinckney, J. (1999) Patch size and spatial patterns of meiobenthic copepods and benthic microalgae in sandy sediments: a microscale approach. *Journal of Sea Research* 41, 179–187.
Schratzberger, M. and Warwick, R.M. (1999) Differential effects of various types of disturbances on the structure of nematode assemblages: an experimental approach. *Marine Ecology Progress Series* 181, 227–236.
Schratzberger, M., Bolam, S., Whomersley, P. and Warr, K. (2006) Differential response of nematode colonist communities to the intertidal placement of dredged material. *Journal of Experimental Marine Biology and Ecology* 334, 244–255.
Schratzberger, M., Warr, K. and Rogers, S.I. (2007) Functional diversity of nematode communities in the southwestern North Sea. *Marine Environmental Research* 63, 368–389.
Short, A.D. (1999) *Beach and Shoreface Morphodynamics*. Wiley, Chichester, West Sussex, 379 pp.
Smith, A., Ajani, P.A. and Roberts, D.E. (1999) Spatial and temporal variation in fish assemblages exposed to sewage and implications for management. *Marine Environmental Research* 47, 241–260.
Somerfield, P.J., Gee, J.M. and Warwick, R.M. (1994) Soft sediment meiofaunal community structure in relation to a long-term heavy metal gradient in the Fal estuary system. *Marine Ecology Progress Series* 105, 79–88.
Somerfield, P.J., Rees, H.L. and Warwick, R.M. (1995) Interrelationships in community structure between shallow-water marine meiofauna and macrofauna in relation to dredging disposal. *Marine Ecology Progress Series* 127, 103–112.
Stacey, B. and Marcotte, B. (1987) Chronic effect of No. 2 fuel oil on population dynamics of harpacticoid copepods in experimental marine mesocosms. *Marine Ecology Progress Series* 40, 61–68.
Steyaert, M., Vanaverbeke, J., Vanreusel, A., Barranguet, C., Lucas, C. and Vincx, M. (2003) The importance of fine-scale, vertical profiles in characterising nematode community structure. *Estuarine, Coastal and Shelf Science* 58, 353–366.
Suderman, K. and Thistle, D. (2003) Spills of fuel oil #6 and Orimulsion can have indistinguishable effects on the benthic meiofauna. *Marine Pollution Bulletin* 46, 49–55.
Sundelin, B. and Elmgren, R. (1991) Meiofauna of an experimental soft bottom ecosystem-effects of macrofauna and cadmium exposure. *Marine Ecology Progress Series* 70, 245–255.
Terlizzi, A., Benedetti-Cecchi, L., Bevilacqua, S., Fraschetti, S., Guidetti, P. and Anderson, M.J. (2005a) Multivariate and univariate asymmetrical analyses in environmental impact

assessment: a case study of Mediterranean subtidal sessile assemblages. *Marine Ecology Progress Series* 289, 27–42.

Terlizzi, A., Scuderi, D., Fraschetti, S., Anderson, M.J. (2005b) Quantifying effects of pollution on biodiversity: a case study of highly diverse molluscan assemblage in the Mediterranean. *Marine Biology* 148, 293–305.

Tietjen, J.H. (1980) Population structure and species distribution of the free-living nematodes inhabiting sands of the New York Bight Apex. *Estuarine Coastal Marine Science* 10, 61–73.

Tsujino, M. (1998) Distribution of free-living nematodes in relation to bottom environments in Hiroshima Bay. *Bulletin of Nansei Natural and Fisheries Research Institute* 31, 25–32.

Van den Brink, P.J. and Ter Braak, C.J.F. (1999) Principal response curves: Analysis of time dependent multivariate responses of biological community to stress. *Environmental Toxicology and Chemistry* 18, 138–148.

Vanreusel, A. (1990) Ecology of the free-living marine nematodes from the Voordelta (Southern Bight of the North Sea). 1. Species composition and structure of the nematode communities. *Cahiers de Biologie Marine* 31 (4), 439–462.

Vidakovic, J. (1983) The influence of row domestic sewage on density and distribution of meiofauna. *Marine Pollution Bulletin* 14, 84–88.

Vincx, M., Meire, P. and Heip, C. (1990) The distribution of nematode communities in the Southern bight of the North Sea. *Cahiers de Biologie Marine* 31, 107–129.

Vincx, M. and Heip, C. (1991) The use of meiobenthos in pollution monitoring studies; a review. In: Rees, H.L., Heip, C., Vincx, M. and Parker, M.M. (eds) *Benthic Communities: Use in Monitoring Point Source Discharges. Techniques in Marine Environmental Sciences*, No. 16. International Council for the Exploration of the Sea, pp. 50–67.

Waldock, R., Paipai, E. and Bolam, S.G. (2003) New land from old, beneficial use of dredged material. *Dredging and Port Construction*. DMG World Media, Southwick, pp. 26–29.

Warwick, R.M. (1981) The nematode/copepod ratio and its use in pollution ecology. *Marine Pollution Bulletin* 12, 329–333.

Warwick, R.M. and Clarke, K.R. (1991) A comparison of some methods for analysing changes in benthic community structure. *Journal of the Marine Biology Association, UK* 71, 225–244.

Wormald, A.P. (1976) Effect of an oil spill of marine diesel oil on meiofauna of a sandy beach at Picnic Bay Hong Kong. *Environmental Pollution* 11, 117–130.

Zhang, J. (1996) Geochemistry of arsenic in the Huanghe (Yellow river) and its delta region: a review of available data. *Aquatic Geochemistry* 1, 241–275.

Zhinan, Z., Hongyue, D. and Zishan, Y. (1993) A study on the meiobenthic community in an organically polluted beach of Qingdao Bay. *Journal of Ocean University of Qingdao/Qingdao Haiyang Daxue Xuebao* 23, 83–91.

Zullini, A. (1976) Nematodes as indicators of river pollution. *Nematologia Mediterranea* 4, 13–22.

Zullini, A. (1988) The ecology of the Lambro river. *Rivista di Idrobiologia* 27, 39–58.

7 Case Studies Using Nematode Assemblage Analysis in Terrestrial Habitats

PÉTER NAGY

Department of Zoology and Animal Ecology, Szent István University, Gödöllő, Hungary

Introduction

The demand to record and understand possible biological responses to environmental disturbances of human origin emerged some decades ago. Within this period, studies on nematodes as biological indicators of environmental pollution have a relatively long history. Among the various pollutants, heavy metals are the most thoroughly explored for their effects on soil nematode assemblages. Therefore this chapter will focus on key studies in this field.

Johnson *et al.* (1974) used a community level approach to study effects of various impacts on nematode communities of forest woodlots. Probably the most important achievement of this work was, in the light of the subsequent studies, stressing the sensitivity of omnivorous and predatory dorylaimid nematodes to environmental disturbances.

Later, Zullini and Peretti (1986) studied the effects of Pb pollution on moss inhabiting nematodes. Although there was no correlation between Pb load and abundance values, samples of increasing Pb content showed a decrease in nematological parameters such as taxon richness, Shannon diversity and nematode biomass. They found members of the suborder Dorylaimina to be the most sensitive to Pb load. Concentrations around 240 mg/kg and 320 mg/kg caused a repeated sharp decline in dorylaimid numbers. Above these values hardly any dorylaimid nematodes appeared in the samples.

Also in the mid-1980s, Dr Dieter Sturhan performed an experiment on the effects of several microelements on soil nematode assemblages. Unfortunately, the results of this very interesting survey have never been published.

The knowledge gathered in the field by that time allowed for drawing important conclusions and outlining a concept of revolutionary consequences for subsequent studies. In 1990, Bongers published his important paper introducing the Maturity Index that became a landmark in environmental nematology (see Ferris and Bongers, Chapter 5, this volume).

Since then a considerable amount of literature has elaborated on effects of both accidental and artificial pollution on nematode community structure. Polluting effects of microelements (mostly heavy metals) on terrestrial nematode assemblages will be examined in the light of practical bioindication. Therefore, where available, the severity of the given contamination will be described in relation to statutory domestic regulation collected from the cited papers.

Key Studies

Weiss and Larink (1991) described an experiment from Braunschweig, Germany in which agricultural plots were treated with sewage sludge containing heavy metals. Metals involved were: Cd, Cr, Cu, Ni, Pb and Zn. The experiment was performed on a loamy sand (luvisol) of pH (H_2O)=6.4. It is particularly difficult to compare the results of this study with those of most others for the following reasons. There were no attempts made to quantify effects of individual metal contaminants. In addition, the feeding type characterizations applied in the paper contradict the generally accepted classification by Yeates *et al.* (1993) for several taxonomic groupings and there is no consideration of nematode life strategies (r-K). These factors together may lead to erroneous or questionable conclusions being drawn on the effects of heavy metals. Probably the most striking is the often cited finding regarding the great increase of predators in the heavy metal treated plots. This is misleading for the following reasons. The only 'predator' that showed a great increase in the heavy metal treated plots was *Pristionchus*, a diplogastrid genus that is a strong r-strategist. In addition, *Pristionchus* is classified as a bacterial feeder *or* predator by Yeates *et al.* (1993). On the contrary, K-strategist dorylaimid nematodes, the real predators or omnivores like *Aporcelaimellus*, *Ecumenicus*, *Mesodorylaimus*, *Paravulvus*, etc. completely disappeared in the heavy metal treated plots. Therefore, in the light of more recent knowledge, it is better not to use this study as a reference for heavy metal tolerance of predatory nematodes. Calculation of the Maturity Index from the data resulted in values of 2.28, 1.62 and 1.44 for the untreated, the sludge treated and the heavy metal contaminated sludge treated plots, respectively. This reveals that the sludge and particularly the heavy metal treatment posed a considerable disturbance on the relevant nematode assemblages, despite the great increase in abundance in values (1729, 4481 and 7216, respectively).

A series of papers from Bardgett *et al.* (1994) and Yeates *et al.* (1994, 1995) reported the effects of a timber preservative based on Cu, Cr and As on soil biota, including nematodes. The experimental site was near Levin, North Island, New Zealand.

Perhaps the most significant and systematic studies on effects of some heavy metals on nematode assemblages are those of Korthals and his co-authors. They published their work principally in the mid to late 1990s. They studied the acute effects of fresh contamination (Korthals *et al.*, 1996a), how these effects were influenced by the presence or absence of vegetation (Korthals *et al.*, 1998) and they also monitored long-term effects (Korthals *et al.*, 1996b). The experimental site was on the Bovenbuurt pastures, near Wageningen in The Netherlands. To study short-term pollution effects, Cu, Ni and Zn were applied in doses of 100, 200, 400, 800 and 1600 mg/kg, while Cd was applied in doses of 10, 20, 40, 80 and 160 mg/kg. Long-term effects of Cu and impact of pH were studied in combinations of four Cu-levels (0, 250, 500 and 750 kg/ha) and pH values of 4.0, 4.7, 5.4 and 6.0. The influence of vegetation on Cu and Zn pollution was studied along a concentration gradient of 0, 25, 50, 100, 200 and 400 mg/kg.

It was also an innovation by Korthals *et al.* (1996c) to propose the use of MI2–5 instead of the original MI. The concept of the new parameter was the elimination of nematodes belonging to group c-p1 from the calculations. These so called 'enrichment opportunist' taxa show a much more sensitive response to the nutrient status of soils than to actual pollution. Therefore, the index value calculated excluding these is a more relevant tool for assessment of pollution effects under certain circumstances.

Somewhat later, Nagy and his co-workers started to publish results of a small-plot agricultural field experiment from the Nagyhörcsök experimental site in the central region of Hungary. The artificial pollution experiment had been started in 1991, 5–6 years before the first nematode samples were taken. This set of studies included a first screening of 13 microelements (Nagy, 1999), a dose-response experiment (Nagy *et al.*, 2004), and also a study of long-term (5–10 years) effects (Bakonyi *et al.*, 2003). Nominal total values of each studied pollutant were set to approx. 0, 30, 90 and 270 mg/kg at the beginning of the experiment. A new approach in these studies was the analysis of diversity profiles, instead of single diversity indices. This was achieved by the use of diversity ordering software developed by Tóthmérész (1993).

Another experiment investigating the effects of Zn was described by Smit *et al.* (2002). This work focused on the nematological effects of a wide range of Zn concentrations: 32–3200 mg/kg (nominal total concentrations).

Georgieva *et al.* (2002) reported results of a long-term experiment in which sewage sludge contaminated with heavy metals was applied to plots in the Gleadthorpe Research Centre in Nottinghamshire, United Kingdom on two separate dates. Nematode samples were taken in 1994, 12 and 8 years after the first and the second application of sludge respectively.

Yeates *et al.* (2003) studied long-term (1–4 years after sludge application) effects of heavy metal pollution on a silt loam soil in New Zealand. They reported effects of Cu, Ni and Zn in low levels (near the NZ statutory soil limit concentrations for sewage sludge application). Total concentrations of Cu, Ni and Zn were between 141–185 mg/kg, 53–58 mg/kg and 266–349 mg/kg, respectively, while activity values in the soil solution ranged from 0.005–2.2, 0.8–8.0 and 1.0–198 mg/kg, respectively. These values showed a decreasing trend during the subsequent years after repeated sludge treatments in 1997 and 1998.

After Weiss and Larink (1991) and the series of papers from New Zealand (Bardgett *et al.*, 1994; Yeates *et al.*, 1994, 1995), some more recent case studies have appeared that focus on complex contaminations, i.e. where the nature of pollution did not allow for separation of individual metal effects. These cases are probably more realistic, but much more difficult to interpret.

Sánchez-Moreno and Navas (2007) published detailed, food-web based nematological results from a survey of the heavy metal polluted soils within the basin of the Guadiamar river in Southern Spain. Their study focused on Cu, Ni, Pb and Zn effects of riparian soils in the area polluted with toxic sludge of a pyrite mine. Soil texture was variable in the experimental area, with a significantly higher sand and lower silt and clay content in the control than in the polluted sites. Soil pH values averaged from 6.7 to 8.0 in the control and 6.0 to 7.5 in the polluted sites. The average values for Cu ranged from 111.0 to 257.1 mg/kg, for Ni from 9.8 to 15.2 mg/kg, for Pb from 207.5 to 329.4 mg/kg and for Zn from 462.9 to 977.3 mg/kg.

The following nematological effects were found as a consequence of this complex pollution. Total nematode abundance, taxon richness and diversity indices (Simpson, Shannon, Margalef) were significantly ($P<0.05$) higher in the non-polluted than in the polluted soils. MI and Σ MI were affected significantly by pollution in only two and one occasions, respectively. From the indices proposed by Ferris *et al.* (2001), Structure Index was always lower (in two cases significantly) and the Channel Index was in most of cases higher in polluted than in control plots. The other indices (Enrichment Index, Basal Index) did not show any clear difference as a response to pollution. SI and CI showed a slightly increasing, while EI and BI a slightly decreasing trend throughout the sampled period.

This paper will further be referred to only when effects of a particular heavy metal on a given nematological parameter are specifically expressed.

Very recently, several case studies have been published from China. Among these, Zhang *et al.* (2007) surveyed responses of nematode assemblages to Cu and Zn pollution on corn fields near a copper smelter in Northeast China. Total Cu and Zn values were between 35.7 and 376.3 mg/kg and between 104.4 and 175.1 mg/kg respectively, while available Cu and Zn values ranged from to 9.0 to 72.1 and from 13.3 to 27.9 mg/kg, respectively. Both total and available concentrations of Cu and Zn showed significant decreases with increasing distance from the smelter. Based on the Chinese soil quality classification, soils of the two sites closer to the smelter were unsuitable for agricultural use (principally due to the high Cu content), whereas soils of the remote sites were suitable for agriculture. Therefore, the conditions of this experiment can be considered as a transition between 'high' and 'medium' load, based on the local classification of soil contamination.

It has to be noted that there are two major ways for quantification of heavy metal content in soil:

1. Measuring 'total' concentration, i.e. the amount of all the metals digestible from the sample. There are simple and straightforward techniques for this, but the outcome is less meaningful in terms of biological effects.

Table 7.1. Summary of key studies regarding use of nematodes as biological indicators in soils.

Country	Soil description	Pollutants tested	Form measured	Analysis	Reference(s)
CN	Cambisol pH = 5.2–5.4	Cu, Zn	Total, available	AAS	Zhang *et al.* (2007)
E	Variable pH = 6.0–8.0	Cu, Ni, Pb, Zn			Sánchez-Moreno and Navas (2007)
GB	Sandy loam pH = 5.8–7.2	Cu, Ni, Zn	Total	ICP-AES	Georgieva *et al.* (2002)
HU	Loamy chernozem pH (KCl) = 7.4	Al, As, Ba, Cd, Cr, Cu, Hg, Mo, Ni, Pb, Se, Sr, Zn	Total, available	ICP	Nagy (1999); Nagy *et al.* (2004)
NL 1	Sandy, acidic pH (KCl) = 4.1	Cd, Cu, Ni, Zn	Total, available	AAS	Korthals *et al.* (1996a); Korthals *et al.* (1998)
NL 2	Sandy, acidic pH (KCl) = 5.2	Zn	Total, $CaCl_2$-exch.		Smit *et al.* (2002)
NZ 1	Silt loam pH = 5.6–6.0	As, Cr, Cu	Total	XRF	Yeates *et al.* (1994); Bardgett *et al.* (1994)
NZ 2	Silt loam pH = 5.5–7.0	Cu, Ni, Zn	Total	XRF	Yeates *et al.* (2003)

CN: China, E: Spain, GB: Great Britain, HU: Hungary, NL: the Netherlands, NZ: New Zealand, $CaCl_2$-exch.: $CaCl_2$-exchangeable, AAS: atomic absorption spectrometry, AES: atomic emission spectrometry, ICP: inductively-coupled plasma spectrometry, XRF: X-ray fluorescence spectrometry.

2. Quantifying the biologically 'available' fraction. This is more informative in explaining possible contaminant effects. However, the techniques used for its estimation are variable worldwide, making it difficult to compare values from different studies.

Table 7.1 summarizes the key methodological features of studies reviewed in this chapter.

Findings of Key Studies

There are certain heavy metals, generally considered to be toxic, that appear to have no effects on nematode assemblages even in concentrations comparable to or slightly higher than statutory limit values. For example, Nagy (1999) showed the following microelements to have no deleterious effects compared to the control. Aluminium (in an 'available' concentration of 66.4 mg/kg), As (41.8 mg/kg), Ba (56.2 mg/kg), Hg (22.8 mg/kg), Mo (8.1 mg/kg), and Sr (132.5 mg/kg). The given load of these elements did not significantly affect total nematode abundance, nematode taxon richness or feeding group distribution. Maturity Index, tested with Kruskal–Wallis ANOVA, showed a significant increase ($P<0.01$) in Ba-, Sr- and ($P<0.05$) in Mo-treated plots. Therefore it can be concluded that the applied doses in the referred experiment were too low to cause long-term toxic effects on the given soil which had a high buffering capacity. For these elements, no or hardly any other interpretable nematological data were found in the literature. Conversely, there are several toxic microelements (Zn, Cu, Ni, Cd, Pb, Cr and Se) for which several data sets exist, and some broad generalizations about the effects of these substances can be drawn.

Zinc

Korthals *et al.* (1996a) found various parameters to decrease significantly with increasing Zn load. The proportion of predators and omnivores appeared to be the most sensitive parameter showing significant decreases at the first contamination level (100 mg/kg). Nematode abundance and Maturity Index showed the first significant reductions at 200 mg/kg and 400 mg/kg, respectively. In the light of the Dutch reference values for contaminated and severely contaminated soils (68 and 340 mg/kg, respectively) all the experimental soils can be classified as contaminated, and the nematological data of Korthals *et al.* (1996a) confirms the validity of these values. Differences were also seen in the relative abundance of certain taxa. For example, the relative abundance of *Pratylenchus* was found to increase significantly compared to the control even when exposed to the highest metal concentrations. This has since been confirmed by Bakonyi *et al.* (2003) who found *Pratylenchus* to significantly increase in abundance over five subsequent years of Zn pollution but under very different soil conditions.

Nagy (1999) found the maximum level of Zn to influence nematode assemblages in several ways. Species richness and Maturity Index both increased significantly. The proportion of fungal feeders decreased, but that of plant feeders, predators/omnivores and c-p3–5 nematodes doubled compared with control plots. However, the soil involved in this experiment was depleted in Zn. Therefore, these nematological reactions were attributed to a microelement completion effect. This proves that nematode assemblages are able to indicate positive microelement effects as well. What is more, from these results it may be concluded that nematode assemblages are not particularly sensitive to Zn load under favourable soil conditions.

Later, Nagy *et al.* (2004), in a study of dose-dependent effects, found Zn to decrease the Structure Index as well as the proportion of the most sensitive predatory and omnivorous taxa. Nematode diversity profiles did not reflect any obvious effect along the gradient of increasing Zn load. The results by Bakonyi *et al.* (2003) confirmed this loss of stimulating effects with increasing time after treatment.

Smit *et al.* (2002) recorded several responses from nematode communities under conditions when the Zn concentrations were below the nominal 'total' value of 560 mg/kg. At this value the $CaCl_2$ exchangeable Zn content did not reach 100 mg/kg. In this study the average nematode abundance did not change, but taxon richness, Shannon–Wiener diversity and the Maturity Index de-creased. (The latter parameter showed a remarkable increase at the two highest concentrations but this was explained by factors other than the treatment.) The critical value where most parameters started to decrease was identified as approximately 10 mg/kg of $CaCl_2$ exchangeable Zn. The Shannon–Wiener index and taxon diversity showed comparable sensitivity and total density was found to be the least sensitive parameter to Zn contamination of soils and these trends were similar throughout the experimental period (3–22 months).

Based on Zn sensitivity responses, three groups of taxa could be established. Zn-sensitive taxa (showing a decline above 56 mg/kg) included *Anaplectus* sp., *Aphelenchus* sp., several cephalobid nematodes (*Cephalobus* and *Eucephalobus* species) and a *Eudorylaimus*-like nematode. The majority of taxa were moderately sensitive to Zn with numbers decreasing at levels between 100 and 560 mg/kg. Tolerant taxa decreased only in levels above 1000 mg/kg. In general, this group comprised *Acrobeloides* and *Aphelenchoides*, but some *Ditylenchus* and rhabditid nematodes appeared to be tolerant too.

In the study of Georgieva *et al.* (2002), Zn significantly increased total nematode abundances, but it decreased taxon richness alone, and in some combinations with Cu. There was a significant positive correlation between Zn concentrations and number of plant-feeding nematodes, due largely to increasing abundance of *Criconemoides* and *Paratylenchus*. On the other hand, abundance of *Trichodorus* decreased significantly in Zn and Cu treated plots. Bacterial feeders increased significantly in some Zn treatments mainly due to an increase of certain rhabditid and cephalobid taxa, although some bacterial feeders, e.g. *Bastiania*, *Alaimus* and *Acrobeles*, showed negative correlations with Zn and Cu concentrations. Similarly, numbers of fungal feeders increased significantly in the low doses of Zn + Ni and Zn + Cu. *Aphelenchoides* seemed to benefit from increasing concentrations in comparison with *Aphelenchus*.

The above findings confirm the importance of articulated resolution of data even within trophic groups. Omnivores and predators, comprising mainly K-strategist mononchid and dorylaimid nematodes, decreased in several Zn and Cu treatments and were completely absent from the highest Zn treatments. Both MI and MI2–5 decreased from the first Zn level onwards. Thus, nematode assemblages showed a sensitive response even to a pollution level of approximately 50% of the EC-value for permissible Zn loads in sewage sludge.

Yeates *et al.* (2003) found that Zn, just like the other heavy metals involved in that study, caused relatively few significant effects on nematological parameters. Fungal feeders were enhanced and Nematode Channel Ratio was decreased in the Zn-treated plots. On the genus level, only *Meloidogyne* juveniles and *Diplogaster* showed a positive, while *Cephalobus* exhibited a negative correlation with the Zn treatment. No significant effects on ΣMI and Shannon–Wiener diversity were found.

Sánchez-Moreno and Navas (2007) found significant negative correlations between zinc concentrations and Margalef's index, taxon richness, SI, MI and Shannon diversity on more than one sampling occasion. Total abundance, ΣMI and Simpson diversity were similarly affected but only on one sampling occasion.

In the study by Zhang *et al.* (2007), the significant difference between soil heavy metal contents at the sampling sites created a gradient of decreasing pollution with increasing distance from the smelter. Correspondingly, significant negative correlations were detected between both total and available Zn and nematode density, proportion of bacterivores, plant feeders and Nematode Channel Ratio ($P<0.01$) as well as ratio of omnivores / predators ($P<0.05$). This concentration gradient also defined a significant increasing trend in taxon richness, Structure Index and MI2–5 with the distance from the smelter.

In conclusion, Zn cannot be classified as obviously harmful to nematode assemblages, as the effects observed largely depend on factors other than the Zn contamination level. Soil conditions (e.g. pH, buffering capacity) and the composition of indigenous nematode fauna appear to be particularly important in this respect. This is probably true for some other metals as well, but more data would be needed to confirm this.

Copper

In the study of Korthals *et al.* (1996a), Cu was applied in doses of 100, 200, 400, 800 and 1600 mg/kg and this caused effects similar to those of the Ni treatments in the same experiment. Even the lowest dose decreased most of the studied parameters (total abundance, Maturity Index and proportion of omnivores and predators) significantly. From these parameters, the proportion of omnivores and predatory nematodes was the most sensitive. Bacterial-feeding r-strategists (particularly Rhabditidae) were quite tolerant with numbers being significantly higher in plots treated with the maximum Cu concentration compared to the control. Based on the estimated EC_{50} values, taxa like *Aporcelaimellus*, *Clarkus* and *Plectus* were the most sensitive to Cu contamination. Sensitivity of *Acrobeles* and *Alaimus* was confirmed by the analyses of Ekschmitt and Korthals (2006).

The Dutch reference value to discriminate between uncontaminated and contaminated status is 20 mg/kg, while the value to define a serious Cu contamination status for that type of soil is 100 mg/kg. The results of this experiment show a very good match with the second value.

In a study focusing on long-term effects of Cu and pH on nematological parameters, Korthals *et al.* (1996b) found Cu effects to enhance as soil pH decreased. Trophic groups reacted to the treatments in a more sensitive way than total abundance values. Bacterial feeders decreased and fungal feeders increased along the concentration gradient. Predatory and, in particular, omnivorous nematodes were highly sensitive to increasing Cu concentrations and decreasing pH. On the genus level, *Acrobeles*, *Acrobeloides*, *Basiria*, *Cervidellus*, *Diphtherophora*, *Merlinius* and *Trichodorus* were reported to decrease significantly, while *Chiloplacus* increased in the sequence of Cu levels.

In order to explore the influence of vegetation on heavy metal effects, Korthals *et al.* (1998) performed an experiment involving ryegrass (*Lolium perenne*). In this experiment, both Cu and Zn were found to cause significant decreases in total abundance and Maturity Index and a non-significant decrease in taxon richness. Trophic structure was also influenced: bacterial and particularly fungal feeders increased, plant feeders decreased in proportion, while predators and omnivores disappeared in the plots of higher doses. However, similar levels of pollutants had less extreme effects on nematological parameters in the presence of vegetation. The authors suggested that this may be a result of plants enriching the soil thus diminishing pollution stress and/or the plants in some way reduced the bioavailability of the pollutant. Both can be responsible for the less harmful pollution effects, making vegetation an important factor in risk assessment studies.

According to Nagy (1999), Cu in a dose of 133 mg/kg did not cause any negative acute impacts on nematodes, though it affected certain nematological parameters 5–6 years after contamination. Total abundance values were the highest among all the heavy metal-treated samples and significantly ($P<0.01$) higher than in the control plots. Fungal feeders (particularly *Aphelenchoides* and *Paraphelenchus*) seemed to benefit from this treatment. The proportion of *Pratylenchus* also increased, while some bacterial-feeding taxa (*Chiloplacus* and *Heterocephalobus*) decreased.

Subsequently, Nagy *et al.* (2004) studied dose-dependent Cu effects on several nematological parameters. No dose related trends were found in terms of nematode abundance, taxon richness and Maturity Index. The most remarkable finding was the sharp decrease of the sensitive predatory and omnivorous nematodes in the second highest treatment. However, in plots with the maximum Cu level, their proportion was comparable to the control. There were significant ($P<0.05$) differences in feeding type distributions in the plots with the two highest Cu doses. The proportion of fungal feeders increased, while those of bacterial feeders and predators plus omnivores decreased. Moreover, diversity profiles showed a clear (although not significant) decrease with increasing Cu load.

Georgieva *et al.* (2002) found Cu alone and in combination with Zn to have a negative effect on various nematological parameters on a sandy loam. Cu and Zn+Cu decreased taxon richness and MI, and affected c-p group and feeding group distributions. The MI2–5 appeared to be more sensitive to Cu contamination than MI showing a significant decrease at a lower concentration. The two concetrations affecting the indices were approximately 24% and 50% above the EC limit value for permissible Cu loads in sewage sludge. (Combined Cu+Zn effects were mentioned in the section on Zn).

Yeates *et al.* (2003) found that their Cu treatments significantly increased numbers of *Heterodera* juveniles but significantly decreased the Shannon–Wiener diversity. No significant effects on the ΣMI and Nematode Channel Ratio were found.

Zhang *et al.* (2007) found a significant negative correlation between total Cu and nematode density, proportions of bacterivores, plant feeders, omnivores / predators and Nematode Channel Ratio ($P<0.01$) and SI ($P<0.05$).

In the Guadiamar river basin case study, Sánchez-Moreno and Navas (2007) found a significant negative correlation between levels of Cu contamination and taxon richness, Margalef's index, SI, MI, Shannon diversity, Simpson diversity and ΣMI.

A very valuable addition to the above detailed field work has been published for acute effects of $CuSO_4$ by Bongers *et al.* (2001). This paper gives sensitivity data for 70 nematode taxa. These results are particularly important since it is possible to use them as reference values in attempts to link community-level effects with generic sensitivity patterns. It is hard to overestimate the significance of this type of data while our knowledge on specific sensitivity of nematodes to individual pollutants is so limited.

In conclusion, Cu effects on nematode assemblages can be quite variable. Below the concentrations destroying all nematodes, some articulated reactions from certain groups can be detected. For example, various feeding types react to moderate Cu levels in different ways: predators and omnivores always drastically decrease, whereas plant feeders decrease in some cases. On the contrary, fungal feeders often show an increase in response to Cu contamination. These findings underline the importance of food-web related nematological studies.

Nickel

Korthals *et al.* (1996a), found the acute effects of Ni to be quite harmful from concentrations of 100 mg/kg and above. A stepwise significant decrease was measured in the following parameters: total nematode abundance, feeding group distribution (with a particular decrease of omnivores and predators) and the MI. *Plectus*, *Clarkus*, *Prismatolaimus* and *Acrobeles* appeared to be the most sensitive genera to Ni contamination. The Dutch reference value separating serious and not serious contamination for Ni on the given type of soil is 80 mg/kg. Thus, even the lowest applied concentration was above this limit value.

In contrast to the above results, Nagy (1999) failed to show any harmful effect on nematological parameters in plots treated with the highest Ni load. Instead, nematological parameters, including total abundance and genus richness, increased although not to a significant extent. Values of genus richness and the proportion of plant-feeding taxa (25%) were the highest throughout the whole experiment for all treatments. The increase of MI was significant ($P<0.01$) compared to the control. Some less common taxa including Alaimidae, *Paraphelenchus*, *Paratylenchus* and Tylenchorchynchidae showed a considerable increase in the Ni treated plots, compared to the control. The taxa comparable to Korthals *et al.* (1996a), *Acrobeles* and *Prismatolaimus*, showed a decrease in this experiment as well. It has to be pointed out, however, that the highest available Ni concentration in this experiment (51.9 mg/kg) was between the 'B' (20 mg/kg) and 'C' (60 mg/kg) values proposed for Hungary (Kadar, 2001). This could explain the above-mentioned contradiction: in Hungary lower doses were applied to a soil with a higher buffering capacity.

In the study by Georgieva *et al.* (2002), one intermediate Ni level (of an average 28.4 mg/kg in total concentration) and the maximum of combined Zn+Ni treatment (141.7 mg/kg and 27.0 mg/kg, respectively) caused a significant increase in nematode abundance compared to the control. The latter treatment increased bacterial-feeding nematodes as well. At the genus level, *Alaimus* seemed to benefit from these treatments. Low concentrations of Ni and Zn+Ni favoured fungal feeders significantly while predators and omnivores were not influenced significantly. Furthermore, the sub-maximum Ni dose (of 42.4 mg/kg in total concentration) significantly increased the MI2–5 value. This Ni value is approx. 30% lower than the EC recommendation for Ni limit in sewage sludge-treated soils. (All the values were expressed in 'total' concentrations in this experiment.)

Yeates *et al.* (2003) found Ni to cause little effect on nematological parameters. Numbers of *Aphelenchus* and *Alaimus* were significantly ($P<0.05$) increased. Shannon–Wiener diversity and Nematode Channel Ratio were affected significantly only in one of the five experimental years. The Ni loads in this experiment were of medium severity (50–58 mg/kg in total concentration).

Sánchez-Moreno and Navas (2007) found that samples with a moderate Ni content showed a significant negative correlation with taxon richness and Margalef's diversity on two occasions and with Structure Index on one occasion. In one case there was a significant positive correlation between Ni and the Enrichment Index.

In conclusion, Ni may have a negative impact on nematode assemblages, but apparently not in concentrations below or around the existing statutory limit values. In lower doses, however, it has an increasing effect on certain nematological parameters. The meta-analyses in Ekschmitt and Korthals (2006) mostly confirmed the above observations on sensitivity (*Prismatolaimus*) and tolerance of certain taxa (Alaimidae, *Paratylenchus* and Tylenchorchynchidae).

Cadmium

Korthals *et al.* (1996a) applied Cd in doses of 10, 20, 40, 80 and 160 mg/kg. The Dutch reference values to determine soil contamination status for this element are 0.5 mg/kg and 8 mg/kg for uncontaminated and severely contaminated soils, respectively. On the applied contamination levels, this heavy metal had no acute effects (nematode abundance, percentage of omnivores and predators, Maturity Index) on nematode assemblages even on a sandy soil with low pH where effects would be expected to be most pronounced.

Other studies indicating the negligible effects of Cd on nematodes include that of Nagy (1999). In this study 'available' Cd concentrations of 190 mg/kg caused no detectable change in community parameters (nematode abundance, taxon richness, feeding type distribution), apart from a significant ($P<0.01$) decrease in the Maturity Index. Furthermore, in a study of dose-dependent effects, Nagy *et al.* (2004) found most nematological values were unaffected within the concentration range of 26–190 mg/kg. The proportion of fungal feeders and the most sensitive predatory and omnivorous taxa increased significantly ($P<0.01$), while proportions of bacterial feeders and plant feeders decreased significantly ($P<0.01$). On the other hand, in the same experiment high Cd doses decreased biomass of various crop plants to a significant extent, even eight years after pollution. The fact that this contamination level hardly caused any nematological effect underlines the tolerance of nematode assemblages to this stressor.

In conclusion, nematode assemblages appear to be fairly insensitive to Cd. The range of soils tested, and the high test doses (compared to statutory limit values) in these experiments give extra confidence in this statement. Furthermore, a number of toxicity test results (Haight *et al.*, 1982; Kammenga *et al.*, 1994; Anderson *et al.*, 2001) confirm this, highlighting how useful laboratory toxicity tests can be in predicting field effects (see Höss and Williams, Chapter 9, this volume and Lagido, Chapter 10, this volume). This is in sharp contrast with our knowledge on the sensitivity of Cd on the majority of soil invertebrates (Van Straalen and Denneman, 1989; Kammenga *et al.*, 2001). These differences should be considered when evaluating the functional consequences of a Cd contamination.

Lead

Apart from the classical work from Zullini and Peretti (1986), very little information is available on the effect of Pb on terrestrial nematodes.

Nagy (1999) showed Pb at a concentration of 188.5 mg/kg had no deleterious effects compared to the control. This conclusion was based on a detailed study that examined total nematode abundance, taxon richness, Maturity Index and feeding group distribution. This level of Pb load can be considered quite high; according to the proposed limit concentration for the NH_4-acetate + EDTA-'available' fraction, it exceeds the level (150 mg/kg) for which compulsory intervention is required for sensitive areas.

According to Sánchez-Moreno and Navas (2007), Pb caused significant negative correlations with: taxon richness and Margalef's index (on all four sampling dates), SI and total abundance (three sampling dates) and MI, ΣMI, Shannon diversity and Simpson diversity on two sampling dates. In contrast, the Basal Index was singificantly enhanced by Pb in one case. However, the (total) Pb levels measured in the exposed areas of the Guadiamar river basin were considerably higher than those in the previous case study.

The above results, in agreement with Zullini and Peretti (1986), underline the possible toxic character of lead for terrestrial nematode assemblages. However, this toxicity appears only above a concentration limit that is: (i) dependent on soil type; and (ii) cannot yet be established based on the rather scarce available data.

Chromium

Hardly any results could be found to show the effects of chromium on soil nematode assemblages. Yeates *et al.* (1994) studied the effects of a pasture contamination with a Cr-containing timber preservative liquid. However, this liquid also contained As and Cu and it is only possible to estimate approximate contamination levels. Shannon–Wiener diversity was increased by the low and medium level of contamination and decreased by the highest level, compared to the control. One of the most striking results was the increase in the proportion of predators with increasing contamination. This is in sharp contrast with our knowledge on the sensitivity of predatory nematodes to heavy metal stress. This contradiction probably arises because in this study *Aporcelaimus*, the more numerous genus of large, K-strategist nematodes, was clearly considered as omnivorous, in contrast to the classification of Yeates *et al.* (1993) where it is classified as a predator *and* an omnivore. The above conflict can be solved if the two most sensitive feeding groups of soil nematodes are handled together, as in Korthals *et al.* (1996a), Nagy (1999) and Nagy *et al.* (2004). Although this technique causes some loss of information, it is made relevant by the taxonomic relationship, the similarities of stress-responses and the overlaps in feeding strategies among the dominant taxa of these groups. Notwithstanding the above, *Mononchus* appeared to tolerate these adverse conditions, as its relative proportion increased with the contamination levels.

Regarding specific harmful effects, Nagy (1999) showed that Cr decreased nematode abundance, taxon richness ($P<0.001$) and MI ($P<0.01$). Some taxa decreased (*Aporcelaimellus, Ecumenicus, Heterocephalobus, Pratylenchus*) or disappeared (*Eudorylaimus, Prismatolaimus, Tylenchorchynchus*) while others (*Acrobeloides, Helicotylenchus, Paraphelenchus*) increased in the plots with an 'available' Cr level of 1.4 mg/kg. In a subsequent study, Nagy *et al.* (2004) found Cr to have clear dose-dependent negative effects on nematological parameters including abundance, taxon richness, Maturity Index and Structure Index. The distribution of c-p groups and feeding types were significantly affected by the increasing Cr loads. The proportion of the most sensitive predatory and omnivorous taxa significantly decreased in these treatments. Diversity ordering techniques comprising several diversity indices also showed the two highest Cr doses (of approx. 0.8 mg/kg and 1.4 mg/kg, respectively) to significantly decrease diversity.

In conclusion, nematode assemblages gave quite articulated sensitive reactions to Cr pollution. In particular, K-strategist dorylaimid nematodes are sensitive to this disturbance. In the case of chromium, the different toxicity of the two ionic forms (CrIII and the more toxic CrVI) may complicate evaluation of the effects. Nevertheless, the above findings underline the sensitivity of nematode community parameters to chromium pollution. It even seems possible that changes in these parameters could be used as an early warning signal for Cr contamination in terrestrial systems.

Selenium

Selenium is a non-metallic microelement about which little is known of its effects on nematodes. Nagy (1999) reported the drastic negative effects of Se in an 'available' concentration of 36 mg/kg on total abundance and taxon richness. It was not possible to determine to what extent this destruction could be attributed to direct or indirect effects of Se, such as a suppression of vegetation in the given plots.

In a more refined study of dose-dependent Se effects along a concentration gradient of 10, 30, 90 and 270 mg/kg (Nagy *et al.*, 2004), Se had clearer effects on nematode assemblages. Nematode density decreased significantly in the third ($P<0.05$) and fourth ($P<0.01$) pollution level. Taxon richness values showed a continuous decline, with significant ($P<0.05$) differences in the third and fourth pollution levels. The MI2–5 and Structure Index values showed a slight increase with increasing Se levels and then decreased sharply. Correspondingly, the c-p group distribution of nematodes differed significantly between the second and the third pollution level ($P<0.01$ and $P<0.001$ respectively). It was not possible to calculate many nematode parameters for the highest Se concentrations because so few nematodes survived. As in the cases of most other pollutants, higher doses of Se completely eradicated predators and omnivores, resulting in increasing proportions of bacterial and fungal feeders.

In Bakonyi *et al.*'s (2003) study of long-term contamination effects, Se became less available over time but the maximum doses were still quite harmful to nematodes. There was still a significant difference in total abundance, c-p group distribution ($P<0.001$) and feeding type distribution ($P<0.001$) between the control and the Se plots 10 years after the treatment. Proportions of plant feeders and fungal feeders (in most cases) were much lower than in the control plots and predators and omnivores were almost absent throughout the entire study period. It is also noteworthy that there was an extremely high yearly taxon turnover in the Se treated plots (over 60% on average, compared with approximately 30% in the other plots).

Conclusions

As can be seen from the above discussion, it is difficult to make many generalizations about the effects of toxic metal pollutants on nematode assemblages in soil. Direct comparison of the above data would make little sense. It is known that ecosystem type, spatial scale (Neher *et al.*, 2005) as well as local characteristics, including pH, vegetation and composition of indigenous nematode fauna (Yeates, 1994), largely influence the outcome of the analyses.

In terms of statutory regulation, it is often difficult to compare actual contamination levels with limit values. Most standards refer to 'total' concentration of pollutants in soil, while the 'available' values (of much higher relevance) are ignored.

As the present chapter illustrates, considerable knowledge has been gathered from case studies on soil heavy metal pollution effects. Although there are still open questions, new research directions should focus on mechanisms behind and consequences of the above patterns, rather than enforcing further descriptive work.

Acknowledgement

Preparation of this work was supported by the Bolyai Scholarship programme of the Hungarian Academy of Sciences.

References

Anderson, G.L., Boyd, W.A. and Williams, P.L. (2001) Assessment of sublethal endpoints for toxicity testing with the nematode *Caenorhabditis elegans*. *Environmental Toxicology and Chemistry* 20, 833–838.

Bakonyi, G., Nagy, P. and Kádár, I. (2003) Long term effects of heavy metals and microelements on nematode assemblage. *Toxicology Letters* 140–141, 391–401.

Bardgett, R.D., Speir, T.W., Ross, D.J., Yeates, G.W. and Kettles, H.A. (1994) Impact of pasture contamination by copper, chromium and arsenic timber preservative on soil microbial properties and nematodes. *Biology and Fertility of Soils* 18, 71–79.

Bongers, T. (1990) The maturity index: an ecological measure of environmental disturbance based on nematode species composition. *Oecologia* 83, 14–19.

Bongers, T., Ilieva-Makulec, K. and Ekschmitt, K. (2001) Acute sensitivity of nematode taxa to $CuSO_4$ and relationships with feeding type and life-history classification. *Environmental Toxicology and Chemistry* 20, 1511–1516.

Ekschmitt, K. and Korthals, G.W. (2006) Nematodes as sentinels of heavy metals and organic toxicants in the soil. *Journal of Nematology* 38, 13–19.

Ferris, H., Bongers, T. and De Goede, R.G.M. (2001) A framework for soil food web diagnostics: extension of the nematode faunal analysis concept. *Applied Soil Ecology* 18, 13–29.

Georgieva, S.S., McGrath S.P., Hooper, D.J. and Chambers, B.S. (2002) Nematode communities under stress: the long-term effects of heavy metals in soil treated with sewage sludge. *Applied Soil Ecology* 20, 27–42.

Haight, M., Mudry, T. and Pasternak, J. (1982) Toxicity of seven heavy metals on *Panagrellus silusiae*: the efficacy of the free-living nematode as an in vivo toxicological bioassay. *Nematologica* 28, 1–11.

Johnson, S.R., Ferris, J.M. and Ferris, V.R. (1974) Nematode community structure of forest woodlots III. Ordinations of taxonomic groups and biomass. *Journal of Nematology* 6, 118–126.

Kadar, I. (2001) [Considerations of soil properties in application of soil contamination limit values II. (Mineral elements)]. Report for Ministry of Environmental Protection, Budapest, pp. 1–30.

Kammenga, J.E., Van Gestel, C.A.M. and Bakker, J. (1994) Patterns of sensitivity to cadmium and pentachlorophenol among nematode species from different taxonomic and ecological groups. *Archives of Environmental Contamination and Toxicology* 27, 88–94.

Kammenga, J.E., Van Gestel, C.A.M. and Hornung, E. (2001) Switching life-history sensitivities to stress in soil invertebrates. *Ecological Applications* 11, 226–238.

Korthals, G.W., Van der Ende, A., Van Megen, H., Lexmond, T.M., Kammenga, J.E. and Bongers, T. (1996a) Short-term effects of cadmium, copper, nickel and zinc on soil nematodes from different feeding and life-history strategy groups. *Applied Soil Ecology* 4, 107–117.

Korthals, G.W., Alexiev, A.D., Lexmond, T.M., Kammenga, J.E. and Bongers, T. (1996b) Long-term effects of copper and pH on the nematode community in an agroecosystem. *Environmental Toxicology and Chemistry* 15, 979–985.

Korthals, G.W., de Goede, R.G.M., Kammenga, J.E. and Bongers, T. (1996c) The Maturity Index as an instrument for risk assessment of soil pollution, 85–94. In: Van Straalen, N.M. and

Krivolutsky, D.A. (eds) *Bioindicator Systems for Soil Pollutions*. Kluwer Academic Publishers, Dordrecht, The Netherlands.

Korthals, G.W., Popovici, I., Iliev, I. and Lexmond, T.M. (1998) Influence of perennial ryegrass on a copper and zinc affected terrestrial nematode community. *Applied Soil Ecology* 10, 73–85.

Nagy, P. (1999) Effects of an artificial metal pollution on nematode assemblage of a calcareous loamy chernozem soil. *Plant and Soil* 212, 35–47.

Nagy, P., Bakonyi, G., Bongers, T., Kádár, I., Fábián, M. and Kiss, I. (2004) Effects of microelements on soil nematode assemblages seven years after contaminating an agricultural field. *Science of the Total Environment* 320, 131–143.

Neher, D.A., Wu, J., Barbercheck, M.E. and Anas, O. (2005) Ecosystem type affects interpretation of soil nematode community measures. *Applied Soil Ecology* 30, 47–64.

Sánchez-Moreno, S. and Navas, A. (2007) Nematode diversity and food web condition in heavy metal polluted soils in a river basin in southern Spain. *European Journal of Soil Biology* 43, 166–179.

Smit, C.E., Schouten, A.J., Van den Brink, P.J., Van Esbroek, M.L.P. and Posthuma, L. (2002) Effects of zinc contamination on a natural nematode community in outdoor soil mesocosms. *Archives of Environmental Contamination and Toxicology* 42, 205–216.

Tóthmérész, B. (1993) DivOrd 1.50. A program for diversity ordering. *Tiscia* 27, 33–44.

Van Straalen, N.M. and Denneman, C.A.J. (1989) Ecotoxicological evaluation of soil quality criteria. *Ecotoxicology and Environmental Safety*, 18, 241–251.

Weiss, B. and Larink, O. (1991) Influence of sewage sludge and heavy metals on nematodes in an arable soil. *Biology and Fertility of Soils* 12, 5–9.

Yeates, G.W. (1994) Modification and qualification of the nematode maturity index. *Pedobiologia* 38, 97–101.

Yeates, G.W., Bongers, T., de Goede, R.G.M., Freckman, D.W. and Georgieva, S.S. (1993) Feeding habits in soil nematode families and genera: an outline for soil ecologists. *Journal of Nematology* 25, 315–331.

Yeates, G.W., Orchard, V.A., Speir, T.W., Hunt, J.L. and Hermans, M.C.C. (1994) Impact of pasture contamination by copper, chromium and arsenic timber preservative on soil biological activity. *Biology and Fertility of Soils* 18, 200–208.

Yeates, G.W., Orchard, V.A. and Speir, T.W. (1995) Reduction in faunal populations and decomposition following pasture contamination by a Cu-Cr-As based timber preservative. *Acta Zoologica Fennica* 196, 297–300.

Yeates, G.W., Percival, H.J. and Parshotam, A. (2003) Soil nematode responses to year-to-year variation of low levels of heavy metals. *Australian Journal of Soil Research* 41, 613–625.

Zhang, W.D., Wang, X.F., Li, Q., Jiang, Y. and Liang, W.J. (2007) Soil nematode responses to heavy metal stress. *Helminthologia* 44, 87–91.

Zullini, A. and Peretti, E. (1986) Lead pollution and moss-inhabiting nematodes of an industrial area. *Water, Air and Soil Pollution* 27, 403–410.

8 Molecular Tools for Analysing Nematode Assemblages

Roy Neilson,[1] Suzanne Donn,[1] Bryan Griffiths,[1] Tim Daniell,[1] Katarzyna D. Rybarczyk,[2] Sven van den Elsen,[2] Paul Mooyman[2] and Johannes Helder[2]

[1]*Environment Plant Interactions Programme, SCRI, Invergowrie, Dundee, United Kingdom;* [2]*Laboratory of Nematology, Department of Plant Sciences, Wageningen University, The Netherlands*

Introduction

Physical, chemical and biological indicators can be used to estimate the condition of terrestrial or freshwater habitats. Biological indicators are highly relevant because they are integrative; they reflect the overall impact of physical and/or chemical and/or biological changes. Furthermore, most biotic indices also rely on indigenous organisms as opposed to exotic test organisms because the diversity and condition of indigenous organisms reflects both acute and chronic effects of disturbances. Chronic effects are especially difficult to assess when standard test organisms are used.

Intensive land use has been implicated in declining soil health, raising concern over the sustainability of agronomic production under current management strategies (Kaiser, 2004). Soil protection is one of the seven themes identified in the European Community's Sixth Environment Action Programme, recognizing the need for a monitoring system (Van Camp *et al.*, 2004). As part of any such monitoring scheme, soil faunal communities may be used as indicators for soil health and can also give an insight into wider soil processes (Kopeszki, 1997; Van Straalen, 1998; Neher, 2001; Schloter *et al.*, 2003; Gormsen *et al.*, 2006). Although no one faunal group may provide a definitive answer, selection and analysis of an appropriate group may yield an insight into general soil conditions, thus avoiding the excessive time and financial constraints involved in assessing the whole soil faunal assemblage.

While the large diversity and short generation time within microfaunal (protozoa and nematodes) groups makes them ideal indicators of changing soil conditions (Ritz and Trudgill, 1999), their inherent diversity also represents a problem in the characterization of assemblages. Notwithstanding that a large proportion of species remain undescribed, those that have, are identified by a declining pool of taxonomists (André *et al.*, 2001). Furthermore, André *et al.* (2002) highlight the need for development and consistency of methods in soil faunal monitoring. Molecular methods provide an alternative to traditional morphological identification for routine assessment of described species. Their application has enabled profiling of environmental samples of soil microbial populations, overcoming the need to culture and identify bacteria and fungi from complex mixtures (Amann *et al.*, 1995) and similarly may reduce the taxonomic expertise currently required to characterize microfaunal communities.

Nematodes are a faunal group that satisfy many of the conditions for suitable indicators of soil and sediment health (see Yeates *et al.*, Chapter 1, this volume). Abundant, diverse and easily extracted from soil (Ritz and Trudgill, 1999), they can be classified into trophic groups (Yeates *et al.*, 1993) functioning at various levels of the soil food web (de Ruiter *et al.*, 1993). Due to their links throughout the food web, analysis of nematode communities may also prove to be a powerful tool allowing extrapolation to associated organisms. For example, Bardgett and Shine (1999) established a link in grassland systems between nematode groups and both resource allocation in plants and dynamics of the bacterial population. Nematodes display a range of responsiveness to toxins and stresses such as desiccation, making them valuable indicators in disturbed systems (Neher, 2001). Univariate indices utilizing nematode assemblage data have been developed for the monitoring of soil and water quality (Beier and Traunspurger, 2003; see also Ferris and Bongers, Chapter 5, this volume).

While some authors (Ritz and Trudgill, 1999) recommend analysis at trophic group level, others emphasize the importance of identifying nematodes at a greater taxonomic resolution e.g. species (Yeates *et al.*, 1993; Bongers and Bongers, 1998; Yeates, 2003). Traditionally, nematode identification relies on high powered light microscopy. However, as with other faunal groups, a declining taxonomic skill base is problematical (Coomans, 2002), notwithstanding the time-consuming nature of the task. Spatially aggregated populations (Boag *et al.*, 1992) and temporal variations in assemblages (Boag *et al.*, 1998) necessitate repeated sampling and so a high throughput method is desirable. Several authors have proposed molecular methods as an alternative to morphological identification of soil nematodes (Floyd *et al.*, 2002; Foucher and Wilson, 2002; Waite *et al.*, 2003; Foucher *et al.*, 2004; Griffiths *et al.*, 2006; Jones *et al.*, 2006). In addition to more rapid identification requiring less specialized skills, molecular techniques may also readily allow identification of cryptic species and juveniles (Blouin, 2002; Powers, 2004).

Anthropogenic impacts such as pollution represent a massive stress factor for both terrestrial and aquatic ecosystems. Recognizing the risks of any given pollution event, and – more in general – for the management of terrestrial and freshwater ecosystems, an adequate biosensor system is needed. Unfortunately, biosensor systems used so far do not meet the criteria set by policy makers and environment management agencies.

Currently available biosensors for pollution monitoring

A range of biosensor systems are currently available. Either single test organisms are used (see example 1 below), or indigenous microbial communities are characterized (see examples 2, 3, and 4 below). For reasons mentioned below, the available tools for ecological assessment of terrestrial or freshwater habitats are unsatisfactory because they provide limited ecological information, and do not make *in situ* use of indigenous species.

1. The luminescent bacterium Vibrio fischeri

Microtox (Coastal Bioanalysts Inc., Gloucester, USA) measures the decrease in respiration, and subsequent light output, of the luminescent bacterium *Vibrio fischeri* as the toxic response. Microtox is a rapid, cost-effective tool in assessing toxicity of effluents, sediments, leachates, soils, sludges, ground water and surface water. However, Microtox does not differentiate between short-term stresses.

2. Multiple substrate-induced respiration

The MicroStation 3E System (Biolog, Hayward, USA) characterizes the microbial assemblage in a sediment sample by making a substrate utilization profile. The 96-well MicroPlate test panels generate 'metabolic fingerprints' characteristic of the assemblage under investigation. Software is provided to translate these fingerprints into information about the composition of the microbial assemblage. Using this Biolog system only culturable bacteria can be identified. However, less than 1% of the microbes present in soil and sediments can be cultured on artificial media (Rozsak and Colwell, 1987). Moreover, isolates from soil and sediment samples are frequently misidentified by the Biolog system (Wünsche and Babel, 1996). This concept has been expanded for use with intact soils (rather than microorganisms extracted from the soils) by adding a similar range of liquid substrates to soil and measuring respiration profiles (Degens and Harris, 1997). A microplate system using this technique has been developed and is commercially available (Campbell *et al.*, 2003).

3. Fatty acid methyl ester (FAME) profiles

Microbial ID Inc (Newark, DE, USA) characterizes microbial communities (aerobic and anaerobic bacteria, yeast and fungi) by gas chromatographic analysis of the whole cell fatty acid content. The extractable FAME and phospholipid fatty acid contents are determined directly, and – as such – FAME profiles are more informative than substrate utilization profiles. However, the resolution of FAME profiles is low. It distinguishes for example, the α-*Proteobacteria* from the γ-*Proteobacteria* and it determines the presence or absence of members of the *Cytophaga–Flexibacter–Bacteroides* group. However, these are huge taxonomic groups. As such, FAME profiles contain too little information to be useful as a biosensor for sediment pollution.

4. Molecular analysis of eubacteria communities

Information about the structural composition of microbial communities can be obtained from analysis of PCR amplified DNA sequences coding for conserved regions in the target gene, commonly the SSU rDNA, and profiled using various methods such as denaturing gradient gel electrophoresis (DGGE; Kowalchuk *et al.*, 1997; Smit *et al.*, 1999) or terminal restriction fragment length polymorphism (T-RFLP; e.g. Liu *et al.*, 1997; Marsh *et al.*, 1998). DNA is generally extracted directly from environmental samples. Broad spectrum bacterial primers are used for non-restricted assemblage analysis whereas specific primers are used for analysing functional groups like ammonium-oxidizing bacteria (Smit *et al.*, 1997). The SSU rDNA-DGGE technique can detect alterations in microbial assemblage composition induced by contaminants in a range of environments and is not dependent on cultivating microorganisms. However, the relationship between observed bands in a DGGE gel or peaks by T-RFLP and the *in situ* dominance of organisms containing that SSU rDNA sequence is still unclear because of, for example, biases associated with DNA extraction and amplification. Thus, this technique should be viewed as a qualitative assessment of microbial assemblage composition and not a quantitative assessment of the relative prevalence of genotypes in samples.

What factors obstruct the use of nematodes as biosensors?

There is no doubt that nematodes have a high potential as an indigenous biosensor for sediment and soil health (Doran *et al.*, 1996; Yeates *et al.*, Chapter 1, this volume, Trett *et al.*, Chapter 12, this volume). In general, pollutants will induce a shift in assemblage structure towards dominance by opportunistic species (reducing the fraction of survivalists). By classifying nematodes by feeding behaviour and life-history strategy, algorithms have been developed that allow for the translation of the composition of the nematode assemblage into a quantitative indicator for ecological condition of habitats (Ferris and Bongers, Chapter 5, this volume).

However, the conserved morphology of nematodes is a major operational reason for the under-exploitation of nematodes as biosensors. Very few experts are able to analyse the nematode assemblage in a given soil or sediment sample (André *et al.*, 2001; Coomans, 2002). In our experience, routine analysis of a single mass-slide by an expert takes on average 2–4 hours (analysing only the first 75–150 individuals!). In such an analysis only adults are considered as most juveniles are discarded because of the lack of useful discriminatory morphological characters. In contrast, a downstream high-throughput molecular approach does not bias towards adult nematodes and in theory could process 96 samples (plate format) in a matter of hours compared to weeks/months for the same sample turnover using classical methodologies. It is therefore concluded that a nematode-based biosensor should, if possible, be based on non-morphological traits.

The Use of Ribosomal DNA Sequences for Nematode Identification

The small subunit (SSU=18S) rRNA is an essential part of the 40S ribosomal subunit. Ribosomes are crucial in protein synthesis and, consequently, the SSU rDNA is relatively resistant to change over time compared to non-coding regions of DNA such as ribosomal ITS regions (Mindell and Honeycutt, 1990). The stem loop structure of the transcribed RNA of this non-protein coding gene has resulted in a patchy but predictable mosaic of more variable and conserved domains. The presence of extremely conserved regions greatly facilitates the amplification of homologous rRNA genes from molecularly uncharacterized members of the phylum Nematoda (actually the majority). The suitability of SSU rDNA for phylogenetic reconstruction (Aleshin *et al.*, 1998; Blaxter *et al.*, 1998; Holterman *et al.*, 2006) and the DNA barcode based identification of nematodes at family, genus and occasionally even at species level has been shown (e.g. Floyd *et al.*, 2002; Blaxter *et al.*, 2005; Holterman *et al.*, 2006; Holterman *et al.*, 2008a).

Currently, approximately 1500 full length nematode SSU rDNA sequences are available from across the phylum including a major part of the terrestrial and freshwater nematode fauna in the temperate climate zone. Marine nematodes are still under-represented although two recent papers from Meldal *et al.* (2007) and Holterman *et al.* (2008b) suggest that this could be increased in the near future. The resulting alignment constitutes an appropriate starting point for the design of taxon-specific primers and/or probes.

In this chapter we will describe two complementary techniques based on nematode SSU to characterize nematode assemblages from environmental samples: the use of real-time quantitative PCR, and the use of terminal restriction fragment length polymorphism analysis of PCR products.

Real-time Quantitative PCR Approaches

Traditionally, end-point PCR amplified, specific DNA fragments, and amplified products are examined by running on agarose gels. In real-time PCR, the accumulation of specific products in the reaction is monitored continuously during cycling by recording changes in fluorescence within the reaction vessel caused by fluorescent dyes incorporated into the reaction. Quantification is achieved by identifying the cycle number (Ct) at which the reporter dye emission intensity rises above a predetermined threshold. The Ct is inversely proportional to the copy number of the target template and quantities of template in a sample can be calculated by direct comparison with Ct values of known standards (Brunborg *et al.*, 2004; Atkins *et al.*, 2005). In order for this technique to be applied to nematodes, specific primers or probes need to be designed for the group (family, genus, or species) under investigation. Once designed, appropriate PCR cycling conditions need to be optimized, and specificity of the primers/probes must be checked prior to using the technique with environmental samples.

Design of taxon-specific primers or probes

For the mining of nematode ribosomal DNA databases various software packages can be used. Microbiologists use large 16S rDNA databases, and within this community a program called ARB ('arbor' = tree) was developed, and is available as Linux-based freeware (Ludwig *et al.*, 2004). ARB software has been used by the authors for the design of family/genus/species-specific PCR primers. In order to facilitate the multiplex detection of nematode genera, primers are designed in such a way that they all have annealing temperatures between 62 and 64°C. The ARB software also identifies close non-targets for the taxa under investigation. It can recognize one to three mismatches, with the risks for false positives being dependent on the position of the mismatches.

ARB interface – practical hints about the use of ARB

Define your target group on the basis of Bayesian inference-based phylogenetic trees (or other robust methods), and preferably not on the Neighbour joining (NJ) tree produced by ARB. Select the individuals for which you would like to produce a primer / probe (press the Mark button at the left side of your window, and click the desired individuals). Click Probes-> Design probes. This will open the Probe design window. Select the appropriate *PT_Server* (click and scroll), and choose for instance the following settings:

Length of Output:	50	
Max non-group hits:	0 (increase if no results are found)	
Max hairpin bonds:	4	
Min group hits:	100% (decrease if no results are found)	
Length of probe:	10	
Temperature	min: 0	max: 100
G+C content	min: 0	max: 100
E. coli position	min: 0	max: 10000
Press GO.		

The mismatch candidates will be shown in yellow, with the actual mismatches shown in red.

If no hits are found, either change the Min group hits, or the Max non-group hits, in small steps, until a reasonable number of probe candidates is found. Note: depending on the *Max non-group hits* and *Min group hits* settings that are needed to find probe candidates altogether, you can also decide to enlarge or decrease the selected group in the *ARB_NT* main screen, by adding or removing species, and recalculate.

Selection of positive targets using ARB

For a 'family-based primer' select a representative number of positive targets with a minimum of one target per genus within this family. With a 'genus-based primer' select a representative number of positive targets with a minimum of one target per species within this genus.

Identification of potential false-positive targets using ARB

A false positive target is not by definition the one closest in the phylogenetic tree. The most likely candidates are selected in ARB for each forward and reverse primer allowing potential false positive sequences to be tested with the primers.

Determination of the optimal annealing temperature (T_a)

For an appropriate estimation of the annealing temperature of a primer, we use MELTING, a program found on the website http://bioweb.pasteur.fr/seqanal/interfaces/melting.html (Le Novère, 2001). Although the annealing temperature predictions of this program are fairly accurate, we always verify the T_a experimentally before further use (Fig. 8.1).

Verification of the specificity of two family-specific primer sets – case study Plectidae

To test the specificity of nematode family-specific primers, bacterial clones harbouring a TOPO TA vector with an SSU rDNA fragment of interest were

Tm (°C)	Ct	Ct	
67.0	25.99	43.17	
66.7	20.01	27.98	
65.8	**14.3**	**17**	**Critical T_a**
64.5	**12.04**	**12.33**	
62.7	11.74	11.67	
61.5	11.74	11.66	
60.6	11.94	11.75	
60.0	12.16	12.08	

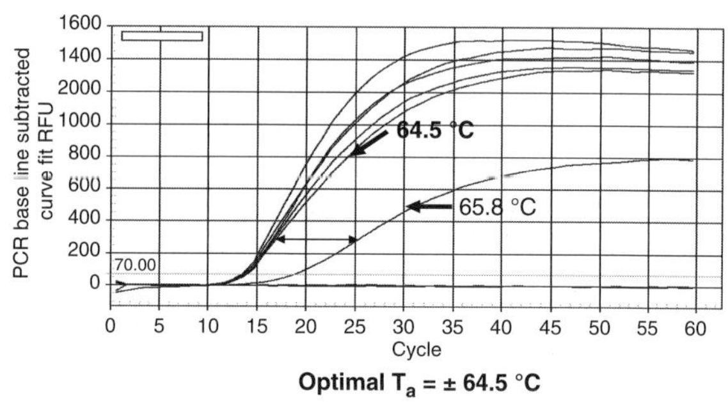

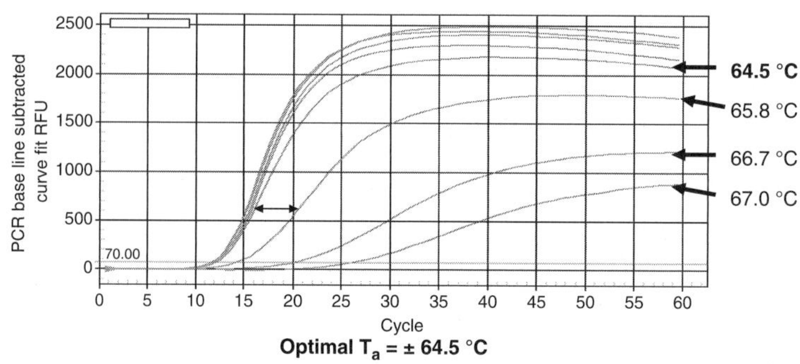

Fig. 8.1. The experimental verification of T_a values for an *in silico* designed Plectidae-specific primer combination.

Primer combination 211/188

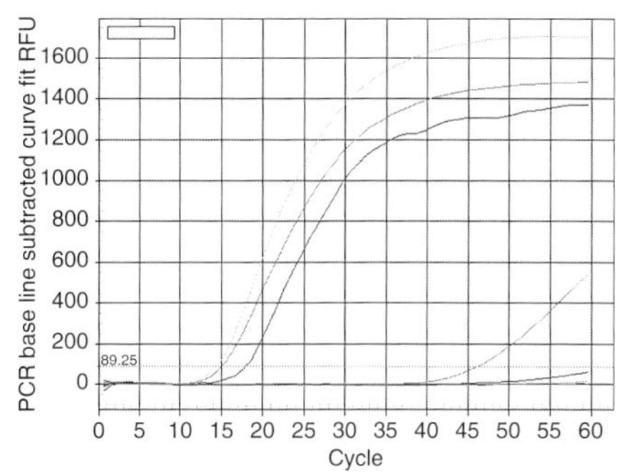

	T_a (°C)	Ct value
Tested positive target		
TyCeAur2 12 Tylocephalus auriculatus 2 P165 (AY284707)	64.0	**17.0**
CeraArm1 365 Ceratoplectus armatus 1 P167 (AY284706)	64.0	**14.7**
PlecRhi1 587 Plectus rhizophilus 1 P170 (AY593928)	64.0	**14.5**
WilsOto2 535 Wilsonema otophorum 2 P168 (AY593927)	64.0	**N/A (*)**
(Possible) False positive target		
MTCeCra2 2 Metateratocephalus crassidens 2 P164 (AY284687)	64.0	**N/A**
AnapGra1 175 Anaplectus grandepapillatus 1 P166 (AY284697)	64.0	**N/A**
MeloMar1 1003 Meloidogyne maritima 1 P171 (EU669944)	64.0	**50.2**
BunoRet3 1006 Bunonema reticulatum 3 P172 (FJ040450)	64.0	**N/A**

Primer combination 212/192

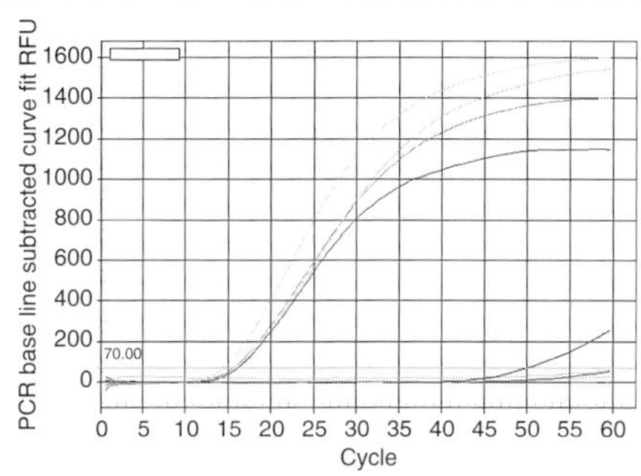

	T_a (°C)	Ct value
Positive target		
TyCeAur2 12 Tylocephalus auriculatus 2 106 (AY284707)	64.0	**16.17**
WilsOto2 Wilsonema otophorum 2 708 (AY284708)	64.0	**15.67**
CeraArm1 365 Ceratoplectus armatus 1 250 (AY284706)	64.0	**16.06**
PlecRhi1 587 Plectus rhizophilus 1 837 (AY593928)	64.0	**14.63**
(Possible) False positive target		
MonhRie1 504 Monhystera riemanni1 696 (AY593938)	64.0	**50.02**
TherAgi1 128 Theristus agilis 1 252 (AY284693)	64.0	**60**
AnapGra1 175 Anaplectus grandepapillatus 1 343 (AY284697)	64.0	**60**
BunoRet3 1006 Bunonema reticulatum 3 1205 (FJ040450)	64.0	**60**
TrisSp 1 58 Trischistoma 1 138 (AY284735)	64.0	**60**
DiGafamil 249 Diplogasteridae 480 (AY284689)	64.0	**60**
MTCeCra2 2 Metateratocephalus crassidens 2 74 (AY284687)	64.0	**60**

Fig. 8.2. Testing the specificity of Plectidae-specific primers.

grown. Plasmid extraction was performed as described in Holterman *et al.* (2008a). DNA concentrations were measured with a NanoDrop spectrophotometer and adjusted to 10 ng/μl. For Q-PCR application 3 μl of 1000 times diluted template was mixed with Plectidae-specific primers and iQ SYBR Green Supermix (Bio-Rad) in a total reaction volume of 25 μl. In Fig. 8.2, results are shown for two primer combinations.

Semi-quantitative detection of individual nematode taxa in complex environmental samples

'Classical' nematode assemblage analysis is based on the counting of individuals, irrespective of whether these individuals are small, e.g. second-stage juveniles, or relatively large (a fourth-stage juvenile or an adult). If the total DNA content increased drastically during nematode development this would seriously hamper the conversion of the DNA signal into an accurate estimation of the number of individuals involved. However, except for their gonads, the hypodermis and the intestinal epithelium, many nematode taxa are eutelic (the cell number and arrangement remains constant during development).

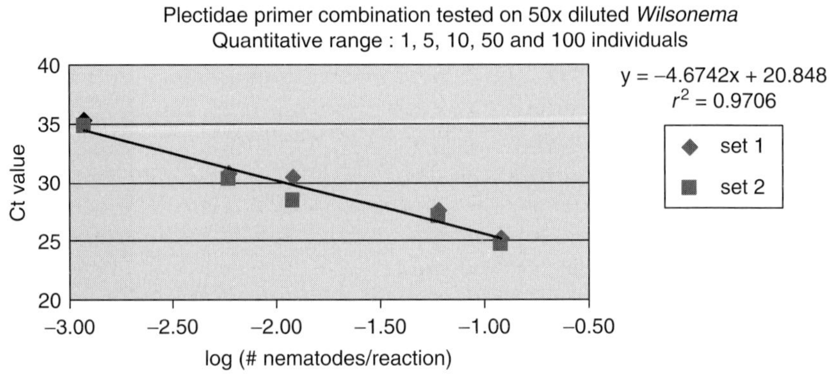

Fig. 8.3. Plectidae-specific primers – relationship between number of nematodes and Ct values.

For example, in the soil nematode *C. elegans*, the cell number only increases from ~550 to ~950 during all of post-embryonic growth (http://www.wormatlas.org/). Hence, the relationship between DNA signal and number of individuals is not exactly linear, but is easily modelled.

This characteristic makes it possible to translate a Q-PCR signal into a fairly accurate estimation of the number of individuals involved. We aim at developing a system that will allow us to quantify nematodes at a logarithmic scale. In other words, such a set up should be able to discriminate between 0, 1, 10, 100, 1000 and 10,000 individuals. For this purpose, we collected quantitative series of individual nematodes from a single genus, and Fig. 8.3 shows an example of the relationship between Ct value and number of individuals. The Ct (cycle threshold) is defined as the number of cycles required for the fluorescent signal to cross the threshold (i.e. exceeds background level). It is noted that the genera *Wilsonema* and *Plectus* both belong to the family Plectidae. It is concluded that it is technically possible to design a relatively simple and affordable system for the quantitative analyses of complex nematode communities. Currently the robustness of this approach is being tested for a range of nematode families under field conditions at the Laboratory of Nematology, Wageningen University.

Terminal-Restriction Fragment Length Polymorphism (T-RFLP) Approaches

T-RFLP has been applied to a wide variety of organisms and questions, since its inception in 1994 as a method for the identification of bacterial species within a mixed community, based on analysis of the small subunit ribosomal RNA (SSU rRNA; Avaniss-Aghajani *et al.*, 1994). The method has most frequently been used for bacterial assemblage profiling (Liu *et al.*, 1997), but has also been applied to archaeal (Chin *et al.*, 1999), ectomycorrhizal (Zhou and

Hogetsu, 2002) and arbuscular mycorrhizal (Vandenkoornhuyse *et al.*, 2003) communities over a range of habitats including polluted soils (Osborn *et al.*, 2000), forest (Zhou and Hogetsu, 2002), wastewater reactors (Guieysse *et al.*, 2001), stream sediments, ocean surface waters (Baldwin *et al.*, 2005), flood plains (Kemnitz *et al.*, 2004), gold mines (Takai *et al.*, 2001), and termite guts (Liu *et al.*, 1997), as well as in several medical applications (Nagashima *et al.*, 2003; Sakamoto *et al.*, 2004; Jernberg *et al.*, 2007). Use of T-RFLP for non-fungal eukaryotic organisms has been more limited but includes analysis of activated sludge assemblages dominated by ciliates (Marsh *et al.*, 1998), marine picoeukaryotes (Diez *et al.*, 2001), protists (Countway *et al.*, 2005) and in an analysis of the blood meals of mosquitos, vertebrates and invertebrates (Meece *et al.*, 2005; Griffiths *et al.*, 2006; Donn *et al.*, 2007). The vast majority of these studies have been based upon restriction of full, or partial length small subunit rRNA gene sequences.

T-RFLP, like all PCR based fingerprinting techniques, is subject to bias including preferential amplification due to primer–target mismatch or secondary structure in some sequences and formation of chimaeras (Kitts, 2001). However, these can be limited by careful primer design and testing, limiting PCR cycle number, optimizing template concentration and annealing temperature. Any analysis based on rDNA must also take into account variation in gene copy number between taxa (Kitts, 2001). Awareness of these potential problems, coupled with careful interpretation of results means PCR-based techniques should be capable of providing useful information. T-RFLP has several advantages over other PCR-based techniques, such as denaturing gradient gel electrophoresis (DGGE), temperature gradient gel electrophoresis (TGGE) and single strand conformation polymorphism (SSCP), which all rely on sequence variation altering the rate of movement of amplicons through a gel matrix (Marsh, 1999), with the exception of the recently developed high throughput methodology using SSCP linked to a capillary sequencer, known as CE-SSCP (Zinger *et al.*, 2007). One major advantage is that T-RFLP is semi-quantitative (Spiegelman *et al.*, 2005), with the relative area of a peak in the profile corresponding to the relative abundance of that terminal restriction fragment (TRF) in the amplicon mixture, and, it is assumed, in the template mixture. This does not translate directly into proportional abundance of taxa, because factors such as gene copy number and amplification bias must be taken into account. However, it allows for cross-sample comparison of relative abundance in addition to presence/absence data.

The ability to analyse data from different sequencer runs together is also an advantage over DGGE, where comparison between gels is difficult (Moeseneder *et al.*, 1999; Nunan *et al.*, 2005). In T-RFLP, the inclusion of a size standard with every sample allows accurate sizing of fragments (Marsh, 1999), and thus comparison across electrophoretic runs. Fragments can be sized accurately up to 700 bp (Osborn *et al.*, 2000), and beyond this a small degree of error may be associated with sizing up to 1000 bp. Automated sample loading with a capillary sequencer reduces variability between runs (Osborn *et al.*, 2000) and the digital output allows rapid and objective data analysis. Unlike DGGE, TGGE or SSCP, some taxonomic information can be inferred from T-RFLP output when compared to database sequence information (Osborn *et al.*, 2000), since *in silico* digestion of sequences with a chosen enzyme is possible using a number of web-based tools designed to estimate likely phylogenetic identity such as TAP-T-RFLP (Marsh *et al.*, 2000). Nevertheless, care needs to be taken in the interpretation of such data. Unlike DGGE, operational taxonomic units (OTUs) cannot be recovered for sequencing. One possible solution is to couple T-RFLP with construction of a clone library providing high quality sequence information to allocate Terminal Restriction Fragments (TRFs) to taxa while maintaining high throughput (Moesender *et al.*, 2001, Scheid *et al.*, 2004). An alternative, directed T-RFLP approach, is to use the sequence information from clone libraries in order to select an enzyme digest that will produce taxonomically informative TRFs (directed T-RFLP approach). Taxonomically similar sequences share the same enzyme cut sizes while taxonomically disparate sequences have distinct cut sites resulting in different TRF sizes.

Requirement for a high-throughput system

Relatively standard equipment is required for the establishment of a high-throughput T-RFLP system. Access to a PCR thermocycler, standard gel electrophoresis equipment and a capillary sequencer are a necessity. For an undirected T-RFLP approach, software available with sequencing machines will allow the processing of raw data to allow downstream analysis. If a directed approach is desired there is a requirement for sequence analysis both in terms of phylogenetic and restriction analysis. There are a number of software packages available that can aid T-RFLP design. The molecular ecology team at SCRI in collaboration with the University of Dundee, is currently testing a package which facilitates the design of such digest strategies. Most restriction enzymes can be purchased from the usual proprietary molecular biology suppliers. Fluorescent primers are also freely available from most oligo houses although some sequencers contain filter sets requiring the use of primers labelled with proprietary dyes available only from the manufacturer.

Extraction of nematode DNA for T-RFLP assemblage analyses

The first step in any molecular analysis is the extraction of representative high-quality nematode DNA. Amounts of less than 1 g of an environmental sample are routinely used for the analysis of bacterial populations (Duarte *et al.*, 1998) but the comparatively low abundance of soil nematodes and their patchy spatial distribution (Webster and Boag, 1992) requires extraction from a larger soil volume to achieve a representative sample. Alternatively, extracting the nematode assemblage from the soil with a subsequent extraction of the nematode assemblage DNA (Foucher and Wilson, 2002; Hübschen *et al.*,

2004) is thus a preferable option. Methods of DNA extraction from individual nematodes include squashing a single nematode in a droplet of water (Powers and Harris, 1993), lysis using sodium hydroxide (Stanton *et al.*, 1998), lysis by the combined use of proteinase K and β-mercaptoethanol (Holterman *et al.*, 2006), or phenol chloroform (Rusin *et al.*, 2003). However, with the latter method used by Foucher and Wilson (2002), it was deemed labour intensive with each sample having to be handled individually, making it impractical for ecological studies requiring large-scale sampling.

Variations of the sodium hydroxide extraction technique (Stanton *et al.*, 1998) have been used on multiple nematodes mainly from laboratory cultures (Floyd *et al.*, 2002; Hübschen *et al.*, 2004) although Hübschen *et al.* (2004) extracted field nematode assemblage DNA using a modification of the method. Zijlstra *et al.* (2004) used a PCR template preparation kit for the extraction of DNA from juveniles of *Meloidogyne* species and previously Griffiths *et al.* (2006) utilized a combination of bead beating and purification through a PCR purification kit. The use of a commercial kit avoids the need for long incubation steps required by sodium hydroxide and most proteinase K lysis methodologies. Recently Donn *et al.* (2007) tested the efficiency of DNA extraction using sodium hydroxide, phenol chloroform and three commercial PCR purification kits namely Purelink, Qiaquick and Wizard and demonstrated differences in consistency among the extraction methods. It is noted that the need for an additional DNA purification step depends on the type of soil under investigation. In the case of soils with a high organic matter content (OM%) purification is certainly advisable (and often even necessary), but in the case of sandy, loamy or clay soils with a relatively low OM%, a simple 100, 500 or 1000 times dilution will in most cases be sufficient to avoid inhibitory effects during PCR amplification (J. Helder, Wageningen, 2007, personal communication).

Resolution of T-RFLP

T-RFLP has often been found to provide either a greater resolution than DGGE (Marsh *et al.*, 1998; Moeseneder *et al.*, 1999) or a similar resolution with an increased capability to compare large numbers of samples (Nunan *et al.*, 2005). This depends on the choice of restriction enzyme as several taxa may share the same restriction site and appear as one TRF. Resolution may be increased by combining information from more than one enzyme (Casamayor *et al.*, 2002) or utilizing information from both forward and reverse labelled primers. The volume of sample required for detection of an OTU with T-RFLP is an order of magnitude lower than that for DGGE (Moeseneder *et al.*, 1999), possibly allowing detection of rarer OTUs, however, artefactual peaks may also contribute to the perceived diversity (Osborne *et al.*, 2006). These artefacts may be PCR-related, such as chimaeras and heteroduplexes (Thompson *et al.*, 2002), and therefore have the same influence on any of the PCR-based methods, or may be digestion artefacts. These can occur for example when digestion is not complete, when single stranded PCR artefacts form secondary structures which can be cut, or when enzyme efficiency is dependent on the sequence surrounding a cut site.

Many of these problems can be avoided by using a directed T-RFLP approach, where TRFs are of expected sizes and thus any artefactual peaks can be ignored. Using a directed approach also means that sharing of TRFs by unrelated taxa can be avoided and that biologically meaningful information can be interpreted from the profile. Care must be taken to exclude the possibility that any TRFs of unexpected length are artefacts and not new OTUs, and therefore clone library construction and screening remains an important element of the approach.

Verification of sequence integrity

The primary objective of a high-throughput molecular technique is to substantially reduce the necessity for resource-hungry classical taxonomy, i.e. microscope work time. However this instigates a question: How do you know what you are sequencing? To determine this one can either: (i) compare their sequence to that published on public databases such as NCBI (http://www.ncbi.nlm.nih.gov/); or (ii) initially identify the specimen using classical techniques prior to sequencing and then subsequently focus on molecular data. The latter method adds a time component to the methodology but guarantees accurate typing of the nematodes and relates specimen to sequence. Utilizing database sequences for comparison can be fraught with dangers as one relies on the taxonomic ability of an individual or individuals with unknown skills. Without question there are incorrect taxonomic names assigned to sequences residing in public databases. However, with diligent analysis of sequence alignments it is usually a straightforward task to identify an 'unusual' sequence or chimaeras which can be omitted from further analysis.

Example of using T-RFLP in different biotopes

New molecular methods such as that described by Donn *et al.* (2007) must demonstrate that they can operate over a range of different habitats. To test this, soil samples were collected from five sites of differing land use and soil type in Scotland (arable soil, dune sand, coniferous forest soil, long-term pasture soil and a moorland soil with a high organic matter). Samples were taken from the top 10 cm. Nematodes were extracted from 4 × 200 g soil using a modified Baermann funnel extraction over 48 h (Brown and Boag, 1988) and allowed to sediment overnight. PCR used a fluorescently labelled primer, VIC-SSU_R81 (Applied Biosystems, California), and 4 μl of the subsequent PCR product was digested for 4 h at 37 °C with 1 unit HinfI (Promega, Madison) in a 6 μl reaction with buffer concentrations manipulated to reflect the recommended conditions listed by the manufacturer.

Digested samples were diluted and analysed on an ABI 3730 capillary sequencer (Fig. 8.4) and assigned to bins using GeneMapper (Applied Biosystems). T-RFLP, which was performed on products amplified from purified DNA extracts, showed that the nematode communities from the five different habitats could be distinguished on the basis of a single enzyme digest (ANOVA PC1, $P<0.001$, Fig. 8.5).

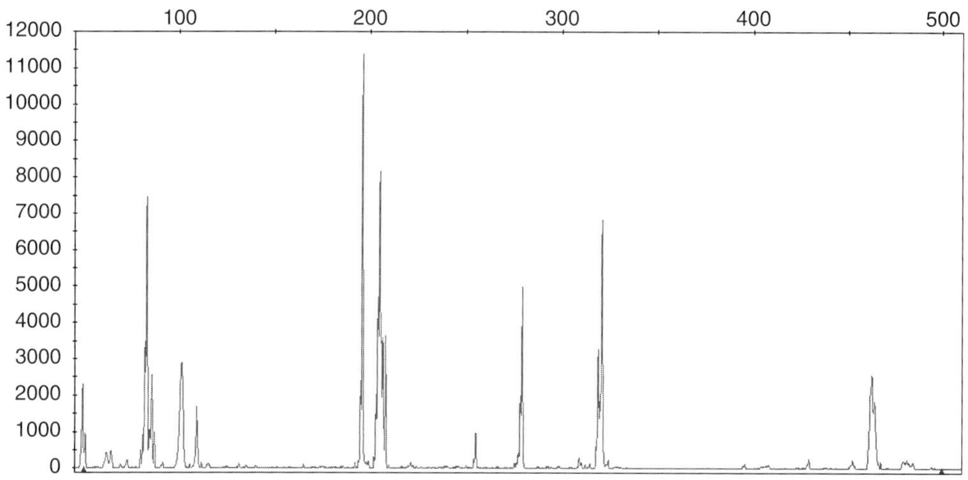

Fig. 8.4. A typical output plot from a capillary sequencer exhibiting Terminal Restriction Fragments with different fragment sizes. Amplitude of the peaks represents fluorescent units. Area under the peaks is a semi-quantitative measure of the abundance of each TRF.

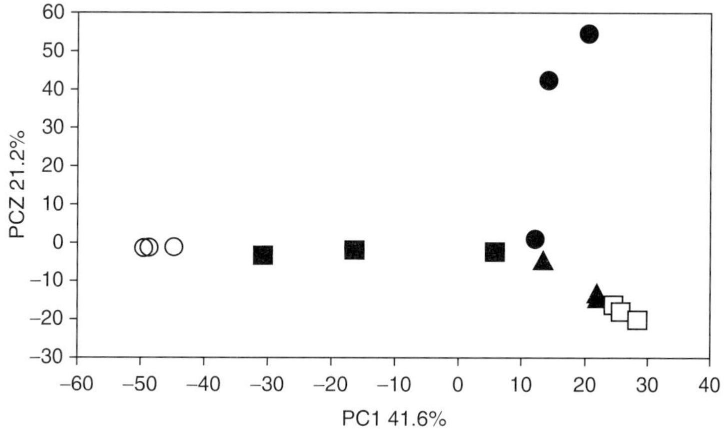

Fig. 8.5. Plot of principal components 1 and 2 of the T-RFLP profile of nematode communities from ● arable, ▲ dune, ❑ coniferous forest, ○ pasture and ■ moorland soil using DNA extracted by physical disruption and purification through the Purelink PCR purification kit. The nematode communities from the five different habitats are distinguished based on a single enzyme digest (ANOVA PC1, $P<0.001$).

Future Prospects

The public availability of SSU and/or LSU ribosomal DNA sequences of many major groups within the Animal Kingdom, and – more specifically – of most major eukaryotes in soil and sediment ecosystems, combined with its remarkable good resolution for most major taxonomic groups within the phylum Nematoda, makes the genes within the rDNA cistron highly useful for the analyses of nematode communities in complex (DNA) backgrounds. Most soil- and sediment-resident nematodes are all small (0.05 to 2 mm), vermiform, and geotropic, and these characteristics allow – contrary to bacteria and fungi – a straightforward and cheap extraction of the whole nematode assemblage.

For the DNA barcode-based analysis of nematode communities two techniques have been discussed in more detail in this chapter, namely quantitative PCR and T-RFLP. The suitability of real-time PCR for the analysis of individual taxa within a complex DNA background was shown by Blgg bv, a Dutch company offering biological soil analysis services. Since 2005 they have analysed tens of thousands of soil samples for the presence or absence of the stem nematode *Ditylenchus dipsaci* based on its DNA sequence signature (http://www.nwo.nl/nwohome.nsf/pages/NWOP_6HQCCT_Eng) and, more recently, a similar product was launched for the quantitative detection of root knot nematodes (*Meloidogyne* spp.). Based on the same technology, researchers at the Laboratory of Nematology, Wageningen University, in summer 2008, performed a pilot experiment in which 21 nematode families were detected and quantified (on a log10 scale) in total nematode extracts from environmental samples (J. Helder, Wageningen, 2008, personal communication). This is the first substantial step that demonstrates the potential of this technology for routine nematode assemblage analysis.

In parallel, T-RFLP was shown to allow assemblage analysis in a 96- or a 384-well plate format. Coupled with the electronic data output and automated fragment calling software means this is a high throughput technique with rapid data analysis. This will permit large-scale temporal and spatial studies with a resolution that hitherto has not been possible and with the replication to lend statistical power to the analyses. It is conceivable that T-RFLPs could be designed and targeted for specific functional groups to answer specific ecosystem level questions. Furthermore some authors advocate the utility of certain nematode groups as indicators, for example, Cephalobidae (Yeates, 2003). At SCRI, researchers are currently testing a directed T-RFLP approach aimed at discriminating nematode assemblages to assess soil quality.

Whatever the DNA barcode analysis platform will look like in the very near future, we foresee routine analyses of nematode assemblages at an unprecedented scale (both in terms of numbers of individuals analysed and in terms of sample number). These tools will help us to understand the ecological story told by a wide range of changes in highly diverse and complex nematode assemblages.

References

Aleshin, V.V., Kedrova, O.S., Milyutina, I.A., Vladychenskaya, N.S. and Petrov, N.B. (1998) Relationships among nematodes based on the analysis of 18S rRNA gene sequences: molecular evidence for monophyly of chromadorian and secernentian nematodes. *Russian Journal of Nematology* 6, 175–184.

Amann, R.I., Ludwig, W. and Schleifer, K.H. (1995) Phylogenetic identification and *in situ* detection of individual microbial-cells without cultivation. *Microbiology Reviews* 59, 143–169.

André, H., Ducarme, X., Anderson, J., Crossley Jr, D., Koehler, H., Paoletti, M., Walter, D. and Lebrun, P. (2001) Skilled eyes are needed to go on studying the richness of the soil. *Nature* 409, 761.

André, H.M., Ducarme, X., and Lebrun, P. (2002) Soil biodiversity: myth, reality or conning? *Oikos* 96, 3–24.

Atkins, S.D., Clark, I.M., Pande, S., Hirsh, P.R. and Kerry, B.R. (2005) The use of real-time PCR and species-specific primers for the identification and monitoring of *Paecilomyces lilacinus. FEMS Microbiology Ecology* 51, 257–264.

Avaniss-Aghajani, E., Jones, K., Chapman, D. and Brunk, C. (1994) A molecular technique for identification of bacteria using small subunit ribosomal RNA sequences. *Biotechniques* 17, 144–149.

Baldwin, A.J., Moss, J.A., Pakulski, J.D., Catala, P., Joux, F. and Jeffrey, W.H. (2005) Microbial diversity in a Pacific Ocean transect from the Arctic to Antarctic circles. *Aquatic Microbial Ecology* 41, 91–102.

Bardgett, R.D. and Shine, A. (1999) Linkages between plant litter diversity, soil microbial biomass and ecosystem function in temperate grasslands. *Soil Biology and Biochemistry* 31, 317–321.

Beier, S. and Traunspurger, W. (2003) Seasonal distribution of free-living nematodes in the Korsch, a coarse grained submountain carbonate stream in southwest Germany. *Nematology* 5, 481–504.

Blaxter, M.L., De Ley, P., Garey, J.R., Liu, L.X., Scheldeman, P., Vierstraete, A., Vanfleteren, J.R., Mackey, L.Y., Dorris, M., Frisse, L.M., Vida, J.T. and Thomas, W.K. (1998) A molecular evolutionary framework for the phylum Nematoda. *Nature*, 392, 71–75.

Blaxter, M., Mann, J., Chapman, T., Thomas, F., Whitton, C., Floyd, R. and Abebe, E. (2005) Defining operational taxonomic units using DNA barcode data. *Philosophical Transactions of the Royal Society* B. 360, 1935–1943.

Blouin, M.S. (2002) Molecular prospecting for cryptic species of nematodes: mitochondrial DNA versus internal transcribed spacer. *International Journal for Parasitology* 32, 527–531.

Boag, B., Neilson, R. and Brown, D.J.F. (1992) Nematode sampling and prediction. *Nematologica* 38, 459–465.

Boag, B., Hebden, P.M., Neilson, R. and Rodger, S.J. (1998) Observations on the effect of different management regimes of set-aside land on nematode community structure. *Applied Soil Ecology* 9, 339–343.

Bongers, T. and Bongers, M. (1998) Functional diversity of nematodes. *Applied Soil Ecology* 10, 239–251.

Brown, D.J.F. and Boag, B. (1988) An examination of methods used to extract virus-vector nematodes (Nematoda: Longidoridae and Trichodoridae) from soil samples. *Nematologia Mediterranea* 16, 93–99.

Brunborg I.M., Moldal, T. and Jonassen, C.M. (2004) Quantitation of porcine circovirus type 2 isolated from serum/plasma and tissue samples of healthy pigs and pigs with postweaning multisystemic wasting syndrome using a TaqMan-based real-time PCR. *Journal of Virology Methods* 122, 171–178.

Campbell, C.D., Chapman, S.J., Cameron, C.M., Davidson, M.S. and Potts, J.M. (2003) A rapid microtitre plate method to measure carbon dioxide evolved from carbon substrate amendments as to determine the physiological profiles of soil microbial communities by using whole soil. *Applied and Environmental Microbiology* 69, 3593–3599.

Casamayor, E.O., Massana, R., Benlloch, S., Ovreas, L., Diez, B., Goddard, V.J., Gasol, J.M., Joint, I., Rodriguez-Valera, F. and Pedros-Alio, C. (2002) Changes in archaeal, bacterial and eukaryal assemblages along a salinity gradient by comparison of genetic fingerprinting methods in a multipond solar saltern. *Environmental Microbiology* 4, 338–348.

Chin, K.J., Lukow, T. and Conrad, R. (1999) Effect of temperature on structure and function of the methanogenic archaeal community in an anoxic rice field soil. *Applied and Environmental Microbiology* 65, 2341–2349.

Coomans, A. (2002) Present status and future of nematode systematics. *Nematology* 4, 573–582.

Countway, P.D., Gast, R.J., Savai, P. and Caron, D.A. (2005) Protistan diversity estimates based on 18S rDNA from seawater incubations in the western North Atlantic. *Journal of Eukaryotic Microbiology* 52, 95–106.

de Ruiter, P.C., Moore, J.C., Zwart, K.B., Bouwman, L.A., Hassink, J., Bloem, J., de Vos, J.A., Marinissen, J.C.Y., Didden, W.A.M., Lebbink, G. and Brussaard, L. (1993) Simulation of nitrogen mineralization in belowground food webs of two winter wheat management practices. *Journal of Applied Ecology*, 30, 95–106.

Degens, B.P. and Harris, J.A. (1997) Development of a physiological approach to measuring the catabolic diversity of soil microbial communities. *Soil Biology and Biochemistry* 29, 1309–1320.

Diez, B., Pedros-Alio, C., Marsh, T.L. and Massana, R., (2001) Application of denaturing gradient gel electrophoresis (DGGE) to study the diversity of marine picoeukaryotic assemblages and comparison of DGGE with other molecular techniques. *Applied and Environmental Microbiology* 67, 2942–2951.

Donn, S., Griffiths, B.S., Neilson, R. and Daniell, T.J. (2007) DNA extraction from soil nematodes for multi-sample community studies. *Applied Soil Ecology* 38, 20–26.

Doran, J.W., Sarrantonio, M. and Liebig, M.A. (1996) Soil health and sustainability. *Advances in Agronomy* 56, 1–54.

Duarte, G.F., Rosado, A.S., Seldin, L., Keijzer-Wolters, A.C. and Van Elsas, J.D. (1998) Extraction of ribosomal RNA and genomic DNA from soil for studying the diversity of the indigenous bacterial community. *Journal of Microbiological Methods* 32, 21–29.

Floyd, R., Abebe, E., Papert, A. and Blaxter, M. (2002) Molecular barcodes for soil nematode identification. *Molecular Ecology* 11, 839–850.

Foucher, A. and Wilson, M. (2002) Development of a polymerase chain reaction-based denaturing gradient gel electrophoresis technique to study nematode species biodiversity using the 18S rDNA gene. *Molecular Ecology Notes* 2, 45–48.

Foucher, A.L.J.L., Bongers, T., Noble, L.R. and Wilson, M.J. (2004) Assessment of nematode biodiversity using DGGE of 18S rDNA following extraction of nematodes from soil. *Soil Biology and Biochemistry* 36, 2027–2032.

Gormsen, D., Hedlund, K. and Wang, H.F. (2006) Diversity of soil mite communities when managing plant communities on set-aside arable land. *Applied Soil Ecology* 31, 147–158.

Griffiths, B.S., Donn, S., Neilson, R. and Daniell, T.J. (2006) Molecular sequencing and morphological analysis of a nematode community. *Applied Soil Ecology* 32, 325–337.

Guieysse, B., Wikstrom, P., Forsman, M. and Mattiasson, B. (2001) Biomonitoring of continuous microbial community adaptation towards more efficient phenol-degradation in a fed-batch bioreactor. *Applied Microbiology and Biotechnology* 56, 780–787.

Holterman, M., Van der Wurff, A., Van den Elsen, S., Van Megen, H., Bongers, T., Holovachov, O., Bakker, J. and Helder, J. (2006) Phylum-wide analysis of SSU rDNA reveals deep phylogenetic relationships among nematodes and accelerated evolution toward crown clades. *Molecular Biology and Evolution* 23, 1792–1800.

Holterman, M.H.M., Rybarczyk, K.D., Elsen, S.J.J., Van den Megen, H.H.B., Van Mooijman, P.J.W., Santiago, R.P., Bongers, A.M.T., Bakker, J. and Helder, J. (2008a) A ribosomal DNA-based framework for the detection and quantification of stress-sensitive nematode families in terrestrial habitats *Molecular Ecology Resources* 8, 23–34.

Holterman, M.H.M., Holovachov, O., Elsen, S.J.J., Van den Megen, H.H.B., Bongers, A.M.T., Bakker, J. and Helder, J. (2008b) Small subunit ribosomal DNA-based phylogeny of basal Chromadoria (Nematoda) suggests that transitions from marine to terrestrial habitats (and vice versa) require relatively simple adaptations. *Molecular Phylogenetics and Evolution* 48, 758–763

Hübschen, J., Kling, L., Ipach, U., Zinkernagel, V., Bosselut, N., Esmenjaud, D., Brown, D.J.F. and Neilson, R. (2004) Validation of the specificity and sensitivity of species-specific primers that provide a reliable molecular diagnostic for *Xiphinema diversicaudatum*, *X. index* and *X. vuittenezi*. *European Journal of Plant Pathology* 110, 779–788.

Jernberg, C., Lofmark, S., Edlund, C. and Jansson, J.K. (2007) Long-term ecological impacts of antibiotic administration on the human intestinal microbiota. *Isme Journal* 1, 56–66.

Jones, K.L., Todd, T.C., Wall-Beam, J.L., Coolon, J.D., Blair, J.M. and Herman, M.A. (2006) Molecular approach for assessing responses of microbial-feeding nematodes to burning and chronic nitrogen enrichment in a native grassland. *Molecular Ecology* 15, 2601–2609.

Kaiser, J. (2004) Wounding earth's fragile skin. *Science* 304, 1616–1618.

Kemnitz, D., Chin, K.J., Bodelier, P. and Conrad, R. (2004) Community analysis of methanogenic Archaea within a riparian flooding gradient. *Environmental Microbiology* 6, 449–461.

Kitts, C.L. (2001) Terminal restriction fragment patterns: a tool for comparing microbial communities and assessing community dynamics. *Current Issues Intestinal Microbiology* 2, 17–25.

Kopeszki, H. (1997) An active bioindication method for the diagnosis of soil properties using *Collembola*. *Pedobiologia* 41, 159–166.

Kowalchuk, G.A., Gerards, S. and Woldendorp, J.W. (1997) Detection and characterization of fungal infections of *Ammophila arenaria* (marram grass) roots by denaturing gradient gel electrophoresis of specifically amplified 18s rDNA. *Applied and Environmental Microbiology* 63, 3858–3865.

Le Novère, N. (2001) MELTING, a free tool to compute the melting temperature of nucleic acid duplex. *Bioinformatics* 17, 1226–1227.

Liu, W.T., Marsh, T.L., Cheng, H. and Forney, L.J. (1997) Characterization of microbial diversity by determining terminal restriction fragment length polymorphisms of genes encoding 16S rRNA. *Applied and Environmental Microbiology* 63, 4516–4522.

Ludwig, W., Strunk, O., Westram, R., Richter, L., Meier, H., Yadhukumar, Buchner, A., Lai, T., Steppi, S., Jobb, G., Förster, W., Brettske, I., Gerber, S., Ginhart, A.W., Gross, O., Grumann, S., Hermann, S., Jost, R., König, A., Liss, T., Lüßmann, R., May, M., Nonhoff, B., Reichel, B., Strehlow, R., Stamatakis, A., Stuckmann, N., Vilbig, A., Lenke, M., Ludwig, T., Bode, A. and Schleifer, K.-H. (2004) ARB: a software environment for sequence data. *Nucleic Acids Research* 32, 1363–1371.

Marsh, T.L. (1999) Terminal restriction fragment length polymorphism (T-RFLP): an emerging method for characterizing diversity among homologous populations of amplification products. *Current Opinions in Microbiology* 2, 323–327.

Marsh, T.L., Liu, W.T., Forney, L.J. and Cheng, H. (1998) Beginning a molecular analysis of the eukaryal community in activated sludge. *Water Science and Technology* 37, 455–460.

Marsh, T.L., Saxman, P., Cole, J. and Tiedje, J. (2000) Terminal restriction length polymorphism analysis program, a web based tool for microbial community analysis. *Applied and Environmental Microbiology* 66, 3616–3620.

Meece, J.K., Reynolds, C.E., Stockwell, P.J., Jenson, T.A., Christensen, J.E. and Reed, K.D. (2005) Identification of mosquito bloodmeal source by terminal restriction fragment length polymorphism profile analysis of the cytochrome B gene. *Journal of Medical Entomology* 42, 657–667.

Meldal, B.H.M., Debenham, N.J., De Ley, P., Tandingan, I., Vanfleteren, J., Vierstraete, A., Bert, W., Borgonie, G., Moens, T., Tyler, P.A., Austen, M.C., Blaxter, M., Rogers, A.D. and Lambshead, P.J.D. (2007) An improved molecular phylogeny of the Nematoda with special emphasis on marine taxa. *Molecular Phylogenetics and Evolution* 42, 622–636.

Mindell, D.P. and Honeycutt, R.L. (1990) Ribosomal RNA in vertebrates: evolution and phylogenetic applications. *Annual Review of Ecology and Systematics* 21, 541–566.

Moeseneder, M.M., Arrieta, J.M., Muyzer, G., Winter, C. and Herndl, G.J. (1999) Optimization of terminal-restriction fragment length polymorphism analysis for complex marine bacterioplankton communities and comparison with denaturing gradient gel electrophoresis. *Applied and Environmental Microbiology* 65, 3518–3525.

Moeseneder, M.M., Winter, C., Arrieta, J.M. and Herndl, G.J. (2001) Terminal restriction length polymorphism (T-RFLP) screening of a marine archaeal clone library to determine the different phylotypes. *Journal of Microbiological Methods* 4, 159–172.

Nagashima, K., Hisada, T., Sato, M. and Mochizuki, J. (2003) Application of new primer–enzyme combinations to terminal restriction fragment length polymorphism profiling of bacterial populations in human faeces. *Applied and Environmental Microbiology* 69, 1251–1262.

Neher, D.A. (2001) Role of nematodes in soil health and their use as indicators. *Journal of Nematology* 33, 161–168.

Nunan, N., Daniell, T.J., Singh, B.K., Papert, A., McNicol, J.W. and Prosser, J.I. (2005) Links between plant and rhizoplane bacterial communities in grassland soils, characterized using molecular techniques. *Applied and Environmental Microbiology* 71, 6784–6792.

Osborn, A.M., Moore, E.R.B. and Timmis, K.N. (2000) An evaluation of terminal-restriction fragment length polymorphism (T-RFLP) analysis for the study of microbial community structure and dynamics. *Environmental Microbiology* 2, 39–50.

Osborne, C.A., Rees, G.N., Bernstein, Y. and Janssen, P.H. (2006) New threshold and confidence estimates for terminal restriction fragment length polymorphism analysis of complex bacterial communities. *Applied and Environmental Microbiology* 72, 1270–1278.

Powers, T. (2004) Nematode molecular diagnostics: from bands to barcodes. *Annual Review of Phytopathology* 42, 367–383.

Powers, T.O. and Harris, T.S. (1993) A polymerase chain-reaction method for identification of 5 major *Meloidogyne* species. *Journal of Nematology* 25, 1–6.

Ritz, K. and Trudgill, D.L. (1999) Utility of nematode community analysis as an integrated measure of the functional state of soils: perspectives and challenges. *Plant and Soil* 212, 1–11.

Rozsak, D.B. and Colwell, R.R. (1987) Survival strategies of bacteria in the natural environment. *Microbiology Review* 51, 365–379.

Rusin, L.Y., Aleshin, V.V., Tchesunov, A.V. and Atrashkevich, G.I. (2003) The 18S ribosomal RNA gene of *Soboliphyme baturini* Petrow, 1930 (Nematoda: Dioctophymida) and its implications for phylogenetic relationships within Dorylaimia. *Nematology* 5, 615–628.

Sakamoto, M., Huang, Y., Ohnishi, M., Umeda, M., Ishikawa, I. and Benno, Y. (2004) Changes in oral microbial profiles after periodontal treatment as determined by molecular analysis of 16S rRNA genes. *Journal of Medical Microbiology* 53, 563–571.

Scheid, D., Stubner, S. and Conrad, R. (2004) Identification of rice root associated nitrate, sulphate and ferric iron reducing bacteria during root decomposition. *FEMS Microbiology Ecology* 50, 101–110.

Schloter, M., Dilly, O. and Munch, J.C. (2003) Indicators for evaluating soil quality. *Agriculture Ecosystems and Environment* 98, 255–262.

Smit, E., Leeflang, P. and Wernars, K. (1997) Detection of shifts in microbial community structure and diversity in soil caused by copper contamination using amplified ribosomal DNA restriction analysis, *FEMS Microbiology Ecology* 23, 249–261.

Smit, E., Leeflang, P., Glandorf, B., Van Elsas, J.D. and Wernars K. (1999) Analysis of fungal diversity in the wheat rhizosphere by sequencing of cloned PCR-amplified genes encoding 18S rRNA and temperature gradient gel electrophoresis. *Applied and Environmental Microbiology* 65, 2614–2621.

Spiegelman, D., Whissell, G. and Greer, C.W. (2005) A survey of the methods for the characterization of microbial consortia and communities. *Canadian Journal of Microbiology* 51, 355–386.

Stanton, J.M., McNicol, C.D. and Steele, V. (1998) Non-manual lysis of second-stage *Meloidogyne* juveniles for identification of pure and mixed samples based on the polymerase chain reaction. *Australian Journal of Plant Pathology* 27, 112–115.

Takai, K., Moser, D.P., DeFlaun, M., Onstott, T.C. and Fredrickson, J.K. (2001) Archaeal diversity in waters from deep South African gold mines. *Applied and Environmental Microbiology* 67, 5750–5760.

Thompson, J.R., Marcelino, L.A. and Polz, M.F. (2002) Heteroduplexes in mixed-template amplifications: formation, consequence and elimination by 'reconditioning PCR'. *Nucleic Acids Research* 30, 2083–2088.

Van Camp, L., Bujarrabal, B., Gentile, A.-R., Jones, R.J.A., Montanarella, L., Olazabal, C. and Selvaradjou, S.-K. (2004) Reports of the Technical Working Groups Established under the Thematic Strategy for Soil Protection. EUR 21319 EN/1. Office for Official Publications of the European Communities, Luxembourg.

Van Straalen, N.M., (1998) Evaluation of bioindicator systems derived from soil arthropod communities. *Applied Soil Ecology* 9, 429–437.

Vandenkoornhuyse, P., Ridgway, K.P., Watson, I.J., Fitter, A.H. and Young, J.P.W. (2003) Co-existing grass species have distinctive arbuscular mycorrhizal communities. *Molecular Ecology* 12, 3085–3095.

Waite, I.S., O'Donnell, A.G., Harrison, A., Davies, J.T., Colvan, S.R., Ekschmitt, K., Dogan, H., Wolters, V., Bongers, T., Bongers, M., Bakonyi, G., Nagy, P., Papatheodorou, E.M., Stamou, G.P. and Bostrom, S. (2003) Design and evaluation of nematode 18S rDNA primers for PCR and denaturing gradient gel electrophoresis (DGGE) of soil community DNA. *Soil Biology and Biochemistry* 35, 1165–1173.

Webster, R. and Boag, B. (1992) Geostatistical analysis of cyst nematodes in soil. *Journal of Soil Science* 43, 583–595.

Wünsche, L. and Babel, W. (1996) The suitability of the Biolog Automated Microbial Identification System for assessing the taxonomical composition of terrestrial bacterial communities. *Microbiological Research* 151, 133–143.

Yeates, G.W. (2003) Nematodes as soil indicators: functional and biodiversity aspects. *Biology and Fertility of Soils* 37, 199–210.

Yeates, G.W., Bongers, T., de Goede, R.G.M., Freckman, D.W. and Georgieva, S.S. (1993) Feeding habits in soil nematode families and genera – an outline for soil ecologists. *Journal of Nematology* 25, 315–331.

Zhou, Z.H. and Hogetsu, T. (2002) Subterranean community structure of ectomycorrhizal fungi under *Suillus grevillei* sporocarps in a *Larix kaempferi* forest. *New Phytologist* 154, 529–539.

Zijlstra, C. Van Hoof, R. and Donkers-Venne, D. (2004) A PCR test to detect the cereal root-knot nematode *Meloidogyne naasi*. *European Journal of Plant Pathology* 110, 855–860.

Zinger, L., Gury, J., Giraud, F., Krivobok, S., Gielly, L., Taberlet, P. and Geremia, R.A. (2007) Improvements of polymerase chain reaction and capillary electrophoresis single-strand conformation polymorphism methods in microbial ecology: toward a high-throughput method for microbial diversity studies in soil. *Microbial Ecology* 54, 203–216.

9 Ecotoxicity Testing with Nematodes

SEBASTIAN HÖSS[1] AND PHILLIP L. WILLIAMS[2]

[1]*Ecossa, Starnberg, Germany;* [2]*College of Public Health, University of Georgia, Athens, USA*

Toxicity Testing in Ecotoxicology

Anthropogenic pollution represents a tremendous stress factor for aquatic and terrestrial ecosystems. By affecting living organisms, pollutants strongly influence the functions of an ecosystem, such as nutrient cycling and overall habitat quality. Moreover, pollutants may accumulate in food webs or migrate to water sources, thus also affecting human health via contaminated food or drinking water. The fact that the hazard of countless anthropogenic substances that are released into the environment (most of them still unknown) can no longer be assessed solely by chemical analysis, has forced scientists and regulators to develop biological methods that allow for the estimation of the bioavailability and toxicity of environmental pollutants (Rand, 2003).

Toxicity tests with whole organisms are important components of an ecotoxicological risk assessment (Suter II, 2003). They provide information about the effects of single substances, as well as of known or unknown chemical mixtures on organisms under standardized laboratory conditions. Depending on the type of organism, various exposure scenarios can be chosen to simulate different uptake routes of substances into the organisms and thus consider their bioavailability. Additionally, the use of various toxicological endpoints provides information about the mode of action of certain toxicants.

The use of single-species bioassays has two major aims: (i) Toxicity assessment (monitoring) of environmental samples containing at least partly unknown mixtures of chemicals, integrating all factors that influence the interaction between a potential toxicant and the organism, such as exposure, bioavailability, mixture effects, and susceptibility; and (ii) Prediction of hazardous effects by studying the bioavailability, toxicity thresholds and modes of action of known substances or substance mixtures on various test organisms in different substrates.

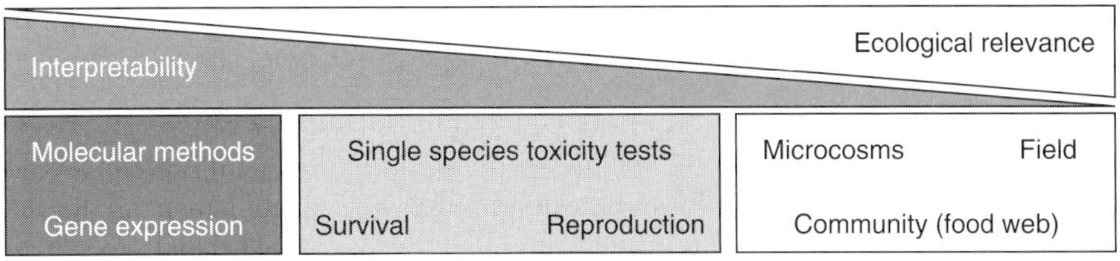

Fig. 9.1. Classification of various ecotoxicological methods regarding two important criteria, interpretability and ecological relevance.

Whole-organism toxicity tests attempt to provide a balance between interpretability and practicability on one hand and ecological relevance on the other hand (Fig. 9.1). The whole organism represents a black box where processes that determine the response to chemicals such as metabolization, detoxification, and elimination, as well as the mode of action, appear as an effect on a certain toxicological endpoint.

Sediment and soil toxicity testing is a challenge for ecotoxicologists. In these complex matrices, the interaction between toxicants, organisms and the environment are not easy to understand, which makes it difficult to interpret toxicity data. The need of infaunal species as test organisms for whole sediment or soil bioassays is beyond doubt (Burton, 1991), and a variety of sediment and soil dwelling organisms have been used for assessments of solid phases (Ingersoll *et al.*, 1995; Keddy *et al.*, 1995; Traunspurger and Drews, 1996). Although nematodes are one of the most important and abundant organisms in sediments and soils, the use of this animal phylum has always been underrepresented in toxicological assessments (see Table 9.1).

Advantages of Nematodes as Test Organisms

For more than three decades, nematodes have been used as test organisms for laboratory bioassays (e.g. Boroditsky and Samoiloff, 1973; Samoiloff *et al.*, 1980; Haight *et al.*, 1982). Various nematode species were used to assess the potential toxicity of various compounds in aqueous media (e.g. Haight *et al.*, 1982; Williams and Dusenbery, 1990b; Traunspurger *et al.*, 1997), on agar (Popham and Webster, 1979; Vranken *et al.*, 1985), as well as in more complex matrices such as sediments and soils (e.g. Donkin and Dusenbery, 1993; Traunspurger *et al.*, 1997). Mainly free-living (i.e., non-parasitic), bacterivorous nematodes were chosen as test organisms: Besides the marine *Monhystera disjuncta* (e.g. Vranken and Heip, 1986) and species of the genus *Panagrellus* (e.g. Samoiloff *et al.*, 1980; Haight *et al.*, 1982; Sherry *et al.*, 1997), ecotoxicology with nematodes has focused on the intensively studied *Caenorhabditis elegans* (see Fig. 9.2; Leung *et al.*, 2008). This soil dwelling species has proven to be an adequate test organism for various substrates, and recently methods have been standardized for the assessment of wastewater (Hitchcock *et al.*, 1997); sediment (Traunspurger *et al.*, 1997; Höss *et al.*, 1999) and soil (Freeman *et al.*, 2000;

Table 9.1. Standards for sediment and soil toxicity tests using invertebrates as test organisms.

Standard	Reference	Test organism	Substrate
ASTM E1367-03e1	ASTM (2003)	*Ampelisca abdita*, *Eohaustorius estuarius*, *Leptocheirus plumulosus*, *Rhepoxynius abronius* (Amphipoda)	Estuarine and marine sediment
ASTM E1611-00(2007)	ASTM (2007)	*Neanthes arenaceodentata* (Annelida)	Estuarine and marine sediment
ASTM E1676-04	ASTM (2004)	*Eisenia fetida*, *Enchytraeus albidus* (Annelida)	Soil
ASTM E1706-05	ASTM (2005)	*Hyallela azteca* (Amphipoda), *Chironomus tentans* (Insecta)	Freshwater sediment
ASTM E2172-01	ASTM (2001)	*Caenorhabditis elegans* (Nematoda)	Soil
ISO 11267	ISO (1999)	*Folsomia candida* (Collembola)	Soil
ISO 11268-1/2	ISO (1993); ISO (1998)	*Eisenia fetida* (Annelida)	Soil
ISO 15952	ISO (2003)	*Helix aspersa* (Mollusca)	Soil
ISO 16387	ISO (2004)	*Enchytraeus albidus* (Annelida)	Soil
ISO 16712	ISO (2005b)	*Corophium volutator* (Amphipoda)	Estuarine and marine sediment
ISO 17512-1	ISO (2008a)	*Eisenia fetida* (Annelida)	Soil
ISO 17512-2	ISO (2008b)	*Folsomia candida* (Collembola)	Soil
ISO 20963	ISO (2005a)	*Oxythyrea funesta* (Coleoptera)	Soil
ISO/DIS 10872	ISO (2009)	*Caenorhabditis elegans* (Nematoda)	Freshwater sediment and soil
OECD 207	OECD (1984)	*Eisenia fetida* (Annelida)	Soil
OECD 220	OECD (2004a)	*Enchytraeus albidus* (Annelida)	Soil
OECD 222	OECD (2004b)	*Eisenia fetida* (Annelida)	Soil
OECD 218/219	OECD (2004c)	*Chironomus riparius* (Insecta)	Freshwater sediment
OECD 225	OECD (2006)	*Lumbriculus variegatus* (Annelida)	Freshwater sediment
OECD 228	OECD (2008a)	*Musca autumnalis*, *Scathophaga stercoraria* (Diptera)	Soil, dung
OECD 226	OECD (2008b)	*Hypoaspis aculeifer* (Acari)	Soil

Peredney and Williams, 2000). For these purposes, a variety of toxicity parameters were studied for *C. elegans*: lethality (Williams and Dusenbery, 1990b; Donkin and Dusenbery, 1993; Tatara *et al.*, 1997); growth (Höss *et al.*, 1997; Traunspurger *et al.*, 1997); reproduction (Traunspurger *et al.*, 1997; Dhawan *et al.*, 1999) and behaviour (Williams and Dusenbery, 1990a; Boyd *et al.*, 2000). Besides classical toxicity tests, *C. elegans* has also been used as a test organism for bioconcentration studies (Haitzer *et al.*, 1999a, b; Jackson *et al.*, 2006).

In this chapter, we provide an overview of the use of nematodes as test organisms for ecotoxicity testing. To do this, we reviewed exemplary studies

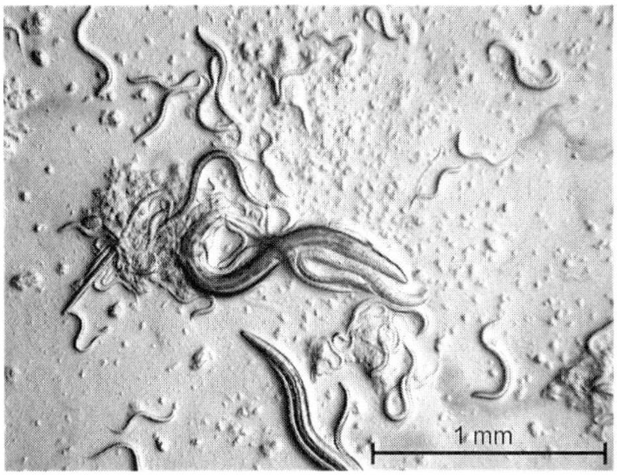

Fig. 9.2. Photograph of *Caenorhabditis elegans*: hermaphrodite (centre) surrounded by offspring in various juvenile stages crawling on agar.

that investigated the toxicity of single substances or environmental samples on nematodes, regarding different nematode species, toxicity endpoints and test substrates. The main focus is on the most frequently used species, *Caenorhabditis elegans*, by presenting examples and case studies, where this test organism has been used in ecotoxicological assessments.

Toxicity of Different Single Substances or Mixtures on Nematodes

One of the major tasks of single-species bioassays is the assessment of the toxicity of potentially harmful chemicals, either as a single substance, or combined with other contaminants in a chemical mixture. Data on the susceptibility of single species towards certain chemicals are necessary for the authorization of chemicals, as well as for a better interpretation of toxicity data of environmental samples. However, it is very important to consider the differences in sensitivity between various species and toxicity parameters, as well as the different exposure routes in various substrates, such as water, sediment or soil. Therefore, this section compares toxicity data between various invertebrate species, between various toxicity parameters of one species, and between different test substrates.

Comparison of different invertebrate species

The extrapolation of toxicity data from lower-tier studies, such as single-species toxicity tests, to higher ecological levels is a crucial issue in environmental risk assessment (Forbes and Calow, 2002). Although it is known that the toxicity of chemicals to various species can vary by several orders of magnitude (Forbes and Calow, 2002), from a practical standpoint, only a limited number of test organisms can be chosen to be used in routine application of toxicity testing. Since the most sensitive relevant species is seldom known, a test organism is usually chosen based on ease of handling and accessibility.

It is often difficult to compare the toxicity data from different studies, as the test conditions (exposure, food, test duration, temperature, etc.) can vary considerably and these differences can significantly influence the outcome of a toxicity test. Thus, only studies that compared the toxicity of chemicals on various species under the same experimental conditions are considered here.

Comparison of various nematode species

Kammenga *et al.* (1994) investigated the acute toxicity of cadmium (Cd) and pentachlorophenol (PCP) on twelve different terrestrial nematode species belonging to different taxonomic and ecological groups: bacterial feeders (*Caenorhabditis elegans*, *Rhabditis* species, *Cephalobus persegnis*, *Plectus acuminatus*, *Acrobeloides buetschlii* and *Diplogasteritus* species), a fungal feeder (*Aphelenchus avenae*), a plant feeder (*Tylenchus elegans*), predators (*Prionchulus punctatus*, *Tobrilus gracilis*) and omnivores (*Dorylaimus stagnalis*, *Aporcelaimellus obtusicaudatus*). These species exhibited large differences in sensitivity. LC50 values (72h) for pentachlorophenol ranged from 0.5 to more than 34.5μmol/l and for cadmium from 29 to more than 800μmol/l. The differences in sensitivity were only partly related to the taxonomic or ecological groups. Species of the group of Chromadorea (*sensu* De Ley and Blaxter, 2002) were less sensitive to pentachlorophenol than species of the group of Enoplea (*sensu* De Ley and Blaxter, 2002), while no differences were observed for cadmium. There appeared to be a correlation between feeding groups and pentachlorophenol sensitivity. Carnivorous, omnivorous, and plant feeding nematodes were relatively sensitive, whereas bacterial and fungal feeders were more tolerant.

Boyd and Williams (2003) compared the sensitivity of three bacterial feeding nematodes, all belonging to the group of Rhabditida (*sensu* De Ley and Blaxter, 2002): *C. elegans* (Rhabditidae), *Panagrellus redivivus* (Panagrolaimidae), and *Pristionchus pacificus* (Neodiplogasteridae) in terms of their sensitivity towards copper (Cu) in water and soil. The LC50 showed a relatively high variation between the three species, with 19mg/l (48mg/kg) for *P. pacificus*, 85mg/l (179mg/kg) for *C. elegans*, and 160mg/l (251mg/kg) for *P. redivivus* for water (soil). Although survival of *C. elegans* was affected at considerably higher concentrations of Cu than the survival of *P. pacificus* (factor 4.5 and 3.7 for water and soil, respectively), sublethal toxicity parameters, such as reproduction and movement, showed no difference in sensitivity between the two species (EC50 reproduction: 2.0 and 2.2mg/l for *C. elegans* and *P. pacificus*, respectively), or even a higher sensitivity of *C. elegans* (EC50 movement: 2.1 and 8.1mg/l for *C. elegans* and *P. pacificus*, respectively). The authors speculated that the Cu-induced changes in movement of *C. elegans* might have protected it from lethal effects, explaining the lower sensitivity of this species in the acute test.

Comparison of nematodes with other standard test organisms

Bezchlebova *et al.* (2007) compared the toxicity of the polychlorinated insecticide toxaphene on various soil invertebrates, including the nematode *C. elegans*, the collembolan *Folsomia candida*, the oligochaetes, *Eisenia fetida*,

Enchytraeus albidus, and *Enchytraeus crypticus*. *F. candida* was the most sensitive with a lowest observed effect concentration (LOEC) of 6.2 mg/kg, followed by *C. elegans*, *E. fetida*, and the *Enchytraeus* species (no observed effect concentration [NOEC]: 69, 496, and >600 mg/kg, respectively). However, it should be noted that, with 48 h, the exposure period for *C. elegans* was much lower than for the other species (28 and 42 d). In a study with aquatic sediments, linear alkylbenzene sulphonates (LAS) had a similar toxicity to *Lumbriculus variegatus* and *C. elegans* in spiked sediments, with LOECs of 82 and 100 mg/kg, respectively (Comber *et al.*, 2006). Using five metallic salts in artificial OECD soil, Peredney and Williams (2000) compared lethality data with *C. elegans* to lethality data with the earthworm *E. fetida*. The *C. elegans* exposure was for 24 hours compared to earthworm exposure of 14 days. *C. elegans* were more sensitive than the earthworm for two of the five metals, Pb and Cd, and less sensitive compared to Cu, Ni, and Zn. Overall, the two species had similar values to the 24 hour nematode data compared to 14 day earthworm data.

Comparison of different toxicity endpoints

In toxicity tests with nematodes and especially with *C. elegans* various toxicological parameters (also called toxicity endpoints) were used. Besides survival, sublethal parameters include growth, fertility, reproduction, feeding, and movement behaviour. Toxicological parameters can be characterized by their sensitivity (sublethal versus lethal), by their ecological relevance (e.g. reproduction as a measure for population growth has a higher ecological relevance than growth), by their specificity in response to certain types of toxicants (e.g. behaviour for neurotoxicants; reproduction for endocrine disruptors), or simply by their reliability and practicability.

Although survival is less sensitive than sublethal toxicity parameters, such as reproduction and behaviour, the discrepancy in sensitivity can vary from species to species. Boyd and Williams (2003) compared the toxicity of Cu on survival, reproduction and movement of two different bacterial feeding species, *C. elegans* and *P. pacificus*. While for *C. elegans*, reproduction and movement were 40 times more sensitive towards Cu than survival (LC50/EC50 = 41.7),[1] reproduction of *P. pacificus* was only nine and movement two times more sensitive than survival (LC50/EC50 = 8.6 and 2.3, respectively).

Anderson *et al.* (2004) showed that movement behaviour might be a specific toxicity parameter for neural toxicants. In 4 h toxicity tests with *C. elegans*, six chemicals of three different chemical classes with known neural toxicity (heavy metals: Pb, Al; organic solvents: ethanol, acetone; organic pesticides: levamisol, chlorpyrifos) had a considerably higher toxicity on locomotion of the nematodes, compared to toxicants of the respective classes without neural toxicity (Cu; dimethylsulfoxide [DMSO]; mebendazol (Fig. 9.3)). The specificity of behavioral toxicity endpoints was already indicated in two earlier studies (Williams and Dusenbery, 1990a; Anderson *et al.*, 2001). However, to detect neural toxicity, the exposure time had to be short (4 h), so

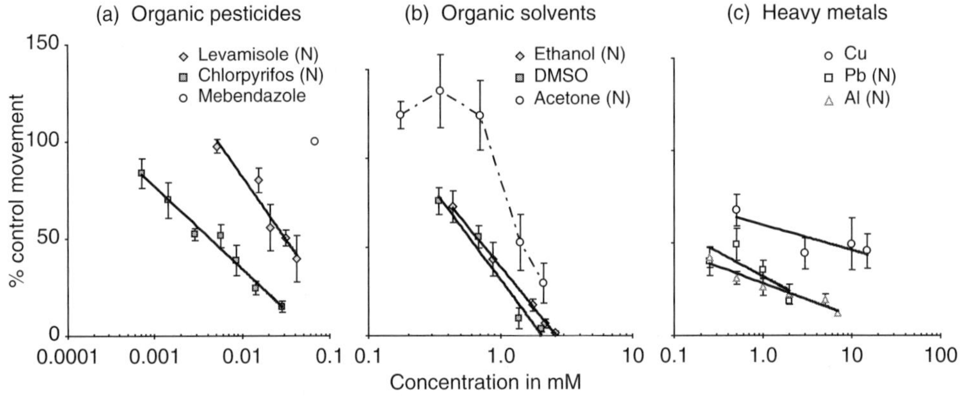

Fig. 9.3. Concentration-response relationship for compounds selected from three chemical classes: (a) organic pesticides, (b) organic solvents, (c) heavy metals. DMSO = dimethylsulfoxide; (N) = chemical with known neural toxicity. Adapted from Anderson *et al.* (2004) with permission from Allen Press Publishing Services; Copyright SETAC, Pensacola, Florida, USA.

that other symptoms of general toxicity, such as starving, could not mask the specific effect on nematode behaviour.

Endocrine disruption is also a specific mode of action that has attracted much attention since the early 1990s. Höss and Weltje (2007) reviewed the literature that dealt with effects of endocrine disruptors on nematodes, and demonstrated with their own data, that endocrine disruptors were able to stimulate the reproduction of *C. elegans*. Although it was not proved that endocrine disruption caused the stimulating effects, studies at the molecular level suggest that nematode reproduction can be influenced by endocrine active substances (Höss and Weltje, 2007).

Comparison of different test substrates

A contaminant can only exhibit a toxic effect on an organism if it is bioavailable. The bioavailability of compounds in a complex matrix, such as soils and sediments, is on one hand determined by the partitioning of chemicals between the particulate and dissolved phase, and on the other hand, dependent on the movement and feeding behaviour of the test organism. In sediments and soils, where nematodes occur, a large proportion of the contaminants may be bound to particles. Moving and feeding in the interstitial space, nematodes are exposed to contaminants via the dissolved and the particulate phase. As it is known that particle-bound contaminants are available to benthic organisms (Landrum *et al.*, 1990; Reinfelder and Fisher, 1991; Leppänen and Kukkonen, 1998; Lee *et al.*, 2000), dissolved and particle-bound contaminants might pose a risk for nematodes.

Cadmium toxicity to *C. elegans* was found to be higher in whole sediment exposure, compared to exposure in porewater that was extracted from the respective sediments (Höss *et al.*, 2001), indicating that particle bound Cd contributed to the toxicity on *C. elegans*. However, the main uptake route for contaminants might still be the dissolved phase, as the toxicity (bioavailability) of chemicals on nematodes is substantially reduced in the presence of sediments or soils. Boyd and Williams (2003) found two times higher LC50s of Cu for three bacterial-feeding species, *C. elegans*, *P. pacificus* and *P. redivivus* in a sandy soil (98% sand, 2% clay, as well as 1.4% organic matter) compared to the respective LC50 in water. This discrepancy can be even higher in soils or sediments with higher contents of fine particles and organic material. Höss *et al.* (2007) compared the toxicity of several metals and organic compounds on *C. elegans* in a test with aqueous medium and natural river sediment that was mainly composed of fine particles (2% sand, 74% silt, 24% clay). Depending on the chemical compound, the toxicity in sediment (based on sediment dry weight) was found to be one to three orders of magnitude lower than in water (Fig. 9.4). The ratio of $EC50_{sed}/EC50_{water}$ for reproduction were 11, 24, 65, 226, 305 and 1326 for dinitrophenol, pentachlorophenol, zinc, nickel, parathion, and fluoranthene, respectively. Similar results were found by Sochova *et al.* (2007) for the toxicity of six organic pollutants in natural soil and water. While toxaphene, chlorinated

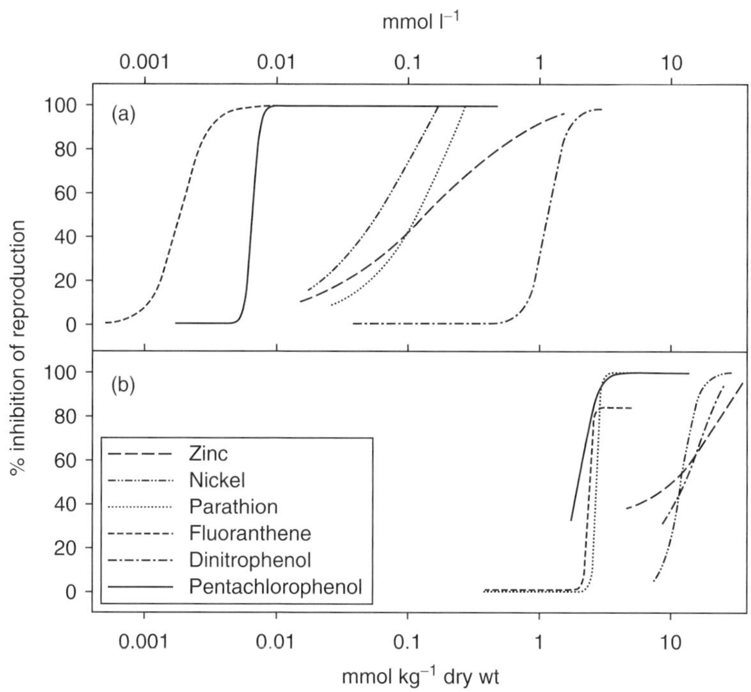

Fig. 9.4. Dose-response curves for the toxicity of six different toxicants on the reproduction of *C. elegans* in (a) water and (b) a natural sediment; dry wt = dry weight; organic substances were dissolved in acetone; for the controls, only acetone without chemicals was added.

paraffin, acridine, phenanthroline, phenanzine, and quinoline showed first lethal effects (LC10) in water at 0.012, 0.3, 3.7, 3.9, 7.1 and 40.3 mg/l, respectively, they showed a considerably lower toxicity in soil, with LC10 of 22, 896, 1788, 17, 1760, and >1000 mg/kg, respectively. The ratios of $LC10_{sed}/LC10_{water}$ for mortality ranged here between 4 and 3000. Thus, the chemical properties as well as the characteristics of the sediment or soil are important for the bioavailability and toxicity for bacterivorous nematodes.

Regardless of the test substrate, the literature supports that the route of exposure for nematodes is the gut, not absorption through the cuticle. In general, it is believed that the cuticle is impervious to most toxicants and this quality may best be demonstrated in the survival of nematodes in such diverse ecosystems and in such varied conditions. Few studies have attempted to assess the actual route of exposure. Jackson *et al.* (2006) used whole body X-ray fluorescence (XRF) to assess metal exposures to *C. elegans* and found all the available metal internal to the organism and none of the metal bound in the region of the cuticle. Furthermore, the maximum particle size that can be ingested by *C. elegans* has been demonstrated using different size fluorescent beads and found to be below 5 μm in diameter (Boyd *et al.*, 2003).

Toxicity Testing of Environmental Samples

Besides studies on the toxicity of known chemical substances, *C. elegans* has been used as a test organism to assess the toxicity of environmental samples, including very complex material, such as freshwater sediments, soils, or wastewater.

Case study with freshwater sediments

Duft (2004) investigated 206 sediment samples taken from 12 large German rivers using a nematode sediment test in order to classify the ecological status of the various sampling sites. Sublethal toxicity parameters, such as growth, fertility, and reproduction were used to rank the samples according to their toxicity. According to the EU-WFD (European Community, 2000), sediments were ranked in five different classes of ecological status, ranging from very good (I) to bad (V). Only 22% of the samples could be assigned to the classes I and II, the highest proportion could be found in class III, and 44 % were placed in the classes IV and V. Compared to the results of the commonly used sediment test with *Chironomus riparius*, that was applied to the same river sediments (Tillmann, 2003), the nematode test was found to be more sensitive. Besides the toxicity data, Duft (2004) also compared some test criteria, such as effort in cultivation, material and test preparation, test duration, required space, flexibility and experience. In most criteria, the nematode test was considered to be equal or even preferable to the chironomid test.

Case study with wastewater

Hitchcock *et al.* (1997) used *C. elegans* to assess the toxicity contribution of various industrial operations to the waste stream of a municipal wastewater treatment plant (WWTP). The water discharge from the WWTP had been found to be above acceptable levels and the source of this problem was believed to be one of the many industrial facilities that fed the plant. Three industrial facilities were identified as the most likely source of the problem. These operations included fibreglass manufacturing, paper packaging, and yarn dyeing facilities. Over a ten month timeframe, 24 hour composite samples were taken monthly from each of these facilities, as well as a common point and discharge point from the WWTP. This approach allowed for the contribution of each industrial operation to be evaluated as well as the combined input from all sources and the treated discharge from the WWTP. Following sample collection from each site, the nematode was used to assess water quality. The samples were also evaluated for standard water quality parameters such as pH and the discharge volumes from each site were assessed. Nematode mortality trends from the testing identified the sources (e.g. industrial facility) for the toxicity in the wastewater and provided the data to assist in controlling the releases. The study demonstrated the ease of use and low cost of this organism in ecotoxicological studies.

Case studies with soils

In January 2000, cyanide and heavy metals were accidently released from a mine waste lagoon in Romania and this initial release was followed in March 2000 by two subsequent mine waste spills. These releases eventually reached the Tisza River that flows through Hungary. Concurrently with the Tisza River contamination, the area received heavy rainfall followed by severe flooding which potentially spread the contamination to soils adjacent to the river. The nematode *C. elegans* was used as a test organism to evaluate soil toxicity following this event and the study demonstrated the benefits of using this organism for soil toxicity studies (Black and Williams, 2001). Soil samples of 100g from each field site were transported back to the laboratory in the USA and triplicate tests were performed with each sample. Using the nematode left enough of the soil sample for use in chemical and physical analyses. A similar approach with earthworms would have required kg samples from each site and added difficulty and expense in both transporting and the eventual disposal of the soil samples. Further, the lethality endpoint could be assessed following 24 hours of exposure compared to 14 days with earthworms. The data were used to provide a preliminary assessment of the metal contamination of the river's floodplain.

In a recent study, *C. elegans* was used to assess the toxicity of soil from fields that were cultivated with transgenic maize (*Bt* maize; MON810). This maize variety is resistant to the European corn borer, because it contains genes for insecticidal δ-endotoxins from the bacterium *Bacillus thuringiensis* (*Bt*). The results showed that growth and reproduction of *C. elegans* were significantly lower in soil from fields with *Bt* maize compared to the isogenic maize and that both toxicological parameters correlated negatively with the toxin (Cry1Ab protein) concentrations in soil (Höss *et al.*, 2008). However, a toxicity test with the pure Cry1Ab protein in aqueous medium revealed toxicity thresholds that were far above the detected soil concentrations (Höss *et al.*, 2008), indicating that the nematodes were not directly affected by the *Bt* toxin. This study demonstrated that *C. elegans* might be a suitable test organism for monitoring the risk of pest-resistant genetically modified plants, as the test is: (i) quick and cost effective; and (ii) can be carried out in aqueous medium (for testing the pure *Bt* toxin solution) and soils (for testing soils containing *Bt* toxin).

Standardization of Toxicity Tests with *C. elegans*

In environmental risk assessment there is a need for standardized tools, to produce reliable data that are able to bear up against legislation. When testing complex matrices, such as sediments and soils, test organisms may be influenced by many different factors present in the matrix. Thus, readily standardized tests are required to differentiate such effects from those of anthropogenic pollution and thus avoid false positive or negative results. For sediment and soil toxicity tests, only a limited number of standards are available for invertebrates (Table 9.1). Among the standardized tests, *Caenorhabditis elegans* is the only representative of the nematodes. However, there are established test protocols using *Plectus accuminatus* (Kammenga *et al.*, 1996) and *Panagrellus redivivus* (Sherry *et al.*, 1997) that can be considered as a suitable basis for future standards. In the USA, Williams and Dusenbery (1990b) developed a bioassay using *C. elegans* for testing aqueous substrates using mortality as the toxicity parameter. In the following years, this test was adapted for testing soils (Donkin and Dusenbery, 1993; Freeman *et al.*, 2000). After confirming the reliability of the test system by testing reference substances (Freeman *et al.*, 1998; Peredney and Williams, 2000) and the tolerance to several abiotic factors (Khanna *et al.*, 1997), an ASTM (American Society for Testing and Materials) standard was developed that:

> Covers procedures for obtaining laboratory data to evaluate the adverse effects of chemicals associated with soil to nematodes from soil toxicity tests. The methods are designed to assess lethal or sublethal toxic effects on nematodes in short-term tests in terrestrial systems. Soils to be tested may be: (i) reference soils or potentially toxic soil sites; (ii) artificial, reference, or site soils spiked with compounds; (iii) site soils diluted with reference soils; or (iv) site or reference soils diluted with artificial soil. (ASTM, 2001).

Based on the work of Williams and his laboratory group, a toxicity test for aqueous medium and freshwater sediments was developed in Europe that used sublethal toxicity parameters, such as growth and reproduction (Traunspurger *et al.*, 1997). The test proved to be suitable for testing aquatic and terrestrial environmental samples (Traunspurger *et al.*, 1997; Höss *et al.*, 1999, 2008; Duft, 2004), as well as the toxicity of chemicals in artificial and natural sediments (Höss *et al.*, 2001; Comber *et al.*, 2006, 2008). Based on these experiences an International Standard was developed that:

> Specifies a method for determining the toxicity of environmental samples on growth, fertility and reproduction of *Caenorhabditis elegans*. The method described is applicable to whole sediment, soil and waste as well as to porewater, elutriates and extracts, that were extracted from sediments and soils. (ISO, 2009).

Conclusions and Outlook

This chapter clearly shows that nematodes are suitable test organisms for ecotoxicity testing. Nematodes have significant ecological relevance for the use in sediment and soil toxicity testing. The ease of culturing and short generation time of some nematode species, such as *Caenorhabditis elegans*, allow for quick and cost-effective tests. Moreover, toxicity tests can be carried out in a variety of test substrates, such as water (wastewater, porewater, eluates, toxicant solutions), sediment, soil, or waste. Thereby, the sensitivity of nematodes (e.g. of *C. elegans*) to various pollutants is within the range of other benthic or soil invertebrates, in spite of the considerably lower exposure time.

In the field of ecotoxicology, toxicity tests take a middle position between highly ecologically relevant community or food web studies that are hard to interpret, and mechanistically informative molecular methods that are difficult to transfer to the field (Fig. 9.1). Toxicity tests with single species are important links in this huge range of ecotoxicological methods that are all necessary to explain and forecast the environmental hazard of chemicals. Nematodes are an ideal organism group to investigate toxic effects consistently from the gene to the community structure, with toxicity testing being only one challenging task for the future.

Note

[1] LC50 (Lethal Concentration 50): concentration at which 50% of test organisms died. EC50 (Effect Concentration 50): concentration at which a sublethal toxicity parameter is inhibited by 50%.

Acknowledgement

Many thanks to Jörg Römbke for completing the list of standards in Table 9.1.

References

Anderson, G.L., Boyd, W.A. and Williams, P.L. (2001) Assessment of sublethal endpoints for toxicity testing with the nematode *Caenorhabditis elegans*. *Environmental Toxicology and Chemistry* 20, 833–838.

Anderson, G.L., Cole, R.D. and Williams, P.L. (2004) Assessing behavioral toxicity with *Caenorhabditis elegans*. *Environmental Toxicology and Chemistry* 23, 1235–1240.

ASTM (2001) Standard guide for conducting laboratory soil toxicity tests with the nematode *Caenorhabditis elegans*. ASTM E2172–01, American Society for Testing and Materials, Westconshohocken, PA, USA.

ASTM (2003) Standard test method for measuring the toxicity of sediment-associated contaminants with estuarine and marine invertebrates. ASTM E1367–03e1, American Society for Testing and Materials, Westconshohocken, PA, USA.

ASTM (2004) Standard guide for conducting laboratory soil toxicity or bioaccumulation tests with the lumbricid earthworm *Eisenia fetida* and the enchytraeid potworm *Enchytraeus albidus*. ASTM E1676–04, American Society for Testing and Materials, Westconshohocken, PA, USA.

ASTM (2005) Standard test method for measuring the toxicity of sediment-associated contaminants with freshwater invertebrates. ASTM E1706–05, American Society for Testing and Materials, Westconshohocken, PA, USA.

ASTM (2007) Standard guide for conducting sediment toxicity tests with polychaetous annelids. ASTM E1611–00(2007), American Society for Testing and Materials, Westconshohocken, PA, USA.

Bezchlebova, J., Cernohlavkova, J., Lana, J., Sochova, I., Kobeticova, K. and Hofman, J. (2007) Effects of toxaphene on soil organisms. *Ecotoxicology and Environmental Safety* 68, 326–334.

Black, M.C. and Williams, P.L. (2001) Preliminary assessment of metal toxicity in the Middle Tisza River (Hungary) flood plain. *Journal of Soils and Sediments* 1, 203–206.

Boroditsky, J.M. and Samoiloff, M.R. (1973) Effects of growth inhibitors on development of the reproductive system of the free-living nematode *Panagrellus redivivus* (Cephalobidae). *Canadian Journal of Zoology* 51, 483–492.

Boyd, W.A. and Williams, P.L. (2003) Comparison of the sensitivity of three nematode species to copper and their utility in aquatic and soil toxicity tests. *Environmental Toxicology and Chemistry* 22, 2768–2774.

Boyd, W.A., Anderson, G.L., Dusenbery, D.B. and Williams, P.L. (2000) Computer tracking method for assessing behavioral changes in the nematode *Caenorhabditis elegans*. In: Price, F.T., Brix, K.V. and Lane, N.K. (eds) *Environmental Toxicology and Risk Assessment: Recent Achivements in Environmental Fate and Transport: Ninth Volume, ASTM STP 1381*. American Society for Testing and Materials, Westconshohocken, PA, USA, pp. 225–238.

Boyd, W.A., Cole, R.D., Anderson, G.L. and Williams, P.L. (2003) The effects of metals and food availability on the behavior of *Caenorhabditis elegans*. *Environmental Toxicology and Chemistry* 22, 3049–3050.

Burton, G.A. (1991) Assessing the toxicity of freshwater sediments. *Environmental Toxicology and Chemistry* 10, 1585–1627.

Comber, S.D.W., Conrad, A.U., Höss, S., Webb, S. and Marshall, S. (2006) Chronic toxicity of sediment-associated linear alkylbenzene sulphonates (LAS) to freshwater benthic organisms. *Environmental Pollution* 144, 661–668.

Comber, S.D.W., Rule, K.L., Conrad, A.U., Höss, S., Webb, S. and Marshall, S. (2008) Bioaccumulation and toxicity of a cationic surfactant (DODMAC) in sediment dwelling freshwater invertebrates. *Environmental Pollution* 153, 184–191.

De Ley, P. and Blaxter, M.L. (2002) Systematic position and phylogeny. In: Lee, D.L. (ed.), *The Biology of Nematodes*. Taylor and Francis, London, pp. 1–30.

Dhawan, R., Dusenbery, D.B. and Williams, P.L. (1999) Comparison of lethality, reproduction, and behavior as toxicological endpoints in the nematode *Caenorhabditis elegans. Journal of Toxicology and Environmental Health* 58, 451–462.

Donkin, S.G. and Dusenbery, D.B. (1993) A soil toxicity test using the nematode *C. elegans* and an effective method of recovery. *Archives of Environmental Contamination and Toxicology* 25, 145–151.

Duft, M. (2004) *Ökologische Sedimentbewertung großer Fliessgewässer mit Nematoden und Gastropoden – vom Biotest zum Freiland*. Cuvillier Verlag, Göttingen, Germany.

European Community (2000) Directive 2000/60/EC of the European Parliament and the council of 23 October 2000 establishing a framework for Community action in the field of water policy. *Official Journal of the European Community* L 327, 1–73.

Forbes, V.E. and Calow, P. (2002) Extrapolation in ecological risk assessment: Balancing pragmatism and precaution in chemical controls legislation. *Bioscience* 52, 249–257.

Freeman, M.N., Marse, T.J. and Williams, P.L. (1998) Establishment of quality assurance procedures for aquatic toxicity testing with the nematode *Caenorhabditis elegans*. In: Little, E.E., DeLonay, A.J. and Greenberg, B.M. (eds) *Environmental Toxicology and Risk Assessment: Seventh Volume, ASTM STP 1333*. American Society for Testing and Materials, Westconshohocken, PA, USA, pp. 45–59.

Freeman, M.N., Peredney, C.L. and Williams, P.L. (2000) A soil bioassay using the nematode *Caenorhabditis elegans*. In: Henshel, D.S., Black, M.C. and Harrass, M.C. (eds) *Environmental Toxicology and Risk Assessment: Standardization of Biomarkers for Endocrine Disruption and Environmental Assessment*. American Society for Testing and Materials, Westconshohocken, PA, USA, pp. 305–318.

Haight, M., Mudry, T. and Pasternak, J. (1982) Toxicity of seven heavy metals on *Panagrellus silusiae*: The efficacy of the free-living nematode as an in vivo toxicological bioassay. *Nematologica* 28, 1–11.

Haitzer, M., Burnison, B.K., Höss, S., Traunspurger, W. and Steinberg, C.E.W. (1999a) Effects of quantity, quality, and contact time of dissolved organic matter on bioconcentration of Benzo[a]pyrene in the nematode *Caenorhabditis elegans. Environmental Toxicology and Chemistry* 18, 459–465.

Haitzer, M., Höss, S., Traunspurger, W. and Steinberg, C. (1999b) Relationship between concentration of dissolved organic matter (DOM) and the effect of DOM on the bioconcentration of benzo[a]pyrene. *Aquatic Toxicology* 45, 147–158.

Hitchcock, D.R., Black, M.C. and Williams, P.L. (1997) Investigations into using the nematode *Caenorhabditis elegans* for municipal and industrial wastewater toxicity testing. *Archives of Environmental Contamination and Toxicology* 33, 252–260.

Höss, S. and Weltje, L. (2007) Endocrine disruption in nematodes – effects and mechanisms. *Ecotoxicology* 16, 15–28.

Höss, S., Haitzer, M., Traunspurger, W., Gratzer, H., Ahlf, W. and Steinberg, C. (1997) Influence of particle size distribution and content of organic matter on the toxicity of copper in bioassays using *Caenorhabditis elegans* (Nematoda). *Water Air and Soil Pollution* 99, 689–695.

Höss, S., Haitzer, M., Traunspurger, W. and Steinberg, C.E.W. (1999) Growth and fertility of *Caenorhabditis elegans* (Nematoda) in unpolluted freshwater sediments – response to particle size distribution and organic content. *Environmental Toxicology and Chemistry* 18, 2921–2925.

Höss, S., Henschel, T., Haitzer, M., Traunspurger, W. and Steinberg, C. (2001) Toxicity of cadmium to *Caenorhabditis elegans* (Nematoda) in whole sediment and porewater – the ambiguous role of organic matter. *Environmental Toxicology and Chemistry* 20, 2794–2801.

Höss, S., Spira, D., Gilbert, D. and Melbye, K. (2007) The SeKT joint research project – TV3: Effects of spiked natural and artificial sediments on the nematode *Caenorhabditis elegans*. 17th Annual Meeting of SETAC Europe, Porto, Portugal, p. 81.

Höss, S., Arndt, M., Baumgarte, S., Tebbe, C., Nguyen-Thu, H. and Jehle, J. (2008) Effects of transgenic corn and Cry1Ab protein on the nematode *Caenorhabditis elegans. Ecotoxicology and Environmental Safety* 70, 334–340.

Ingersoll, C.G., Ankley, G.T., Benoit, D.A., Brunson, E.L., Burton, G.A., Dwyer, F.J., Hoke, R.A., Landrum, P.F., Norberg-King, T.J. and Winger, P.V. (1995) Toxicity and bioaccumulation of sediment-associated contaminants using freshwater invertebrates: A review of methods and applications. *Environmental Toxicology and Chemistry* 14, 1885–1894.

ISO (1993) Soil quality – Effects of pollutants on earthworms (*Eisenia fetida*) – Part 1: Determination of acute toxicity using artificial soil substrate. ISO 11268–1, International Organization for Standardization, Geneva.

ISO (1998) Soil quality – Effects of pollutants on earthworms (*Eisenia fetida*) – Part 2: Determination of effects on reproduction. ISO 11268–2, International Organization for Standardization, Geneva.

ISO (1999) Soil quality – Inhibition of reproduction of Collembola (*Folsomia candida*) by soil pollutants. ISO 11267:1999, International Organization for Standardization, Geneva.

ISO (2003) Soil quality – Determination of the effects of pollutants on juvenile land snails (*Helix aspersa*) – Determination of the effects on growth by soil contamination. ISO 15952, International Organization for Standardization, Geneva.

ISO (2004) Soil quality – Effects of pollutants on Enchytraeidae (*Enchytraeus* sp.) – Determination of effects on reproduction and survival ISO 16387:2004, International Organization for Standardization, Geneva.

ISO (2005a) Soil quality – Effects of pollutants on insect larvae (*Oxythyrea funesta*) – Determination of acute toxicity using artificial soil substrate. ISO 20963, International Organization for Standardization, Geneva.

ISO (2005b) Water quality – Determination of acute toxicity of marine or estuarine sediment to amphipods. ISO 16712:2005, International Organization for Standardization, Geneva.

ISO (2008a) Soil quality – Avoidance test for evaluating the quality of soils and the toxicity of chemicals. Test with earthworms (*Eisenia fetida/andrei*). ISO 17512–1, International Organization for Standardization, Geneva.

ISO (2008b) Soil quality – Avoidance test for evaluating the quality of soils and the toxicity of chemicals. Test with collembolans (*Folsomia candida*). ISO/CD 17512–2, International Organization for Standardization, Geneva.

ISO (2009) Water quality – Determination of the toxic effect of sediment and soil samples on growth, fertility and reproduction of *Caenorhabditis elegans* (Nematoda). ISO/DIS 10872, International Organization for Standardization, Geneva.

Jackson, B.P., Williams, P.L., Lanzirotti, A. and Bertsch, P.M. (2006) Evidence for biogenic pyromorphite formation by the nematode *Caenorhabditis elegans. Environmental Science and Technology* 39, 5620–5625.

Kammenga, J.E., Van Gestel, C.A.M., and Bakker, J. (1994) Patterns of sensitivity to cadmium and pentachlorophenol among nematode species from different taxonomic and ecological groups. *Archives of Environmental Contamination and Toxicology* 27, 88–94.

Kammenga, J.E., Van Koert, P.H.G., Riksen, J.A.G., Korthals, G.W. and Bakker, J. (1996) A toxicity test in artificial soil based on the life history strategy of the nematode *Plectus acuminatus. Environmental Toxicology and Chemistry* 15, 722–725.

Keddy, C.J., Greene, J.C. and Bonnell, M.A. (1995) Review of whole-organism bioassays: Soil, freshwater sediment, and freshwater assessment in Canada. *Ecotoxicology and Environmental Safety* 30, 221–251.

Khanna, N., Cressman III, C.P., Tatara, C.P., Williams, P.L. (1997) Tolerance of the nematode *Caenorhabditis elegans* to pH, salinity, and hardness in aquatic media. *Archives of Environmental Contamination and Toxicology* 32, 110–114.

Landrum, P.F., Robbins, J.A., Baudo, R., Giesy, J. and Muntau, H. (1990) Bioavailability of sediment associated contaminants to benthic invertebrates. *Sediments: Chemistry and Toxicity of In-Place Pollutants*. Lewis Publisher, Ann Arbor, Michigan, pp. 237–263.

Lee, B.G., Griscom, S.B., Lee, J.S., Choi, H.J., Koh, C.H., Luoma, S.N. and Fisher, N.S. (2000) Influences of dietary uptake and reactive sulfides on metal bioavailability from aquatic sediments. *Science* 287, 282–284.

Leppänen, M.T. and Kukkonen, J. (1998) Relative importance of ingested sediment and pore water as bioaccumulation routes for pyrene to oligochaete (*Lumbriculus variegatus*, Müller). *Environmental Science and Technology* 32, 1503–1508.

Leung, M.C.K., Williams, P.L., Benedetto, A., Au, C., Helmcke, K., Aschner, M. and Meyer, J.N. (2008) *Caenorhabditis elegans*: An emerging model in biomedical and environmental toxicology. *Toxicological Sciences* 106, 5–28.

OECD (1984) Earthworm acute toxicity test. No. 207, Organisation for Economic Cooperation and Development, Paris.

OECD (2004a) Enchytraeidae reproduction test. No. 220, Organisation for Economic Cooperation and Development, Paris.

OECD (2004b) Earthworm reproduction test. No. 222, Organisation for Economic Cooperation and Development, Paris.

OECD (2004c) Sediment-water chironomid toxicity test using spiked sediment. No. 218, Organisation for Economic Cooperation and Development, Paris.

OECD (2006) Sediment – water *Lumbriculus* toxicity test using spiked sediment. No. 225, Organisation for Economic Cooperation and Development, Paris.

OECD (2008a) Determination of developmental toxicity of a test chemical to dipteran dung flies (*Scathophaga stercoraria* L. (Scathophagidae), *Musca autumnalis* De Geer (Muscidae)). No. 228, Organisation for Economic Cooperation and Development, Paris.

OECD (2007b) Predatory mite reproduction test in soil (*Hypoaspis* (*Geolaelaps*) *aculeifer*). No. 226, Organisation for Economic Cooperation and Development, Paris.

Peredney, C.L. and Williams, P.L. (2000) Utility of *Caenorhabditis elegans* for assessing heavy metal contamination in artificial soil. *Archives of Environmental Contamination and Toxicology* 39, 113–118.

Popham, J.D. and Webster, J.M. (1979) Cadmium toxicity in the freeliving nematode, *Caenorhabditis elegans*. *Environmental Research* 20, 183–191.

Rand, G.M. (2003) *Fundamentals of Aquatic Toxicology: Effects, Environmental Fate, and Risk Assessment*. Taylor and Francis, New York.

Reinfelder, J.R. and Fisher, N.S. (1991) The assimilation of elements ingested by marine copepods. *Science* 251, 794–798.

Samoiloff, M.R., Schulz, S., Jordan, Y., Denich, K. and Arnott, E. (1980) A rapid simple long-term toxicity assay for aquatic contaminants using the nematode *Panagrellus redivivus*. *Canadian Journal of Fisheries and Aquatic Science* 37, 1167–1174.

Sherry, J., Scott, B. and Dutka, B. (1997) Use of various acute, sublethal and early life stage tests to evaluate the toxicity of refinery effluents. *Environmental Toxicology and Chemistry* 16, 2249–2257.

Sochova, I., Hofman, J. and Holoubek, I. (2007) Effects of seven organic pollutants on soil nematode *Caenorhabditis elegans*. *Environment International* 33, 798–804.

Suter II, G.W. (2003) Introduction to ecological risk assessment for aquatic toxic effects. In: Rand, G.M. (ed.) *Fundamentals of Aquatic Toxicology: Effects, Environmental Fate, and Risk Assessment*. Taylor and Francis, New York, pp. 803–816.

Tatara, C.P., Newman, M.C., McCloskey, J.T. and Williams, P.L. (1997) Predicting relative metal toxicity with ion characteristics: *Caenorhabditis elegans* LC50. *Aquatic Toxicology* 39, 279–290.

Tillmann, M. (2003) Sedimenttoxikologische Untersuchungen mit Gastropoden und Insekten unter besonderer Berücksichtigung endokrin wirksamer Substanzen. PhD thesis, J.-W. Goethe Universität, Frankfurt/Main, Germany.

Traunspurger, W. and Drews, C. (1996) Toxicity analysis of freshwater and marine sediments with meio- and macrobenthic organisms: a review. *Hydrobiologia* 328, 215–261.

Traunspurger, W., Haitzer, M., Höss, S., Beier, S., Ahlf, W. and Steinberg, C. (1997) Ecotoxicological assessment of aquatic sediments with *Caenorhabditis elegans* (Nematoda) – A method for testing in liquid medium and whole sediment samples. *Environmental Toxicology and Chemistry* 16, 245–250.

Vranken, G. and Heip, C. (1986) Toxicity of copper, mercury and lead to a marine nematode. *Marine Pollution Bulletin* 17, 453–457.

Vranken, G., Vanderhaeghen, R. and Heip, C. (1985) Toxicity of cadmium to free-living marine and brackish water nematodes (*Monhystera microphtalma, Monhystera disjuncta, Pellioditis marina*). *Diseases of Aquatic Organisms* 1, 49–58.

Williams, P.L. and Dusenbery, D.B. (1990a) A promising indicator of neurobehavioral toxicity using the nematode *Caenorhabditis elegans* and computer tracking. *Toxicology and Industrial Health* 6, 425–440.

Williams, P.L. and Dusenbery, D.B. (1990b) Aquatic toxicity testing using the nematode *Caenorhabditis elegans. Environmental Toxicology and Chemistry* 9, 1285–1290.

10 Transgenic *Caenorhabditis elegans* as Biosensors

Cristina Lagido

School of Medical Sciences, University of Aberdeen, UK

Introduction

The free-living nematode *Caenorhabditis elegans* is one of the best characterized metazoans in terms of its genetics, development, anatomy and behaviour. Its amenity for genetic modification has been exploited to generate transgenic biosensor strains that are increasingly sensitive and informative about the mechanisms of toxicity. A large percentage of the *C. elegans* genes have been conserved in evolution and have counterparts in vertebrates (Culetto and Sattelle, 2000). This can be used in biomonitoring to develop *C. elegans* strains that report on the activation of specific genetic pathways implicated in human diseases, providing a fast prediction of the threat to human health inflicted by the combination of xenobiotics present in an environmental sample. Combined with the ease of culturing in the laboratory, this is what gives *C. elegans* the edge as a sentinel organism.

Toxicity studies involving the wild-type strains have already been reviewed by Höss and Williams, in Chapter 9, this volume. Here I will focus on transgenic *C. elegans* developed for ecotoxicology purposes to the end of 2007. These fall broadly into three categories or approaches, in terms of the information generated:

1. A strain may express a marker gene simply to facilitate its identification, as exemplified by strain CB5584 that expresses the green fluorescent protein (GFP) constitutively in the pharynx. Graves *et al.* (2005) validated the use of this strain for toxicity testing of field collected soil samples. CB5584 was easier to score and more efficiently recovered from samples containing indigenous nematodes than the wild-type and responded in the same way when exposed to copper (Graves *et al.*, 2005).
2. Strains that carry reporter genes under control of a regulatory region from a stress inducible gene and thus report on the cellular and molecular defence responses caused by environmental stress. Examples include the extensively researched transgenic strains CB4027 and PC72. These strains have respectively *Drosophila* and *C. elegans* heat-shock responsive promoters driving the

(eds M.J. Wilson and T. Kakouli-Duarte)

expression of bacterial *lacZ* gene in response to a variety of stresses. Other promoters that respond to specific stresses, such as transition metals (Cioci *et al.*, 2000; Liao *et al.*, 2002) or PCB exposure (Menzel *et al.*, 2001, 2007) have been used more recently. An important requirement in validating this type of biosensor is that conditions that lead to expression of the surrogate reporter gene will also induce the endogenous genes encoding stress proteins (Jones *et al.*, 1996; Link *et al.*, 1999; Menzel *et al.*, 2001, 2007; Roh *et al.*, 2006).

3. Transgenic strains that report on the perturbation of energy balance caused by stress. Metabolic sensor strains have been generated by constitutive expression of the firefly luciferase gene, *luc* (Lagido *et al.*, 2001). This enzyme catalyses the oxidation of an exogenous substrate, luciferin, in a reaction that is powered by endogenous ATP and produces light and AMP (de Wet *et al.*, 1987). When the other substrates are in excess, the amount of light emitted provides a relative measure of the cellular ATP levels and, more generally, of the metabolic state of the organism. As ATP tends to fall rapidly in response to various stresses (Corton *et al.*, 1994; Lagido *et al.*, 2008), luminometry allows for quantification of the degree of stress experienced by an organism.

The levels of stress/toxins that elicit a measurable response in transgenic strains are generally much lower than those leading to lethality. Additionally, results can be obtained very quickly after an initial exposure to a contaminated environmental sample, within minutes for metabolic biosensors or hours for biosensors that report on the induction of stress responses. Below I will explain how transgenic *C. elegans* strains are generated and review the literature concerning their use in biomonitoring. Developments in the understanding of stress responses will also be explored briefly as they have the potential to be translated into novel *C. elegans* biosensor strains. Standard *C. elegans* genetic nomenclature is used throughout (http://www.wormbase.org/wiki/index.php/Nomenclature).

How to Generate Transgenic *C. elegans*

C. elegans transformation techniques are well established, allowing for routine construction of transgenic animals carrying reporter genes. Detection of reporter gene expression does not require dissection because the worm is transparent. This is a strong advantage and enables *in vivo* analysis. Here, I will present an overview of common reporter genes and will explain how transgenic strains are generated. The *Caenorhabditis* Genetics Center (http://biosci.umn.edu/CGC/strains/) maintains a collection of strains that are available to researchers.

Reporter genes

β-galactosidase

The *E. coli* β-galactosidase gene, *lacZ*, is a sensitive reporter and no background activity is detected in wild-type worms grown on *E. coli* lacking the *lacZ* gene

(*E. coli* P90C). β-galactosidase levels are determined by cleavage of the histochemical stain X-gal (5-bromo-4-chloro-3-indolyl-β-D-galactopyranoside) with formation of an insoluble blue precipitate quantifiable spectrophotometrically at 615 nm. Alternative substrates have been developed that allow for different methods of quantification, including fluorometry (Stringham and Candido, 1994; Daniells *et al.*, 1998; David *et al.*, 2003). A drawback of *lacZ* is that the procedures required to promote diffusion of substrates into the worm are lethal and on this account lost favour to the extensively used *in vivo* GFP reporter (Chalfie *et al.*, 1994). Nevertheless, *lacZ* provides a cheaper methodology that may be useful in environmental monitoring on a global scale.

Fluorescent proteins

The green fluorescent protein (GFP), originally from the jellyfish *Aequorea victoria*, has become the most popular reporter in *C. elegans* research. As soon as the polypeptide folds to form the mature fluorophore, it emits green light intrinsically (peak at 209 nm with a shoulder at 540 nm) upon illumination with a long wave ultraviolet source. GFP conveys an intensely green colour to the cells and tissues and does not interfere with growth or function thus allowing for *in vivo* visualization of gene expression (Chalfie *et al.*, 1994). Some autofluorescence can be observed in wild-type worms, in particular the gut and in hypodermal cells, but this is easily distinguished from the GFP fluorescence displayed by transgenic animals.

Several GFP variants have been developed for expression in *C. elegans*. Two mutations, S65T and S65C (mutation of serine-65 to threonine and cysteine respectively) result in faster folding, stronger fluorescence and improved resistance to fading. Furthermore, GFP colour variants with non-overlapping emission spectra, such as the cyan-coloured derivative CFP and the yellow shifted derivative YFP, enable simultaneous analysis of gene expression driven by different promoters (Miller *et al.*, 1999). Another fluorescent reporter is the red fluorescent protein from *Discoma sp.* reef coral: DsRed and DsRed2 (Clontech). CFP, YFP, DsRed and DsRed2 have been successfully used in combination to label different classes of *C. elegans* neurons (Hutter, 2003). A limitation of GFP and other fluorescent proteins is that gene expression cannot be studied in real-time due to the time required for fluorophore maturation (typically one hour for the S65T variant, as determined in an *E. coli* expression system (Heim *et al.*, 1995)). GFP (S65T), CFP and YFP appear to mature at the same rate in the worm, whereas DsRed variants mature more slowly, however new DsRed variants are being developed to mature faster (Shaner *et al.*, 2004).

Luciferase

The American firefly (*Photinus pyralis*) *luc* gene encodes the enzyme firefly luciferase that is used widely to measure *in vitro* ATP levels. It catalyses the oxidation of luciferin in a reaction that consumes ATP and generates yellow-green light (λ_{max} ≈560 nm) and AMP (de Wet *et al.*, 1987). Luciferase-expressing *C. elegans* emits light when provided with exogenous D-luciferin (Lagido *et al.*, 2001). Light emitted can be easily captured with a sensitive CCD camera or quantified by luminometry because the worms are transparent.

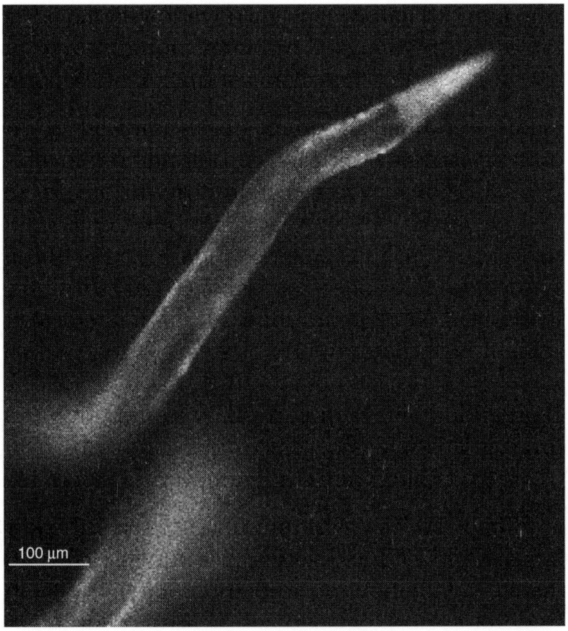

Fig. 10.1. A *C. elegans* strain marked with the *luc+* gene shows widespread *in vivo* luminescence. Image was captured in complete darkness using a Photometrics CASCADE II 512 air cooled (–70°C) CCD sensor (10s integration, 18min after adding 0.1mM luciferin). Reproduced from Lagido *et al.* (2008) with permission from the authors.

The wild-type enzyme is targeted at peroxisomes (Keller *et al.*, 1987). However, modification of the gene has enabled cytoplasmic expression of the enzyme and increased its stability (Kogai *et al.*, 2003), a distinct advantage for *in vivo* ATP measurements.

Stringham and Candido (1994) were the first to suggest using luciferase as a marker gene, but with the newly cloned GFP proving adequate for most purposes, progress was delayed until Lagido *et al.* (2001) produced the first such strains. Unlike other exogenous substrates such as X-gal, permeability of luciferin was not found to be an issue when provided in a buffer containing 1% DMSO and 0.05 % triton-X (Lagido *et al.* 2001). Captured images of bioluminescence revealed widespread luminescence in the worm's tissues (Fig. 10.1), indicating that luciferin permeated through the worm. A unique application of luciferase is to report on ATP levels *in vivo* and in real-time (Maechler *et al.*, 1998; Lagido *et al.*, 2001; Schneider and Gourse, 2004).

Reporter constructs

Standard cloning techniques are used to assemble the constructs for genetic transformation of *C. elegans* (Boulin *et al.*, 2006). Two types of fusions are commonly used for expression in *C. elegans*, transcriptional fusions (putative regulatory region of stress responsive gene fused to the reporter gene) and translational fusions (reporter genes fused to the coding sequence of a stress responsive gene). Translational fusions are more representative of a gene's expression pattern. The 'Fire vector kit' (http://www.addgene.org/Andrew_Fire) contains a large collection of vectors for genetic transformation of *C. elegans* which are based on the pUC19 plasmid and include the *E. coli* ampicillin resistance gene (AmpR) and origin of replication (Fire *et al.*, 1990). These vectors may contain: promoter regions of genes such as the heat-shock genes; various reporter genes; localization signals for the reporter genes, for example the nuclear localization peptide (NLS) from the SV40 virus; and also multiple cloning sites (MCS) to enable insertion of DNA sequences. Genomic DNA preparations, worm lysates and cosmids or other sources available in the *C. elegans* research community can be used as a DNA template for PCR amplification of any promoter/gene. Protocols for generation of reporter gene fusions are provided elsewhere (Boulin et al., 2006; Dolphin and Hope, 2006). The 'assembled' vector containing the promoter and reporter gene(s) can then be used for genetic modification as described below.

Strain construction

Generation of transgenic *C. elegans* can be achieved by microinjection of the recombinant DNA into the gonad or, relatively more recently, by gene bombardment. Microinjection is an established technique that involves injection into the distal arm of the gonad (refer to Mello and Fire (1995) and http://www.wormbook.org for protocols) which results in a large number of oocytes containing the DNA of interest. The outcome in most cases is transient expression in the F1 generation, with DNA being absent in future generations. Occasionally, the injected DNA concatemerizes *in vivo*, to form long tandem extrachromosomal arrays containing several hundred copies of the injected plasmid DNA (Stinchcomb *et al.*, 1985). These are inherited by a fraction of the progeny at each subsequent generation and can easily be selected for by co-injection of a phenotypic marker such as *rol-6 (su1006)* which causes transformed animals to move abnormally by rolling along their axis ('roller' phenotype) (Kramer *et al.*, 1990). Alternatively, animals with stunted growth (dumpy phenotype) are injected with the DNA of interest plus a plasmid carrying a functional copy of the gene that will rescue the dumpy phenotype. In the latter case marked individuals will display a wild-type phenotype, whereas non-marked animals will remain dumpy.

Extrachromosomal DNA arrays are unstable and can be lost from transformed lines without frequent selection. If transgenes are integrated into chromosomes this results in 100% inheritance and less variability in expression. While this happens spontaneously with very low frequency (Mello and Fire, 1995) it can be encouraged using gamma irradiation, UV irradiation or chemical mutagenesis (EMS) (Evans, 2006). In these cases the integrated line should subsequently be out-crossed several times with wild-type males to eliminate random mutations.

Microparticle bombardment is an important development in nematode transgenics as it enables the direct production of integrated transgenic lines. With this technique DNA is bound onto gold particles which are then introduced into worms using a 'gene gun' (Praitis *et al.*, 2001). Particles that enter nuclei may give rise to integrated lines containing low transgene copy number. Such strains do not undergo silencing of the transgene in the germline, unlike strains derived from microinjection (Kelly and Fire, 1998). In addition variability in expression is reduced in low copy integrants.

Transgenic *C. elegans* Biosensor Strains and the Range of Toxicants that Elicit a Response

Biosensor strains reporting on the induction of the heat-shock response

Stress-induced protein denaturation elicits a universal cellular response, known as the heat-shock response, involving rapid and coordinated induction of heat-shock proteins (HSP) (Westerheide and Morimoto, 2005).

There has been much interest in monitoring the expression of HSPs, particularly HSP70 and HSP16 of 70 and 16 kDa mass respectively, as reporters of environmental contamination. HSP70s act as molecular chaperones, promoting folding and minimizing protein denaturation and aggregation. HSP16 proteins are less well characterized and appear to be strictly stress-inducible.

C. elegans shows a classic heat-shock response when exposed to temperatures of 31–33 °C (its maximum growth temperature is 25 °C), with synthesis of stress proteins homologous to the *Drosophila* HSP90, HSP70 and small HSP families (Snutch and Baillie, 1984). The best known *C. elegans hsp-16* genes share common promoters in pairs: *hsp-16.1/hsp-48* and *hsp-16.2/hsp-16.41* (Stringham and Candido, 1994; Shim *et al.*, 2003). Several transgenic strains have been constructed with the *C. elegans hsp-16* or the *Drosophila hsp-70* promoters driving the expression of various reporter genes (see Table 10.1). These respond to a variety of stressful conditions.

Direct visualization of GFP and β-galactosidase expression allows for observation of the stress response in single worms. All somatic cells have the ability to transcribe the *hsp-16* driven reporter genes in response to heat-shock or arsenite exposure, a classic inducer of the heat-shock response (Stringham and Candido, 1994). However, expression is more restricted with toxicants such as metals and fungicides (Stringham and Candido, 1994; Candido and Jones, 1996; Jones *et al.*, 1996). For example, cadmium resulted in reporter expression in the intestine and/or the pharynx, depending on whether food was present (Stringham and Candido, 1994; Candido and Jones, 1996; Chu and Chow, 2002; David *et al.*, 2003). Similarly, the *hsp-70* reporter shows a range of responses in relation to different stressors, from pharyngeal to generalized expression (Guven *et al.*, 1994, 1995, 1999;

Table 10.1. Transgenic strains that report on induction of the heat-shock response.

Strain	Promoter	Reporter gene/fusion	Expression of reporter	Stress	Medium
PC72 PC71 PC73	*hsp-16.1* *hsp-16.1* *hsp-16.48*	*lacZ* (plus SV40 NLS)/ translational	Nuclei of cells. Heat-shock, arsenite (AsO_2^-): generalized; other stresses: differential expression e.g. gut (Hg^{2+}, Cd^{2+}), anterior pharynx (Cu^{2+}, captan, captafol), pharynx (Cd^{2+}, folpet), pharynx and nerve ring (Hg^{2+}), posterior pharynx (Pb^{2+}), hypodermal cells (Zn^{2+}).[a,b]	Heat, Cd^{2+}, Cu^{2+}, Hg^{2+}, Zn^{2+}, Pb^{2+}, AsO_2^-, fungicides (captan, captafol, folpet, mancozeb) herbicide (paraquat);[a,b,c] nematicide (pyrantel), insecticide (Deltamethrin);[c] metal polluted river water;[d] metal mixtures (Cd^{2+}, Cu^{2+}, and Cd^{2+} + Zn^{2+}).[e] Non-inducers: phthalimide, tetrahydrophthalimide, N-phthaloyl-L-glutamic acid.[b] Poor inducers: fungicides (maneb, carbendazim), herbicide (Carbetamex), insecticide (dimethoate);[c] Mn^{2+}[f]	Aqueous: distilled water;[a,d] buffered K medium;[b] K medium,[c,f] extracted soil water,[c,e], polluted river water[d] Soil (Lufa 2.2)[e]
PC161	*hsp-16.1*	*lacZ::gfp* (plus SV40 NLS)/ translational	Heat-shock: generalized; Cd^{2+}: pharynx[g]	Heat (30°C), Cd^{2+}[g]	Aqueous: distilled water[g]
CB4027	*hsp70*	*lacZ*/ transcriptional	Pharynx to general expression depending on stress	Heat (32°C), Cd^{2+}, Hg^{2+}, Zn^{2+}, Ag^{+}, Sn^{2+}, Mn^{2+}, TBT, lindane;[h] thiabendazole, ivermectin;[c] metal polluted river water;[d] Cd^{2+} + Ca^{2+} and Cd^{2+} + Zn^{2+}[i]	Aqueous: K medium;[h,i] distilled water, polluted river water[d]
	hsp-16.2 *hsp-16.48*	*gfp*/translational	Generalized	Heat, EtOH, NaN_3, Cd^{2+};[j,k] weak induction by di(2-ethylhexyl)phthalate (DEHP).[l] Non-inducers:[k] Pb^{2+}, $Cr_2O_7^{2-}$, AsO_2^-	Aqueous: M9;[j] K medium;[K] DEHP contaminated soils[l]
KC136[m]	*hsp-16.2*	*gfp*/ transcriptional	Pharynx and intestine	Cd^{2+}, Hg^{2+}, metal mixtures (Cd^{2+}+ Cu^{2+}, Cd^{2+}+ Cu^{2+}+ $Cr_2O_7^{2-}$, Cd^{2+}+ Cu^{2+}+ Hg^{2+}). Non-inducer: Zn^{2+}	Aqueous: K medium
KC334[n]	*hsp-16.2*	*gfp*/ transcriptional	Pharynx and intestine (Cd^{2+}); pharynx (extracted field sample)	Cd^{2+}; field sample extracted from river sediments, diluted with K medium	Aqueous: K medium
CL2070[o] CL2071[o]	*hsp-16.2*	*gfp*/ transcriptional	Pharynx	Heat (35°C), juglone, paraquat, plumbagin	NGM plates; aqueous: liquid survival medium

[a]Stringham and Candido, 1994; [b]Jones *et al.*, 1996; [c]Guven *et al.*, 1999; [d]Mutwakil *et al.*, 1997; [e]Power and Pomerai, 1999; [f]Dennis *et al.*, 1997; [g]David *et al.*, 2003; [h]Guven *et al.*, 1994; [i]Guven *et al.*, 1995; [j]Hong *et al.*, 2004; [K]Roh *et al.*, 2006; [l]Roh *et al.*, 2007;[m]Chu and Chow, 2002; [n]strain genetic background: *daf-16(m26); unc-75(e950)* Chu *et al.*, 2005; [o]Link *et al.*, 1999.

Mutwakil *et al.*, 1997). The diverse pattern of expression may result from increased exposure of certain tissues to the toxicants due to the route of entry and/or tissue-specific concentration/adsorption of the chemical. It may also reflect tissue-specific susceptibility to the toxicants (Cioci *et al.*, 2000).

Hsp-16 based sensors have been favoured over the *hsp-70* strain, because detection of toxicants can be carried out at the normal growth temperature of 20 °C, whereas the *hsp-70* strain requires combined heat-shock and toxin exposure for maximum sensitivity (Guven *et al.*, 1994; Mutwakil *et al.*, 1997).

Biosensor strains reporting on the induction of metallothioneins

Metallothioneins (MTs) are small (≈ 60 amino acid residues), cystein-rich proteins that are ubiquitous amongst eukaryote organisms (Kagi, 1991). They complex metals, reducing their bioavailability. MTs are induced by physiological concentrations of essential metals such as zinc and copper, but also strongly induced by nonessential metals, e.g. cadmium and mercury, suggesting that they have a dual role in metal homeostasis and detoxification (Palmiter, 1998). Additionally MTs also respond to heat-shock (Freedman *et al.*, 1993).

C. elegans has two metallothionein genes, *mtl-1* and *mtl-2*, encoding polypeptides that are respectively 75 and 63 amino acids long. The importance of *C. elegans* metallothionein genes for protection against cadmium has been demonstrated experimentally: reduction in brood size and lifespan caused by cadmium was magnified when the *mtl-1* or *mtl-2* genes were knocked-down by RNAi or when these genes were deleted (Swain *et al.*, 2004; Hughes and Sturzenbaum, 2007).

The expression of *mtl-1* and *mtl-2* genes has been characterized using surrogate *lacZ* and GFP reporters (Freedman *et al.*, 1993; Swain *et al.*, 2004) (Table 10.2). Expression driven by the *mtl-1* promoter is constitutive in the posterior bulb of the pharynx but extends to the intestine in the presence of cadmium and is stronger in larvae than adults. In contrast, the cadmium induced expression of *mtl-2::lacZ or mtl2::gfp* is not influenced by developmental stage and occurs exclusively in intestinal cells. Cadmium, nickel, or zinc all induced the *lacZ* strains whereas cadmium but not zinc was a strong inducer of the GFP marked strains. Although this is an apparent discrepancy, the *lacZ* and *gfp* strains are not directly comparable as the amplified promoter regions are of different size, with the *lacZ* strains containing a nuclear localization signal for improved reporter visibility (Cioci *et al.*, 2000; Swain *et al.*, 2004).

The applicability of MT-based biomonitors should be evaluated for each toxicant under consideration since, for example, cobalt does not induce MT expression or the JF2.1 reporter (Sunderman *et al.*, 1995; Cioci *et al.*, 2000).

Strain JF2.1 resulted from the chromosomal integration of the *mtl-2::lacZ* reporter (Fig. 10.2). JF2.1 is a sensitive monitor for metal exposure, including cadmium, zinc, mercury, nickel, copper, lead and silver (Cioci *et al.*, 2000).

Table 10.2. Transgenic strains that report on stress responses other than the heat-shock response. All of these strains are transcriptional reporters.

Strain	Promoter	Reporter gene	Expression of reporter	Stress	Medium
	mtl-1[a]	*lacZ* (plus SV40 NLS)	Intestinal cells of larvae; markedly reduced in adults. Un-induced controls: constitutive expression in pharynx	Cd^{2+}	Aqueous: S medium
JF2.1[a,b]	*mtl-2*	*lacZ* (plus SV40 NLS)	Nuclei of intestinal cells of postembryonic stages	Cd^{2+}, Hg^{2+}, Ni^{2+} and Zn^{2+}	Aqueous: S medium;[a] S basal pH 6.0 + cholesterol (exposure to Cd^{2+}, Hg^{2+});[b] K medium (Ni^{2+} Zn^{2+})[b]
JF9[c]	*cdr-1*	*lacZ* (plus SV40 NLS)	Nuclei of intestinal cells	Cd^{2+} Non-inducers: Pb^{2+}, Hg^{2+}, Cu^{2+}, Zn^{2+}, paraquat, juglone, Heat shock (33°C)	Aqueous: K medium
	mtl-1[d] *mtl-2*[d]	*gfp*	As described for *mtl-1::lacZ*[a] and *mtl-2::lacZ*[a]	Cd^{2+} Poor inducers: Cu^{2+} and Zn^{2+}	NGM agar with *E. coli*
	CYP35A2[e]	*gfp*	Whole intestine Un-induced controls: anterior and posterior intestine	PCB52	Aqueous: S complete with *E. coli*
	cyp-35A3[f] *dhs-23*[f] *cyp-14A3*[f]	*gfp*	Intestinal cells (and for *cyp-14A3* also hypodermis and head neurons) un-induced controls: vestigial	PCB52	NGM agar with *E. coli*
PE39[g]	*let-858*	*luc* (no peroxisome targeting sequence)	Constitutive in most cells	3,5-DCP,Pb^{2+}, Cu^{2+}, heat. Response: perturbation of energy balance	Aqueous: ddH_2O

[a]Freedman *et al.*, 1993; [b]Cioci *et al.*, 2000; [c]Liao *et al.*, 2002; [d]Swain *et al.*, 2004; [e]Menzel *et al.*, 2001; [f]Menzel *et al.*, 2007; [g]Lagido *et al.*, 2001.

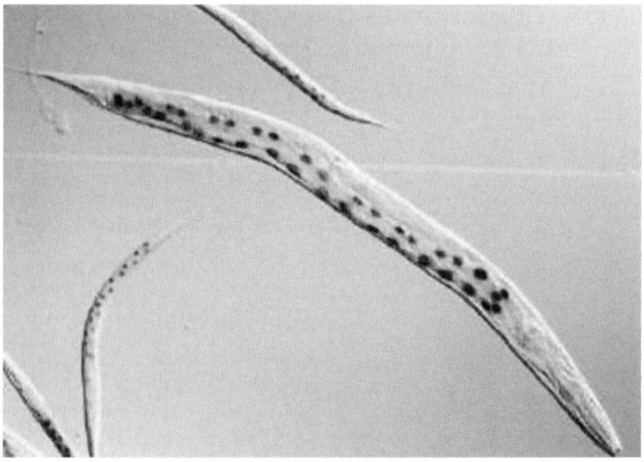

Fig. 10.2. A *C. elegans* strain that contains the *mtl-2::lacZ* transgene displays β-galactosidase activity in intestinal cell nuclei, following treatment with 100 μM $CdCl_2$ for 24 h. Reproduced from Cioci *et al.* (2000) with permission from Allen Press Publishing Services and Jonathan Freedman.

Biosensors reporting on the induction of novel stress promoters

Microarray technology (see Menzel *et al.*, Chapter 11, this volume) has facilitated the identification of genes which exhibit altered expression following xenobiotic exposure and, therefore, are thought to play a role in the response to the stress under investigation. Nevertheless, the significance of stress-induced changes in gene expression will depend on the downstream biological effects elicited (Feder and Walser, 2005). The biological function of stress responsive genes has been tested using loss of function mutants and RNAi (Menzel *et al.*, 2005, 2007; Cui *et al.*, 2007). Once a physiological role has been established, transgenic strains can be generated containing reporter genes under control of the regulatory regions of stress inducible genes.

The cadmium responsive gene *cdr-1* encodes a predicted 32 KDa integral membrane protein targeted to lysosomes, whose strength and sensitivity of induction is comparable to that of the *C. elegans* metallothioneins (Liao *et al.*, 2002). Inhibition of *cdr-1* by RNAi has resulted in impairment of growth and reproduction in the presence of cadmium, demonstrating its role in protection against cadmium. A transgenic strain (JF9) has been constructed where the *cdr-1* regulatory region drives the expression of a β-galactosidase fusion protein that accumulates in the nuclei (Liao *et al.*, 2002) (see Table 10.2). Following cadmium exposure, the fusion protein is transcribed exclusively in intestinal cells of all postembryonic stages of development but not in developing embryos. This expression pattern is in agreement with that observed for reporters of *hsp-16* or metallothionein expression following cadmium exposure. The gut is one of the main routes for exposure to toxicants in the worm and has been suggested to be the equivalent of the mammalian liver in terms of its function in detoxification mechanisms (Custodia *et al.*, 2001; An and Blackwell, 2003; Novillo *et al.*, 2005; Antebi, 2006). The JF9 transgenic strain was unique amongst strains based on metal responsive genes in that it was not induced by other common stresses such as transition metals (lead, mercury, copper and zinc), heat or oxidative stress (paraquat, juglone). Thus, JF9 appears to be exclusively a reporter for cadmium contamination.

Another recently identified cadmium responsive gene, *numr-1*, is induced by exposure to various metals but not by heat exposure, and confers resistance to metal stress (B.E. Tvermoes and J.F. Freedman, USA, personal communication). A *numr-1::gfp* translational reporter shows a basal level of constitutive expression and a dramatic increase in fluorescence in intestinal and pharyngeal nuclei in response to metal exposure.

Menzel *et al.* (2001) focused attention on the xenobiotically induced cytochrome P450 (CYP) genes. CYP genes encode heme-containing NADPH-dependent monooxygenases which catalyse the oxidative metabolism of many exogenous and endogenous compounds in other organisms (Nelson *et al.*, 1996; Nelson, 1999). Using known inducers of CYP genes, including an *ortho*-substituted, 2,2',5,5'-tetrachlorobiphenyl (PCB52), Lansoprazol and β-naphthoflavone, Menzel *et al.* (2001) observed strong induction of the subfamilies CYP35A, 35C and 29A. CYP35A2 was chosen for construction of a transgenic strain containing the whole promoter directly in front of *gfp*. Prior to induction by β-naphthoflavone, GFP fluorescence was limited to small areas of the anterior and posterior intestine. After induction, strong fluorescence was seen along the whole intestine in all larval stages as well as in adults, with at least a 10-fold increase over controls (Fig. 10.3).

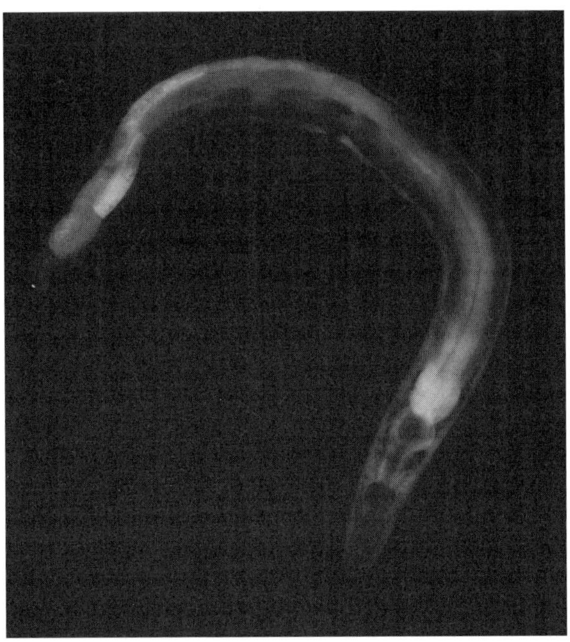

Fig. 10.3. *CYP-35A2::gfp* reporter strain displays intense green fluorescence when exposed to 1 mg L^{-1} β-naphthoflavone for 24 h and observed under fluorescence microscopy. Photograph courtesy of Ralph Menzel.

The polychlorinated biphenyl PCB52 induced three classes of genes strongly: CYP genes, short-chain dehydrogenases/reductases (SDR), and lipases. RNAi confirmed a biological effect of short-chain dehydrogenase genes, *dhs-2* and *dhs-23,* and of *cyp-14A3* or of *cyp-35A* genes, in terms of the brood size in presence of PCB52. The *cyp-35A3, cyp-14A3* and *dhs-23* promoters were selected for construction of *gfp* reporter strains. PCB52 elicited GFP expression in intestinal cells in these strains, and also in the hypodermis and head neurons of the *cyp-14A3::gfp* strain, (Menzel *et al.*, 2007). GFP expression was vestigial in all three strains prior to induction.

The above strains all exploit the induction of gene expression by specific stressors and this class of strains has much potential in terms of biomonitoring.

Biosensors reporting on perturbation of energy balance

An alternative approach to biomonitoring relies on the assessment of stress induced metabolic perturbation (Hollis *et al.* 2000; Lagido *et al.*, 2001; Shao *et al.*, 2002). In healthy cells, ATP consumption and generation are tightly coupled and cells strive to maintain stable ratios of ATP:ADP and of ATP: AMP (Hardie and Hawley, 2001). However, many forms of stress, including heat, metabolic poisoning, anoxia, starvation and oxidative stress perturb the energy balance in a variety of organisms, generally lowering cellular ATP levels (Findley *et al.*, 1983; Takeuchi and Kishimoto, 1983; Gasbarrini *et al.*, 1992; Houthoofd *et al.*, 2002; Tiwari *et al.*, 2002; Apfeld *et al.*, 2004). This triggers the activation of compensatory mechanisms in an attempt to restore the energy balance (Corton *et al.*, 1994; Zhang and Haldenwang, 2005). The type and severity of stress experienced will determine whether such balance is re-established and at what cost to the organism.

Metabolic effects occur prior to other effects that impact on growth, reproduction or survival, thus providing an early warning of hazards. Lagido *et al.* (2001) exploited the ability of luciferase to report on ATP levels at the whole multicellular organism level. The *luc* gene was placed under control of the constitutive *let-858* promoter and co-injected with the *rol-6* marker to produce a spontaneously integrated line, designated strain PE39. Toxicant and heat-stress induced changes in the metabolic state of *C. elegans* were measured *in vivo* by luminescence, after exogenous addition of luciferin in a suitable buffer (see 'Reporter genes' section). Luminescence provided a fast assay for sub-lethal effects of toxicity of heat stress, copper, and 3,5-DCP, as well as for acute toxicity of lead.

Expression levels of luciferase in PE39 are low, requiring that high worm densities are used in assays (1×10^4 worms/ml). The high worm density is a source of stress and may reduce bioavailability of toxins due to adsorption on to the worms' surface. New strains with increased luciferase expression contain a transcriptional fusion of luciferase and *gfp*, under control of the constitutive, generalized expression promoter, *sur-5* (Lagido *et al.*, 2008). Two lines were obtained after integration with EMS; one where the transgene integrated into the X chromosome, outcrossed 8 times with the wild-type

(PE255); and another into chromosome V, outcrossed 10 times (PE254). The luminescence of these two strains is considerably higher than that of PE39 and assays can be carried out in multiwell plates with a density of 300 worms/ml. The bioluminescence response of the PE254 and the PE255 strains following environmental stress has been characterized for cadmium. A 19 to 24h exposure to cadmium at sublethal concentrations (up to 30 μM) leads to decreased luminescence, with a 50% reduction observed at approximately 20–25 μM cadmium (Lagido *et al.*, unpublished data). Bioluminescence in response to (and during recovery from) the metabolic inhibitor sodium azide, specifically demonstrated the link between intracellular ATP levels and luminescence (Lagido *et al.*, 2008). Luminescence gives a rapid, real-time indication of metabolic status and is a technology that is suitable for the high-throughput testing required for biomonitoring. However these strains are not diagnostic of the particular stress encountered.

Characterization of the Response of Transgenic Biosensors

Detection limit, sensitivity and comparison with other endpoints

The usefulness of transgenic strains in biomonitoring is best evaluated by comparison with other physiological parameters commonly used as descriptors of toxicity in *C. elegans* (see Höss and Williams, Chapter 9, this volume). Table 10.3 gives an overview of the detection limits obtained with transgenic strains as compared to lethality, described as LC50. It can be seen from Table 10.3 that transgenic strains enable more sensitive and often faster stress detection than conventional LC50 testing. They respond to sublethal levels of stress which affect sensitive endpoints such as feeding, development and reproduction.

The most sensitive strains are those that report on induction of stressor specific responses. For example, the JF2.1 (*mtl2::lacZ*) strain responds to metal concentrations that are several orders of magnitude lower than the LC50 values (Cioci *et al.*, 2000), and the PCB52 and β-naphthoflavone biosensor strains respond to sublethal concentrations causing a 10 to 20 % reduction in reproduction (Menzel *et al.*, 2001, 2007).

Although 10–200 times less sensitive than *mtl-* (strain JF2.1) and *cdr-* (strain JF9) reporters (Cioci *et al.*, 2000; Liao *et al.*, 2002), the heat-shock response based systems show increased sensitivity over lethality testing. One extensively researched strain is PC72. Induction of β-galactosidase by metals, arsenite and fungicides was observed at sublethal concentrations affecting feeding and growth (Stringham and Candido, 1994; Candido and Jones, 1996; Jones *et al.*, 1996). Another *hsp* strain, CB4027 appears to have good sensitivity at 32 °C with reporter induction at toxicant concentrations that cause less than 10% lethality, however in the absence of heat-stress, PC72 performs better (Guven *et al.*, 1994; Mutwakil *et al.*, 1997).

Strain PE39, the bioluminescent metabolic reporter, also showed an advantage over lethality tests. Copper or lead concentrations that halved luminescence in 2h were respectively 2–13 and 4.5 times lower than the 24h

Table 10.3. Minimum stress level that elicits a response in a transgenic strain (detection limit). Lethality data is expressed as LC50 for comparison (wild-type strain unless otherwise stated).

Stress	Strain	Detection limit[a]	LC50[b]
Cadmium ($CdCl_2 \cdot 2.5H_2O$)	JF2.1	0.1 μM (24 h, 20°C, mixed stages)[c]	8042 μM[d]
	JF9	1 μM (24 h, 20°C, mixed stages)[g]	
	mtl-1::gfp *mtl-2::gfp*	2.5 μM (24 h) or >20 μM (1 h)[h]	8390 μM[e]
	CB4027	4.5 μM (7 h, 32°C, adults)[i]	
	PC71/PC72/PC73	25 μM (16 h, ?°C, L2/L3)[f]	
	PC161	43.8 μM (GFP expression) or 21.9 μM (*lacZ* expression) (16h, 24°C, ? stage)[j]	185 μM, PC71 strain[f]
	KC136	272.8 μM (5 h, 22°C, L3)[k]	
	KC334	54.6 μM (5 h, 22°C, L3)[l]	
	hsp-16.48::gfp	463.7 μM (24 h, 20°C, young adults)[m]	
Zinc ($ZnSO_4 \cdot 7H_2O$)	JF2.1	10 μM (24 h)[c]	
	CB4027	30.6 μM (7 h, 32°C, adults)[i]	
$ZnCl_2$			3089.2 μM[d] 2169 μM[e]
Mercury ($HgCl_2$)	JF2.1	0.2 μM (24 h, 20°C, mixed stages)[c]	49.9 μM[d]
	CB4027	2.5 μM (7 h, 32°C, adults)[i]	72 μM (24 h)[e] Approx. 47.9 μM[f]
	PC71/PC72/PC73	3.7 μM (24 h, ?°C, L2/L3)[f]	
	KC136	15 μM (5 h, 22°C, L3)[k]	
Copper ($CuCl_2$)	PC71/PC72/PC73	74.4 μM (8 h, ?°C, L2/L3)[f]	100.4 μM;[f] 346.2 μM[d] 1420 μM[e]
$CuSO_4 \cdot 5H_2O$	PE39	80 μM (2 h, 25°C, mixed stages)[n]	
	KC136	100 μM (5 h, 22°C, L3)[k]	100 μM (48 h)[k]
Lead ($Pb(NO_3)_2$)	PC71/PC72/PC73	30.2 μM (16 h, ?°C, L2/L3)[f]	622.6 μM;[d] 260 μM[e]
	PE39	50 μM (2 h, 25°C, mixed stages)[n]	
Arsenite ($NaAsO_2$)	PC71/PC72/PC73	50 μM (8 h, ?°C, L2/L3)[f]	2429.2[d];> 200 μM;[e] 170 μM (48 h) PC72[f]
Silver ($AgNO_3$)	CB4027	1.4/4.6 μM (7 h, 32°C, adults)[i]	46.4 μM[d]
Nickel ($NiSO_4 \cdot 6H_2O$)	JF2.1	10 μM (24 h, 20°C, mixed stages)[c]	$NiCl_2$: 49682 μM;[d] 63400 μM[e]
Captan	PC72	10 mg/l (48 h, 15°C, L3)[o]	
Captafol	PC72	1 mg/l (48 h, 15°C, L3)[o]	
Folpet	PC72	10 mg/l (48 h, 15°C, L3)[o]	

[a]Nominal concentrations, exposure conditions in brackets; [b]Wild-type strain and 24 h exposure, unless otherwise stated; [c]Cioci *et al.*, 2000; [d]Williams and Dusenberry, 1990; [e]Tatara *et al.*, 1997; [f]Stringham and Candido, 1994; [g]Liao *et al.*, 2002; [h]Swain *et al.*, 2004; [i]Guven *et al.*, 1994; [j]David *et al.*, 2003; [k]Chu and Chow, 2002; [l]Chu *et al.*, 2005; [m]Roh *et al.*, 2006; [n]Lagido *et al.*, 2001; [o]Jones *et al.*, 1996.

LC50 (Williams and Dusenberry, 1988, 1990; Tatara *et al.*, 1997; Lagido *et al.*, 2001). This was the quickest strain to respond, with a measurable response in as little as 5 min (Hollis *et al.*, 2001). Only one other strain came close, PC161, that could be sensitized through heat-shock to respond optimally after 2.5 h exposure to stress (David *et al.*, 2003).

The luciferase and *hsp*-reporter strains can potentially be made more sensitive by incorporation of specific mutations that decrease stress resistance (Broeks *et al.*, 1996; Koga *et al.*, 2000; Villanueva *et al.*, 2001; Vatamaniuk *et al.*, 2001, 2005; Liao *et al.*, 2002; Swain *et al.*, 2004; Chu *et al.*, 2005). As an example of this, the sensitivity of a heat shock response reporter strain was increased five times by incorporation of the *daf-16(m26); unc-75(e950)* double mutation (Chu *et al.*, 2005).

Concentration-effect and strength of response

In order to assess bioavailability of toxicants it is important to determine the range of linearity of the response to each toxicant for each particular biosensor strain.

Cadmium is generally a strong inducer of both *hsp* based strains and the *mtl* (JF2.1) and *cdr* (JF9) reporters. The response of the CB4027 (*hsp-70::lacZ)* strain to cadmium was concentration-dependent in the 1–16 mg/l range, at 32°C. Cadmium was a strong inducer whereas zinc, mercury, lindane, tin, TBT and manganese were all weak inducers of this strain (Guven *et al.*, 1994). PC72 (*hsp16.1::lacZ*) showed significant concentration-dependent 7 h response to 0–15 mg/l cadmium (Dennis *et al.*, 1997), also observed after 24 h exposure to cadmium spiked soil samples at concentrations up to 500 μg/g and in extracted soil water samples (Power and de Pomerai, 1999). PC161, another *hsp-16.1* based reporter strain expressing *lacZ::gfp*, displayed very significant linear trends between intensity of GFP fluorescence/ β-galactosidase activity and cadmium ($CdCl_2$) concentration in the 0–100 mg/l range, after 16 h (David *et al.*, 2003). The 24 h response of strain JF.9 (*cdr-1::lacZ*) to cadmium was concentration dependent up to 25 μM (Liao *et al.*, 2002) and of strain JF2.1 (*mtl-2::lacZ*) up to 100 μM (Cioci *et al.*, 2000).

Other metals elicit strong concentration-effect relationships in biosensor strains. Examples include: 0.5–50 mg/l silver in the CB4027 strain (Guven *et al.*, 1994); 0–5 mg/l mercury, 0–15 mg/l copper and 0–50 mg/l manganese or zinc in the PC72 strain (Dennis *et al.*, 1997); and 10–50 μM nickel, 10–200 μM zinc and up to 1 μM mercury in the JF2.1 strain (Cioci *et al.*, 2000).

Similar results have been found for non-metal stressors. The fungicides captan, captafol and folpet induced concentration dependent responses up to 50, 20 and 100 mg/l respectively, as assessed by the number of PC72 larvae that expressed β-galactosidase after 48 h (Jones *et al.*, 1996). Above the concentrations indicated the fungicides formed large crystals which could not be ingested. Up to 500 μg/ml, the fungicides mancozeb and maneb elicited concentration dependent induction of β-galactosidase in the strain PC72, with mancozeb resulting in a stronger response (Guven *et al.*, 1999).

Generally, strong inducers of the transgenes appear to give rise to a concentration-dependent response up to certain concentrations, above which expression of reporter gene declines (Guven *et al.*, 1994, 1999; Stringham and Candido, 1994; Cioci *et al.*, 2000). This may be accounted for by animals dying before a stress response is mounted, as was suggested for exposure of PC72 worms to 100 mg/l lead (Stringham and Candido, 1994) or to 5000 mg/l mancozeb or maneb (Guven *et al.*, 1999). Substantially lower than expected β-galactosidase activity was also detected with lindane and TBT, under conditions of high lethality (>20%) (Guven *et al.*, 1994). However, transgene expression decreased sharply from high at 50 mg/l Ag^+ to zero expression at 100 mg/l even though worms survived (Guven *et al.*, 1994). Possible explanations in this case include metal inhibition of enzymatic activity or metal inhibition of transcription/translation apparatus resulting in lower transgene expression (Mazidji *et al.*, 1992). These observations suggest that transgene induction assays should preferably be carried out under conditions of low lethality (e.g. less than 10–20%).

The luminescence responses of metabolic sensor strains are usually concentration-dependent and report both on sublethal effects and lethality. Following a 2 h exposure, bioluminescence was inversely related to 3,5-DCP concentration in the sublethal range. Copper (up to 1.6 mM) also elicited a concentration-effect relationship measured as a drop in bioluminescence. Although, unlike 3,5-DCP, copper caused both a partial inhibition of enzyme activity and lethality, these factors did not explain the observed reduction in luminescence which was due to a concentration-dependent decrease in the metabolic state of surviving worms (Lagido *et al.*, 2001). The decrease in bioluminescence following exposure to up to 0.5 mM lead was mainly due to increased lethality.

Variability in response

Variability in transgene induction

For consistent results in biomonitoring standardized experimental conditions need to be established. Even then, variability in the induction of the transgene was reported for the response of PC72 strain (Stringham *et al.*, 1992). Heat shock lead to a display of β-galactosidase activity in all worms, but the proportion varied in response to other forms of stress and was dependent on the duration of exposure and the concentration of the toxicant. After 96 h exposure to 50–200 μM arsenite, a classical inducer of the heat-shock response, the maximum number of worms that showed induction was 13% of the total. The maximum percentages of worms displaying β-galactosidase activity after exposure to transition metals were as follows (metal concentration and the exposure time that elicited the response are indicated in brackets): cadmium 27% (8 h, 100 μM), copper 38% (24 h, 50 mg/l), mercury 43% (48 h, 5 mg/l) and lead 16% (16 h, 10 mg/l) (Stringham and Candido, 1994). Feeding inhibition by metals was suggested as an explanation for the differences between heat and metal exposure but this was not supported by experimental data. Results may indicate that responses to heat and metals occur by different mechanisms, with HSPs other than HSP-16 preferentially induced by metals, or the primary mechanism for response to metals involving metallothioneins rather than the heat-shock response.

Individual variability in the intensity of the heat-shock response has also been reported. Isogenic heat-shocked worms of a strain containing the *hsp-16.2* promoter coupled to *gfp*, grown under conditions of minimum environmental heterogeneity, showed a wide, normally distributed variation in individual GFP fluorescence which was not heritable (Rea *et al.*, 2005). This was attributed to the inherent molecular variability amongst genetically identical individuals and its effects on the activation of the heat shock response.

Variability in the cells that showed induction was displayed by strains carrying *mtl-2::lacZ* as an extrachromosomal array, where only some of the intestinal cell nuclei contained β-galactosidase (mosaic pattern of expression). When the array was integrated, more than 90% of the resulting strain, JF2.1, expressed the reporter in all intestinal nuclei following stress (Cioci *et al.*, 2000).

Assay reproducibility

It is not surprising that stress exposure assays have a degree of inherent variability. Stress responses have evolved to cope with environmental change, and unnoticeable or small differences in the treatment of worm batches may affect the subsequent response to a tested stress. In assays with the PC161 or the PC72 strains, the variability in amount of reporter product detected between different bioassays is 20–40% (Dennis *et al.*, 1997; David *et al.*, 2003). In order to quantify variability, some studies have included a positive control consisting of 30 °C exposure and the results were expressed as a percentage of the positive control (Guven *et al.*, 1994, 1995; Mutwakil *et al.*, 1997). Slight variations in exposure temperature or worm density can elicit a partial heat-shock response and affect the degree of transgene expression in negative controls (Dennis *et al.*, 1997; Mutwakil *et al.*, 1997).

In order to deal with the inherent variability of induction bioassays, sets of multiple replicates within a single experiment are preferred to several repeat runs using a few replicates for each test condition (Dennis *et al.*, 1997). These considerations also apply to bioassays involving luminescence. In any given experiment using multiple replicates (e.g. n=8), variability in results as assessed by the standard error of the mean, is generally within 10% of the mean value, but occasionally higher (<20%). It is not uncommon for luminescence, expressed as a percentage of the control values, to vary by 20% between repeat experiments, although similar trends of concentration-response are observed.

As previously mentioned, standardization of assay conditions is essential to obtain meaningful results with transgenic biosensor strains. One important biological aspect that influences outcome is the developmental stage of the test population of nematodes, as different larval stages may vary in their stress response (Guven *et al.*, 1994; Stringham and Candido, 1994; Jones *et al.*, 1996). Nevertheless, mixed populations of worms including several developmental stages are easier to obtain and many studies have opted for these (Freedman *et al.*, 1993; Mutwakil *et al.*, 1997; Lagido *et al.*, 2001), at the cost of increasing variability between experiments.

Testing Environmental Samples

Published work with transgenic *C. elegans* as environmental monitors consists of proof of concept studies, focusing on single toxicants or simple mixtures. Yet environmental samples are often complex mixtures that are poorly characterized in terms of their individual components. Only a few pioneering studies have applied transgenic strains to these type of samples.

Strains PC72 and CB4027 have been used to assay water samples from Carnon river basin (England), an area polluted by heavy metals due to past mining activities (Mutwakil *et al.*, 1997). Water samples were taken from five locations ranging from a relatively unpolluted stream to heavily polluted mine-water. Toxicity of samples was ranked on the basis of metal concentrations and macroinvertebrate diversity. The relative toxicity of the water samples was reflected in the strength of induction of β-galactosidase. The least polluted sample induced a clear response even though it contained low metal concentrations that in isolation would not elicit a response (less than 2 μM arsenic, cadmium, aluminium, copper, manganese, zinc and iron). This synergistic effect of mixtures of toxicants was also observed in other studies (Dennis *et al.*, 1997; Power and de Pomerai, 1999; Chu and Chow, 2002), indicating that biosensor strains may be more sensitive monitors in field conditions than predicted from laboratory studies based on single stressors.

Sediment samples from another metal polluted river, the Shing Mun River (China) were tested using a sensitized biosensor strain (KC334) (Chu *et al.*, 2005). Various dilutions of the interstitial water extracted from sediment field samples resulted in pharyngeal *gfp* expression significantly higher than blank controls, giving a good indication of the toxicity of the river sediments.

Toxicity of soil samples has been addressed using the PC72 strain and the loamy sand Lufa 2.2 spiked with toxicants. Higher concentrations of cadmium are required to elicit a response in soil than in aqueous media e.g. 100 μg/g versus 8–16 μg/ml (Guven *et al.*, 1994; Power and de Pomerai, 1999), probably due to reduced bioavailability of cadmium in soil through binding to organic matter. When copper and cadmium are combined in soil toxicity assays, worms exhibited a stronger response than to cadmium alone, whereas no response was detectable for up to 250 μg/g copper. This was possibly due to copper displacing cadmium from soil binding sites resulting in increased cadmium availability in soil pore water. Interestingly, adding zinc and cadmium together had the opposite effect and reduced toxicity of cadmium (Power and de Pomerai, 1999). The authors suggested that this may have resulted from competition for entry, with zinc and cadmium mixtures reducing reporter responses and metal accumulation to 50% of the levels observed with either metal. Availability of the fungicide mancozeb was also reduced in soil, but the fraction entering the soil water compartment was sufficient to induce a mild stress response in PC72 (Guven *et al.*, 1999). These studies highlight the complex bioavailability issues that may arise in soil samples and the need for further studies with different soil types.

Another recent study has applied strains *hsp16.2::gfp* and *hsp16.48::gfp* to the detection of di(2-ethylhexyl)phthalate (DEHP). Although these strains are not very sensitive, the fluorescence signals from both strains increased noticeably with 24 h exposure to DEHP contaminated soils from landfill sites (Roh *et al.*, 2007).

The application of transgenic strains to biomonitoring under field conditions is now attracting increased attention, a trend that is set to continue.

Perspectives and Challenges

Exploiting available transgenic biosensors

The potential of available biosensor strains in real-life monitoring situations remains largely unexploited. As discussed above, only a few studies have attempted to use biosensors to assay field collected samples of soil, water or sediments containing complex mixtures of toxicants. The enhanced toxicity of mixtures of xenobiotics has not been assessed thoroughly. Additionally, the response of different biosensor strains under the same experimental conditions has not been addressed in the literature.

A systematic comparison of the performance of a range of transgenic biosensor strains under a variety of conditions, relative to other monitoring approaches, is required as a first step towards their use for routine monitoring of the environment. Another issue that remains to be addressed is the degree to which 'normal' natural environment conditions stress the biosensor strains. Can some of the most sensitive strains be induced under conditions naturally found in the environment? The distinction between response to natural variation or to hazardous conditions may be possible based on two criteria: the strength of the response in individual worms and the proportion of worms that show a response. There are good indications that these are generally dependent on the severity of the stress (see 'concentration-effect and strength of response' section). A high level of sensitivity is an advantage in monitoring, especially for toxicants that have the potential to bioaccumulate along the food chain. Situations in which direct applicability of transgenic strains is envisaged include the monitoring of potential sources of environmental contamination, such as discharges from industrial wastewater, sewage treatment plants, agricultural runoff, etc., or routine screening of drinking water supplies and of extracted water from soils. Biosensors can also be applied to soil samples as efficient extraction methods are available (Donkin and Dusenberry, 1993).

At present, the available transgenic biosensors include reporters of general stress and of specific stresses. The former include HSP promoter-based reporters and metabolic reporter strains, whereas the latter include many strains that exploit the activation of stress specific promoters such as metallotheionein promoters, CYP promoters, etc. It is unlikely that one particular strain will prove superior to all others in every situation, nevertheless if metal contamination is the major problem in a particular area, the use of a biosensor reporting on metallotheionein induction is likely to be the most appropriate. Perhaps the simultaneous use of two types of sensors will inform on the degree of stress experienced by living organisms, as well as the specific stress that is being inflicted upon them. For instance, the CYP reporter strain may indicate that an organic xenobiotic such as PCB52 is present in a sample, whereas a luciferase-based metabolic reporter may provide a direct measurement of the degree of stress the organism is under in real-time.

Cost-effectiveness is another important aspect in monitoring. This may be achieved through sophisticated technology, with automation of assays and response quantification. The COPAS (*C*omplex *O*bject *P*arametric *A*nalyser and *S*orter) BIOSORT dispenses exact numbers of *C. elegans*, at specific developmental stages, into a microplate format. Amongst other measurements it is able to determine growth rates, distribution of developmental stages, and is capable of performing the functions of a fluorometer, thus reporting on the level of reporter gene expression (*gfp* or *lacZ*). It can also be used in conjunction with a luminometer for luciferase-based assays.

Development of novel transgenic biosensors

C. elegans is an ideal model system to understand and exploit mechanisms of toxicity at a molecular level. Many useful transgenic strains may already be available amongst the *C. elegans* community as GFP fusions are routinely generated to describe the expression pattern of a gene. Indeed this analysis is now being carried out on a large scale, with transcriptional fusions being constructed for as many genes as possible (McKay *et al.*, 2003; http://elegans.bcgsc.bc.ca). Information about transgenic strains can be researched in wormbase (http://www.wormbase.org) under reagents for a particular gene. Most of these strains are not routinely tested under a range of environmental conditions and therefore remain untested as biosensors. Their suitability would largely depend on whether increased expression of the reporter gene, or a change in localization of the reporter protein, is detected under conditions of stress. Below, I will provide examples of stress mechanisms of direct relevance to toxicologists. Some of the novel strains being developed are shown in Table 10.4. An exhaustive exploration of this topic is beyond the scope of this chapter but it is hoped that this section will motivate readers to find out more.

A feature of stress responses is that they tend to be suppressed under normal growth conditions (Solomon *et al.*, 2004; Westerheide and Morimoto, 2005; Baumeister *et al.*, 2006). Depletion of insulin signalling (e.g. by lack of food), leads to activation of stress responses through the translocation of the

Table 10.4. Novel transgenic strains with potential value as biosensors. Biological processes/ genetic pathways where the respective genes are involved. Response to stress is also shown where reported.

Transgene	Biological process /genetic pathway	Response (stress)
daf-16::gfp[a]	Stress response, development, ageing/ IGF-like signalling; JNK signalling; TGF-ß signalling	Fluorescence localizes to nuclei (35°C/200μM juglone)
sod-3::gfp[b]	Oxidative stress response/DAF-16 target	↑ fluorescence (2.5 mM paraquat)
jnk-1::gfp[c]	Stress response, body movement coordination/JNK signalling	Not reported
mek-1::lacZ[d] *mek-1::gfp*[d]	Stress response, feeding/JNK signalling	Not reported
pmk-1::gfp[e]	Immunity, oxidative stress response/p 38 MAPK pathway	Not reported
skn-1::gfp[f,g]	Development, oxidative stress response/p 38 MAPK pathway	Fluorescence localizes to nuclei of intestinal cells (6.2 mM t-butyl hydroperoxide, 5 mM arsenite)
gcs-1::gfp[f,g,h]	Oxidative stress response/p 38 MAPK pathway, SKN-1 target	↑ fluorescence (6.2 mM t-butyl hydroperoxide, 5 mM arsenite)
nhr-8::gfp[i]	Stress response: Nuclear receptor	
mrp-2:: gfp[j]	Stress response: ABC transporter	
gst-1::gfp[e]	Oxidative stress response/ DAF-16 target gene	
sip-1::gfp[e]	Heat shock response/DAF-16 target gene	

[a]Henderson and Johnson, 2001; [b]Essers *et al.*, 2005; [c]Oh *et al.*, 2005; [d]Koga *et al.*, 2000; [e]strains being developed as part of the US National Toxicology Programme (J.F. Freedman, USA, 2007, personal communication); [f]An *et al.* 2005 ; [g]Inoue *et al.*, 2005; [h]An and Blackwell, 2003; [i]Lindblom *et al.*, 2001; [j]Dupuy *et al.*, 2007.

unphosphorylated DAF-16 transcription factor to the nucleus where it activates transcription of stress response genes and represses other genes involved in growth and reproduction. Regulatory regions of DAF-16 target genes are good candidates for construction of biosensors (cf. McElwee *et al.*, 2003; Murphy *et al.*, 2003). One example of this is a transgenic *sod-3::gfp* strain which displayed a strong increase in fluorescence when worms were exposed to paraquat (Essers *et al.*, 2005). As part of the US National Toxicology Program into alternative animal testing and toxicity monitoring, several *C. elegans* transgenic strains are being created that are based on promoters of DAF-16 dependent stress responsive genes, such as metallotheioneins, various *hsp* genes (e.g. small heat shock protein *sip-1*) and gluthatione S-transferase (*gst-1*) gene (J.F. Freedman, USA, 2007, personal communication). A *daf-16::gfp* strain exists that displays cytoplasmic GFP expression under normal growth, or nuclear localization when insulin signalling is off (Lee *et al.*, 2001) and during stress exposure (Henderson and Johnson, 2001). These strains may have much potential in terms of biomonitoring.

Other interesting strains are based on *mek-1* and *pmk-1*, which are part of MAPK signalling pathways activated by stress. MEK-1 functions upstream of JNK-1 which phosphorylates DAF-16 thus modulating its nuclear translocation (Oh *et al.*, 2005). Loss of function of *mek-1* or of *jnk-1* leads to hypersensitivity to copper and cadmium (Koga *et al.*, 2000; Villanueva *et al.*, 2001). More recently a novel signalling protein, KEL-8, has been identified that protects *C. elegans* from cadmium toxicity in a *mek-1* dependent manner (Cui *et al.*, 2007). To my knowledge no *kel-8* based biosensor strains have been constructed so far. Reporter genes have been placed under control of the *jnk-1* and *mek-1* promoters (Koga *et al.*, 2000; Oh *et al.*, 2005) but these stains have not yet been tested in biomonitoring.

The *pmk-1* gene operates in the p38 MAPK pathway, thought to be the ancestral immune pathway in insects, nematodes and vertebrates. More recently the role of *pmk-1* in oxidative stress has been discovered, via the nuclear localization of the transcription factor, SNK-1, in intestinal cells in response to arsenite (Inoue *et al.*, 2005). SKN-1 controls the expression of *gcs-1* in the intestine. This gene encodes a phase II detoxification enzyme involved in the biosynthesis of the critical reducing agent glutathione (GSH) (An *et al.*, 2005). A strain in which the *gcs-1* promoter region was linked to *gfp* was expressed at high levels in the intestine in response to oxidative stress, but at low levels under normal conditions (An and Blackwell, 2003). Oxidative stress is a major etiological factor involved in numerous human diseases, including diabetes, atherosclerosis, cancer, and ageing. Therefore, the *skn-1::gfp*, *gcs-1::gfp* and *pmk-1::gfp* reporter strains are potentially very interesting biosensors.

Nuclear receptors (NR) are transcription factors involved in the response to xenobiotics. The *C. elegans* nuclear hormone receptor, *nhr-8*, is proposed to function in xenobiotic metabolism since loss of function *nhr-8(ok186)* mutants are more sensitive than wild-type to the toxins colchicine and chloroquine (Lindblom *et al.*, 2001). A *nhr-8::gfp* reporter strain indicated that *nhr-8* expression was exclusively intestinal, as observed with many other genes involved in stress responses. The use of this strain as a biosensor has not been reported. The function of NRs is as yet not well characterized in *C. elegans*. Nevertheless, expression of 25 out of 284 *C. elegans* NRs were altered after exposure to the steroids progesterone, estradiol, or cholesterol, with each steroid affecting a different subset of NR genes (Novillo *et al.*, 2005). Therefore NR based biosensors may find an application in the detection of endocrine disrupting chemicals in the environment.

The above are examples of transgenic strains which may prove valuable in environmental monitoring and toxicology. The identification of novel stress responsive genes by microarray studies, as well as RNAi screens and classical genetics approaches, will lead to the construction of novel biosensor strains. We are currently investigating the combined approach of RNAi, stress bioassays and assessment of worm physiology by luminescence as a means of probing gene function in the stress response.

Transgenic *C. elegans* strains are unique biosensors that will bridge different levels of biological information: not only will transgenic *C. elegans* report on whole organism effects as representatives of an ecologically important phylum, they can also be used to detect environmental insults that activate genetic pathways associated with human diseases. Interesting times are ahead in the field.

Acknowledgements

Many thanks to: King L. Chow (Hong Kong University of Science and Technology) for useful input at the early stages of this project; Anne L. Glover, Debbie McLaggan and Valerie Undrill (University of Aberdeen, UK) for useful comments and proof reading of the manuscript and helpful discussions.

References

An, J.H. and Blackwell, T.K. (2003) SKN-1 links *C. elegans* mesendodermal specification to a conserved oxidative stress response. *Genes and Development* 17, 1882–1893.

An, J.H., Vranas, K., Lucke, M., Inoue, H., Hisamoto, N., Matsumoto, K. and Blackwell, T.K. (2005) Regulation of the *Caenorhabditis elegans* oxidative stress defense protein SKN-1 by glycogen synthase kinase-3. *Proceedings of the National Academy of Sciences of the U S A* 102, 16275–16280.

Antebi, A. (2006) Nuclear hormone receptors in *C. elegans*. The *C. elegans* Research Community, *Wormbook*, doi/10.1895/wormbook.1.64.1, http://www.wormbook.org (accessed 20 October 2008).

Apfeld, J., O'Connor, G., McDonagh, T., DiStefano, P.S. and Curtis, R. (2004) The AMP-activated protein kinase AAK-2 links energy levels and insulin-like signals to lifespan in *C. elegans*. *Genes and Development* 18, 3004–3009.

Baumeister, R., Schaffitzel, E. and Hertweck, M. (2006) Endocrine signaling in *Caenorhabditis elegans* controls stress response and longevity. *Journal of Endocrinology* 190, 191–202.

Boulin, T., Etchberger, J.F., Obert, O. and Hughes, H. Reporter gene fusions (April 5, 2006), Wormbook, ed. The *C. elegans* Research Community, Wormbook, doi/10.1895/wormbook.1.106.1, http://www.wormbook.org (accessed 20 October 2008).

Broeks, A., Gerrard, B., Allikmets, R., Dean, M. and Plasterk, R.H. (1996) Homologues of the human multidrug resistance genes MRP and MDR contribute to heavy metal resistance in the soil nematode *Caenorhabditis elegans*. *Embo Journal* 15, 6132–6143.

Candido, E.P. and Jones, D. (1996) Transgenic *Caenorhabditis elegans* strains as biosensors. *Trends in Biotechnology* 14, 125–129.

Chalfie, M., Tu, Y., Euskirchen, G., Ward, W.W. and Prasher, D.C. (1994) Green fluorescent protein as a marker for gene expression. *Science* 263, 802–805.

Chu, K.W. and Chow, K.L. (2002) Synergistic toxicity of multiple heavy metals is revealed by a biological assay using a nematode and its transgenic derivative. *Aquatic Toxicology* 61, 53–64.

Chu, K.W., Chan, S.K. and Chow, K.L. (2005) Improvement of heavy metal stress and toxicity assays by coupling a transgenic reporter in a mutant nematode strain. *Aquatic Toxicology* 74, 320–332.

Cioci, L.K., Qiu, L. and Freedman, J.H. (2000) Transgenic strains of the nematode *Caenorhabditis elegans* as biomonitors of metal contamination. *Environmental Toxicology and Chemistry* 19, 2122–2129.

Corton, J.M., Gillespie, J.G. and Hardie, D.G. (1994) Role of the AMP-activated protein kinase in the cellular stress response. *Current Biology* 4, 315–324.

Cui, Y., McBride, S.J., Boyd, W.A., Alper, S. and Freedman, J.H. (2007) Toxicogenomic analysis of *Caenorhabditis elegans* reveals novel genes and pathways involved in the resistance to cadmium toxicity. *Genome Biology* 8, R122.

Culetto, E. and Sattelle, D.B. (2000) A role for *Caenorhabditis elegans* in understanding the function and interactions of human disease genes. *Human Molecular Genetics* 9, 869–877.

Custodia, N., Won, S.J., Novillo, A., Wieland, M., Li, C. and Callard, I.P. (2001) *Caenorhabditis elegans* as an environmental monitor using DNA microarray analysis. *Annals of the New York Academy of Sciences* 948, 32–42.

Daniells, C., Duce, I., Thomas, D., Sewell, P., Tattersall, J. and de Pomerai, D. (1998) Transgenic nematodes as biomonitors of microwave-induced stress. *Mutation Research* 399, 55–64.

David, H.E., Dawe, A.S., de Pomerai, D.I., Jones, D., Candido, E.P. and Daniells, C. (2003) Construction and evaluation of a transgenic *hsp16-GFP-lacZ Caenorhabditis elegans* strain for environmental monitoring. *Environmental Toxicology and Chemistry* 22, 111–118.

de Wet, J.R., Wood, K.V., DeLuca, M., Helinski, D.R. and Subramani, S. (1987) Firefly luciferase gene: structure and expression in mammalian cells. *Molecular and Cellular Biology* 7, 725–737.

Dennis, J.L., Mutwakil, M.H.A.Z., Lowe, K.L. and de Pomerai, D.I. (1997) Effects of metal ions in combination with a non-ionic surfactant on stress responses in a transgenic nematode. *Aquatic Toxicology* 40, 37–50.

Dolphin, C.T. and Hope, I.A. (2006) *Caenorhabditis elegans* reporter fusion genes generated by seamless modification of large genomic DNA clones. *Nucleic Acids Research* 34, e72.

Donkin, S.G. and Dusenbery, D.B. (1993) A soil toxicity test using the nematode *Caenorhabditis elegans* and an effective method of recovery. *Archives of Environmental Contamination and Toxicology* 25, 145–151.

Dupuy, D., Bertin, N., Hidalgo, C.A., Venkatesan, K., Tu, D., Lee, D., Rosenberg, J., Svrzikapa, N., Blanc, A., Carnec, A., Carvunis, A.R., Pulak, R., Shingles, J., Reece-Hoyes, J., Hunt-Newbury, R., Viveiros, R., Mohler, W.A., Tasan, M., Roth, F.P., Le Peuch, C., Hope, I.A., Johnsen, R., Moerman, D.G., Barabasi, A.L., Baillie, D. and Vidal, M. (2007) Genome-scale analysis of *in vivo* spatiotemporal promoter activity in *Caenorhabditis elegans. Nature Biotechnology* 25, 663–668.

Essers, M.A., de Vries-Smits, L.M., Barker, N., Polderman, P.E., Burgering, B.M. and Korswagen, H.C. (2005) Functional interaction between beta-catenin and FOXO in oxidative stress signaling. *Science* 308, 1181–1184.

Evans, T.C. (ed.) (2006) Transformation and microinjection. The *C. elegans* Research Community, *Wormbook*, doi/10.1895/wormbook.1.108.1, http://www.wormbook.org. (accessed 20 October 2008).

Feder, M.E. and Walser, J.C. (2005) The biological implications of transcriptomics in elucidating stress and stress responses. *Journal of Evolutionary Biology* 18, 901–910.

Findley, R.C., Gillies, R.J. and Schulman, R.G. (1983) *In vivo* phosphorus-31 nuclear magnetic resonance reveals lowered ATP during heat shock of *Tetrahymena. Science* 219, 1223–1225.

Fire, A. (1986) Integrative transformation of *Carnorhabditis elegans. EMBO Journal* 5, 2673–2680.

Fire, A., Harrison, S.W. and Dixon, D. (1990) A modular set of *lacZ* fusion vectors for studying gene expression in *Caenorhabditis elegans. Gene* 93, 189–198.

Freedman, J.H., Slice, L.W., Dixon, D., Fire, A. and Rubin, C.S. (1993) The novel metallothionein genes of *Caenorhabditis elegans*. Structural organization and inducible, cell-specific expression. *Journal of Biological Chemistry* 268, 2554–2564.

Gasbarrini, A., Borle, A.B., Farghali, H., Bender, C., Francavilla, A. and Van Thiel, D. (1992) Effect of anoxia on intracellular ATP, Na+i, Ca2+i, Mg2+i, and cytotoxicity in rat hepatocytes. *Journal of Biological Chemistry* 267, 6654–6663.

Graves, A.L., Boyd, W.A. and Williams, P.L. (2005) Using transgenic *Caenorhabditis elegans* in soil toxicity testing. *Archives of Environmental Contamination and Toxicology* 48, 490–494.

Guven, K., Duce, J.A. and de Pomerai, D.I. (1994) Evaluation of a stress-inducible transgenic nematode strain for rapid aquatic toxicity testing. *Aquatic Toxicology* 29, 119–137.

Guven, K., Duce, J.A. and de Pomerai, D.I. (1995) Calcium moderation of cadmium stress explored using a stress-inducible transgenic strain of *Caenorhabditis elegans. Comparative Biochemistry and Physiology* 110C, 61–70.

Guven, K., Power, R.S., Avramides, S., Allender, R. and de Pomerai, D.I. (1999) The toxicity of dithiocarbamate fungicides to soil nematodes, assessed using a stress-inducible transgenic strain of *Caenorhabditis elegans. Journal of Biochemical and Molecular Toxico*logy 13, 324–333.

Hardie, D.G. and Hawley, S.A. (2001) AMP-activated protein kinase: the energy charge hypothesis revisited. *Bioessays* 23, 1112–1119.

Heim, R., Cubitt, A.B. and Tsien, R.Y. (1995) Improved green fluorescence. *Nature* 373, 663–664.

Henderson, S.T. and Johnson, T.E. (2001) daf-16 integrates developmental and environmental inputs to mediate aging in the nematode *Caenorhabditis elegans. Current Biology* 11, 1975–1980.

Hollis, R.P., Killham, K. and Glover, L.A. (2000) Design and application of a biosensor for monitoring toxicity of compounds to eukaryotes. *Applied and Environmental Microbiology* 66, 1676–1679.

Hollis, R.P., Lagido, C., Pettitt, J., Porter, A.J., Killham, K., Paton, G.I. and Glover, L.A. (2001) Toxicity of the bacterial luciferase substrate, n-decyl aldehyde, to *Saccharomyces cerevisiae* and *Caenorhabditis elegans. FEBS Letters* 506, 140–142.

Hong, M., Kwon, J.Y., Shim, J. and Lee, J. (2004) Differential hypoxia response of hsp-16 genes in the nematode. *Journal of Molecular Biology* 344, 369–381.

Houthoofd, K., Braeckman, B.P., Lenaerts, I., Brys, K., De Vreese, A., Van Eygen, S. and Vanfleteren, J.R. (2002) No reduction of metabolic rate in food restricted *Caenorhabditis elegans. Experimental Gerontology* 37, 1359–1369.

Hughes, S. and Sturzenbaum, S.R. (2007) Single and double metallothionein knockout in the nematode *C. elegans* reveals cadmium dependent and independent toxic effects on life history traits. *Environmental Pollution* 145, 395–400.

Hutter, H. (2003) Extracellular cues and pioneers act together to guide axons in the ventral cord of *C. elegans. Development* 130, 5307–5318.

Inoue, H., Hisamoto, N., An, J.H., Oliveira, R.P., Nishida, E., Blackwell, T.K. and Matsumoto, K. (2005) The C. elegans p38 MAPK pathway regulates nuclear localization of the transcription factor SKN-1 in oxidative stress response. *Genes and Development* 19, 2278–2283.

Jones, D., Stringham, E.G., Babich, S.L. and Candido, E.P. (1996) Transgenic strains of the nematode *C. elegans* in biomonitoring and toxicology: effects of captan and related compounds on the stress response. *Toxicology* 109, 119–127.

Kagi, J.H. (1991) Overview of metallothionein. *Methods in Enzymology* 205, 613–26.

Keller, G.A., Gould, S., Deluca, M. and Subramani, S. (1987) Firefly luciferase is targeted to peroxisomes in mammalian cells. *Proceedings of the National Academy of Sciences of the USA* 84, 3264–3268.

Kelly, W.G. and Fire, A. (1998) Chromatin silencing and the maintenance of a functional germline in *Caenorhabditis elegans. Development* 125, 2451–2456.

Koga, M., Zwaal, R., Guan, K.L., Avery, L. and Ohshima, Y. (2000) A *Caenorhabditis elegans* MAP kinase kinase, MEK-1, is involved in stress responses. *Embo Journal* 19, 5148–5156.

Kogai, T., Kanamoto, Y. and Brent, G.A. (2003) The modified firefly luciferase reporter gene (*luc+*) but not *Renilla* luciferase is induced by all-trans retinoic acid in MCF-7 breast cancer cells. *Breast Cancer Research and Treatment* 78, 119–126.

Kramer, J.M., French, R.P., Park, E.C. and Johnson, J.J. (1990) The *Caenorhabditis elegans rol-6* gene, which interacts with the *sqt-1* collagen gene to determine organismal morphology, encodes a collagen. *Molecular and Cellular Biology* 10, 2081–2089.

Lagido, C., Pettitt, J., Porter, A.J., Paton, G.I. and Glover, L.A. (2001) Development and application of bioluminescent *Caenorhabditis elegans* as multicellular eukaryotic biosensors. *FEBS Letters* 493, 36–39.

Lagido, C., Pettitt, J., Flett, A. and Glover, L.A. (2008) Bridging the phenotipic gap: real-time assessment of mitochondrial function and metabolism of the nematode *Caenorhabditis elegans*. *BMC Physiology* 8.

Lee, R.Y., Hench, J. and Ruvkun, G. (2001) Regulation of *C. elegans* DAF-16 and its human ortholog FKHRL1 by the *daf-2* insulin-like signaling pathway. *Current Biology* 11, 1950–1957.

Liao, V.H., Dong, J. and Freedman, J.H. (2002) Molecular characterization of a novel, cadmium-inducible gene from the nematode *Caenorhabditis elegans*. A new gene that contributes to the resistance to cadmium toxicity. *Journal of Biological Chemistry* 277, 42049–42059.

Lindblom, T.H., Pierce, G.J. and Sluder, A.E. (2001) A *C. elegans* orphan nuclear receptor contributes to xenobiotic resistance. *Current Biology* 11, 864–868.

Link, C.D., Cypser, J.R., Johnson, C.J. and Johnson, T.E. (1999) Direct observation of stress response in *Caenorhabditis elegans* using a reporter transgene. *Cell Stress and Chaperones* 4, 235–242.

Maechler, P., Wang, H. and Wollheim, C.B. (1998) Continuous monitoring of ATP levels in living insulin secreting cells expressing cytosolic firefly luciferase. *FEBS Letters* 422, 328–332.

Mazidji, C.N., Koopman, B., Bitton, G. and Neita, D. (1992) Distinction between heavy metal and organic toxicity using EDTA chelation and microbial assays. *Environmental Toxicology and Water Quality* 7, 339–353.

McElwee, J., Bubb, K. and Thomas, J.H. (2003) Transcriptional outputs of the *Caenorhabditis elegans* forkhead protein DAF-16. *Aging Cell* 2, 111–121.

McKay, S.J., Johnsen, R., Khattra, J., Asano, J., Baillie, D.L., Chan, S., Dube, N., Fang, L., Goszczynski, B., Ha, E., Halfnight, E., Hollebakken, R., Huang, P., Hung, K., Jensen, V., Jones, S.J., Kai, H., Li, D., Mah, A., Marra, M., McGhee, J., Newbury, R., Pouzyrev, A., Riddle, D.L., Sonnhammer, E., Tian, H., Tu, D., Tyson, J.R., Vatcher, G., Warner, A., Wong, K., Zhao, Z. and Moerman, D.G. (2003) Gene expression profiling of cells, tissues, and developmental stages of the nematode *C. elegans*. *Cold Spring Harbor Symposium on Quantitative Biology* 68, 159–169.

Mello, C. and Fire, A. (1995) DNA transformation. In: Epstein, H.F. and Shakes, D.C. (ed.) *Caenorhabditis elegans: Modern Biological Analysis of an Organism*. Academic Press, San Diego, California, pp. 451–482.

Menzel, R., Bogaert, T. and Achazi, R. (2001) A systematic gene expression screen of *Caenorhabditis elegans* cytochrome P450 genes reveals CYP35 as strongly xenobiotic inducible. *Archives of Biochemistry and Biophysics* 395, 158–168.

Menzel, R., Rodel, M., Kulas, J. and Steinberg, C.E. (2005) CYP35: xenobiotically induced gene expression in the nematode *Caenorhabditis elegans*. *Archives of Biochemistry and Biophysics* 438, 93–102.

Menzel, R., Yeo, H.L., Rienau, S., Li, S., Steinberg, C.E. and Sturzenbaum, S.R. (2007) Cytochrome P450s and short-chain dehydrogenases mediate the toxicogenomic response of PCB52 in the nematode *Caenorhabditis elegans*. *Journal of Molecular Biology* 370, 1–13.

Miller, D.M., III, Desai, N.S., Hardin, D.C., Piston, D.W., Patterson, G.H., Fleenor, J., Xu, S. and Fire, A. (1999) Two-color GFP expression system for *C. elegans*. *Biotechniques* 26, 914–918.

Murphy, C.T., McCarroll, S.A., Bargmann, C.I., Fraser, A., Kamath, R.S., Ahringer, J., Li, H. and Kenyon, C. (2003) Genes that act downstream of DAF-16 to influence the lifespan of *Caenorhabditis elegans*. *Nature* 424, 277–283.

Mutwakil, M.H., Reader, J.P., Holdich, D.M., Smithurst, P.R., Candido, E.P.M., Jones, D., Stringham, E.G. and de Pomerai, D.I. (1997) Use of stress-inducible transgenic nematodes as biomarkers of heavy metal pollution in water samples from an English river system. *Archives of Environmental Contamination and Toxicology* 32, 146–153.

Nelson, D.R. (1999) Cytochrome P450 and the individuality of species. *Archives of Biochemistry and Biophysics* 369, 1–10.

Nelson, D.R., Koymans, L., Kamataki, T., Stegeman, J.J., Feyereisen, R., Waxman, D.J., Waterman, M.R., Gotoh, O., Coon, M.J., Estabrook, R.W., Gunsalus, I.C. and Nebert, D.W. (1996) P450 superfamily: update on new sequences, gene mapping, accession numbers and nomenclature. *Pharmacogenetics* 6, 1–42.

Novillo, A., Won, S.J., Li, C. and Callard, I.P. (2005) Changes in nuclear receptor and vitellogenin gene expression in response to steroids and heavy metal in *Caenorhabditis elegans. Integrative and Comparative Biology* 45, 61–71.

Oh, S.W., Mukhopadhyay, A., Svrzikapa, N., Jiang, F., Davis, R.J. and Tissenbaum, H.A. (2005) JNK regulates lifespan in *Caenorhabditis elegans* by modulating nuclear translocation of forkhead transcription factor/DAF-16. *Proceedings of the National Academy of Sciences of the USA* 102, 4494–4499.

Palmiter, R.D. (1998) The elusive function of metallothioneins. *Proceedings of the National Academy of Sciences of the USA* 95, 8428–8430.

Power, R.S. and de Pomerai, D.I. (1999) Effect of single and paired metal inputs in soil on a stress-inducible transgenic nematode. *Archives of Environmental Contamination and Toxicology* 37, 503–511.

Praitis, V., Casey, E., Collar, D. and Austin, J. (2001) Creation of low-copy integrated transgenic lines in *Caenorhabditis elegans. Genetics* 157, 1217–1226.

Rea, S.L., Wu, D., Cypser, J.R., Vaupel, J.W. and Johnson, T.E. (2005) A stress-sensitive reporter predicts longevity in isogenic populations of *Caenorhabditis elegans. Nature Genetics* 37, 894–898.

Roh, J.Y., Lee, J. and Choi, J. (2006) Assessment of stress-related gene expression in the heavy metal-exposed nematode *Caenorhabditis elegans*: a potential biomarker for metal-induced toxicity monitoring and environmental risk assessment. *Environmental Toxicology and Chemistry* 25, 2946–2956.

Roh, J.Y., Jung, I.H., Lee, J.Y. and Choi, J. (2007) Toxic effects of di(2-ethylhexyl)phthalate on mortality, growth, reproduction and stress-related gene expression in the soil nematode *Caenorhabditis elegans. Toxicology* 237, 126–133.

Schneider, D.A. and Gourse, R.L. (2004) Relationship between growth rate and ATP concentration in *Escherichia coli*: a bioassay for available cellular ATP. *Journal of Biological Chemistry* 279, 8262–8268.

Shaner, N.C., Campbell, R.E., Steinbach, P.A., Giepmans, B.N., Palmer, A.E. and Tsien, R.Y. (2004) Improved monomeric red, orange and yellow fluorescent proteins derived from *Discosoma* sp. red fluorescent protein. *Nature Biotechnology* 22, 1567–1572.

Shao, C.Y., Howe, C.J., Porter, A.J. and Glover, L.A. (2002) Novel cyanobacterial biosensor for detection of herbicides. *Applied and Environmental Microbiology* 68, 5026–5033.

Shim, J., Im, S.H. and Lee, J. (2003) Tissue-specific expression, heat inducibility, and biological roles of two *hsp16* genes in *Caenorhabditis elegans. FEBS Letters* 537, 139–145.

Snutch, T.P. and Baillie, D.L. (1984) A high degree of DNA strain polymorphism associated with the major heat shock gene in *Caenorhabditis elegans. Molecular and General Genetics* 195, 329–335.

Solomon, A., Bandhakavi, S., Jabbar, S., Shah, R., Beitel, G.J. and Morimoto, R.I. (2004) *Caenorhabditis elegans* OSR-1 regulates behavioral and physiological responses to hyperosmotic environments. *Genetics* 167, 161–170.

Stinchcomb, D.T., Shaw, J.E., Carr, S.H. and Hirsh, D. (1985) Extrachromosomal DNA transformation of *Caenorhabditis elegans. Molecular and Cellular Biology* 5, 3484–3496.

Stringham, E.G. and Candido, E.P. (1994) Transgenic *hsp16-lacZ* strains of the soil nematode *Caenorhabditis elegans* as biological monitors of environmental stress. *Environmental Toxicology and Chemistry* 13, 1211–1220.

Stringham, E.G., Dixon, D.K., Jones, D. and Candido, E.P. (1992) Temporal and spatial expression patterns of the small heat shock (*hsp16*) genes in transgenic *Caenorhabditis elegans. Molecular Biology of the Cell* 3, 221–233.

Sunderman, F.W., Jr., Plowman, M.C., Kroftova, O.S., Grbac-Ivankovic, S., Foglia, L. and Crivello, J.F. (1995) Effects of teratogenic exposures to Zn2+, Cd2+, Ni2+, Co2+, and Cu2+ on metallothionein and metallothionein-mRNA contents of *Xenopus* embryos. *Pharmacology and Toxicology* 76, 178–184.

Swain, S.C., Keusekotten, K., Baumeister, R. and Sturzenbaum, S.R. (2004) C. elegans metallothioneins: new insights into the phenotypic effects of cadmium toxicosis. *Journal of Molecular Biology* 341, 951–959.

Takeuchi, Y. and Kishimoto, U. (1983) Changes of adenine nucleotide levels in *Chara* internodes during metabolic inhibition. *Plant and Cell Physiology* 24, 1401–1409.

Tatara, C.P., Newman, M.C., McCloskey, J.T. and Williams, P.L. (1997) Predicting relative metal toxicity with ion characteristics: *Caenorhabditis elegans* LC50. *Aquatic Toxicology* 39, 279–290.

Tiwari, B.S., Belenghi, B. and Levine, A. (2002) Oxidative stress increased respiration and generation of reactive oxygen species, resulting in ATP depletion, opening of mitochondrial permeability transition, and programmed cell death. *Plant Physiology* 128, 1271–1281.

Vatamaniuk, O.K., Bucher, E.A., Ward, J.T. and Rea, P.A. (2001) A new pathway for heavy metal detoxification in animals. Phytochelatin synthase is required for cadmium tolerance in *Caenorhabditis elegans. Journal of Biological Chemistry* 276, 20817–20820.

Vatamaniuk, O.K., Bucher, E.A., Sundaram, M.V. and Rea, P.A. (2005) CeHMT-1, a putative phytochelatin transporter, is required for cadmium tolerance in *Caenorhabditis elegans. Journal of Biological Chemistry* 280, 23684–23690.

Villanueva, A., Lozano, J., Morales, A., Lin, X., Deng, X., Hengartner, M.O. and Kolesnick, R.N. (2001) *jkk-1* and *mek-1* regulate body movement coordination and response to heavy metals through *jnk-1* in *Caenorhabditis elegans. Embo Journal* 20, 5114–5128.

Westerheide, S.D. and Morimoto, R.I. (2005) Heat shock response modulators as therapeutic tools for diseases of protein conformation. *Journal of Biological Chemistry* 280, 33097–33100.

Williams, P.L. and Dusenbery, D.B. (1988) Using the nematode *Caenorhabditis elegans* to predict mammalian acute lethality to metallic salts. *Toxicology and Industrial Health* 4, 469–478.

Williams, P.L. and Dusenbery, D.B. (1990) A promising indicator of neurobehavioral toxicity using the nematode *Caenorhabditis elegans* and computer tracking. *Toxicology and Industrial Health* 6, 425–440.

Zhang, S. and Haldenwang, W.G. (2005) Contributions of ATP, GTP, and redox state to nutritional stress activation of the *Bacillus subtilis* sigmaB transcription factor. *Journal of Bacteriology* 187, 7554–7560.

11 *Caenorhabditis elegans* and Chemical Environmental Stressors: from DNA Microarrays and Data Analysis to Genomic Data of True Ecotoxicological Relevance

RALPH MENZEL,[1] CHRISTIAN E.W. STEINBERG[1] AND STEPHEN R. STÜRZENBAUM[2]

[1]*Department of Biology, Freshwater and Stress Ecology, Humboldt University, Berlin, Germany;* [2]*School of Biomedical and Health Sciences, King's College, London, UK*

Introduction

One principle focus of the post-genomic era is to translate the vast amounts of information that have been generated by large-scale sequencing projects into units that aid our understanding of genome function and biological processes in general and in relation to environmental triggers and/or particular stressors. High-throughput technologies, such as DNA microarrays, are able to profile the expression of thousands of genes in a single experiment and in consequence, identify transcripts that are significantly up- or down-regulated. In recent years DNA microarray technology has been used to explore changes in gene expression following exposure to environmental pollutants and natural environmental chemical stressors. This allows a greater understanding of whole genome expression in response to chemical stressors and, in turn, of ecological and toxicological modes of action. The terms 'ecogenomics' (Chapman, 2001) and 'ecotoxicogenomics' (Snape *et al.*, 2004) in their broadest sense, encompass not only transcript profiling (*transcriptomics*), but also determine protein composition (*proteomics*) and metabolic constituents (*metabolomics*) of a cell. This chapter will focus on profiling genome-wide changes in gene expression following exposure to environmental chemical stressors using DNA microarrays. Throughout the chapter the terms 'ecotoxicological' and 'ecotoxicology' comprise both man-made and natural chemical environmental stressors. It may be trivial, but is certainly worth mentioning: the presence of natural endogenic and exogenic chemical stressors has been instrumental for, and in fact driven the development of stress defence systems, such as the antioxidant or biotransformation systems, expression of stress proteins or metal-binding proteins. Consequently, anthropogenic chemical stress, though sometimes severe or even lethal, is one of several stressors that impact on organisms. The purpose of ecogenomics is to provide an insight into the physiological status of organisms and decipher responses and interactions of organisms with the environment and to one another. Ultimately, it is assumed that the interactions between species and their environments can be understood in a similar fashion as the complex interactions of genes and proteins at the cellular level (Chapman, 2001).

Transcription is the initial step in gene expression, thus, a transcriptional response can give an indication of cellular mechanisms that are affected by a pollutant and in consequence provide a sensitive starting point to assess ecological and toxicological (=ecotoxicological) responses. However, each application depends on basic assumptions (Oberemm *et al.*, 2005), including that all ecotoxicologically relevant effects are indeed accompanied by alterations in gene expression. Likewise, the specific pattern of expression changes induced by a chemical stress is expected to be similar for all chemicals with similar toxic modes of action, and thus can be used to categorize any kind of individual chemical and mixtures into different modes of action. This notion forms the basis for predictive ecotoxicology. Several studies have validated classification systems (Burczynski *et al.*, 2000) and large amounts of gene expression data are accumulating in dedicated databases, such as Chemical Effects in Biological Systems (CEBS) (Waters *et al.*, 2003).

Gene expression measured by DNA microarrays is believed to provide a more comprehensive, sensitive and characteristic insight into toxicity than typical toxicological parameters such as morphological changes, altered reproductive capacity or mortality (Hamadeh *et al.*, 2001; Menzel *et al.*, 2005a). In addition to these classical ecotoxicological parameters, ecotoxicogenomics is a powerful tool that unravels mechanistic processes, reveals novel modes of action, and provides the opportunity to get a dynamic picture of biological systems. This in turn offers the ability to comprehensively dissect different states of biological activities in cells, tissues or whole organisms. Changes in gene expression typically occur very rapidly following exposure. Ecotoxicogenomics aims to develop new predictive models for identifying environmental or human health hazards and precise and fast molecular biomarkers of exposure to natural as well as man-made chemical stress. In contrast, changes in phenotype or life cycle parameters may take days, weeks or even months to develop.

Key aims of ecotoxicogenomics include:

- The characterization of the modes of action of known chemical environmental stresses.
- The identification of the modes of action of previously uncharacterized chemical stresses with the intention to correlate the observed molecular signatures with well-characterized pollutant-induced expression patterns.

- The assessment of stress-specific gene expression patterns as a biomarker of chemical exposure.
- The identification of chemical specific, species overlapping gene expression patterns.
- The characterization of complex mixtures of natural and anthropogenic chemicals.
- The examination of chronic versus acute effects as well as low-concentration versus high-concentration exposures.

Transcript Profiling using DNA Microarrays

Several sequential steps are needed for the successful profiling of DNA transcripts exploiting microarray technology (see Fig. 11.1). The underlying concept has been reviewed numerous times and readers interested in a more detailed explanation are referred to Yue *et al.* (2001) or Allison *et al.* (2006). Rather than recap the literature, the following paragraphs aim to summarize the most crucial points in designing and evaluating DNA microarray experiments.

Experimental design

Good experimental design and implementation are probably the single most important steps in any microarray experiment. Rigour and consistency in data analysis and output format enables the building of reference data sets ultimately facilitating the comparison of microarray output across different

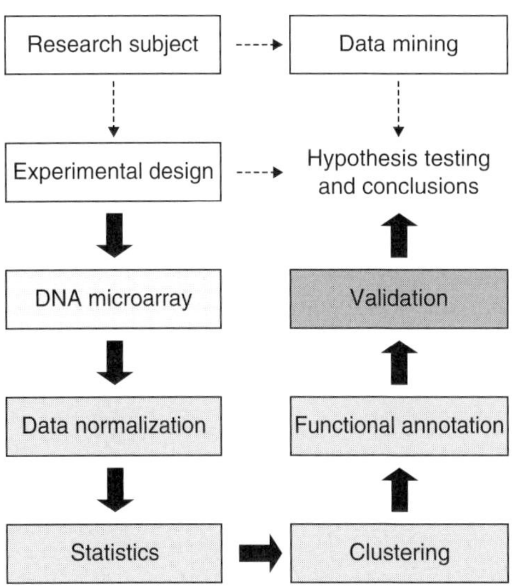

Fig. 11.1. Main steps involved in a DNA microarray analysis.

laboratories and possibly platforms and taxa. The emphasis should always be on the quality, not quantity of data, a challenge that is easily neglected. Replications are the key to reliable results and, depending on the experimental platform and sample availability and variability, at least five experimental repeats are desirable. Similarly, gene expression can be sensitive to environmental change; therefore test organisms should be maintained under well-controlled, homogenous laboratory conditions. In addition to this preconditioning issue, it might be necessary to analyse several time points or developmental stages before reaching a conclusion that is sound and statistically robust. Taking these points into account will avoid, or at least minimize, the interference and variability caused by physical parameters (e.g. temperature, light, medium structure), during development or effects that are dependent on ploidy, gender, age or nutritional status. One of the most challenging issues is the identification of changes in gene expression that represent an adaptive response to external stimuli. These may have no ecotoxicological significance or increased risk and are distinct from early stages of disease progression (Pennie *et al.*, 2000). For example, some changes in gene expression might be a reflection of a non-specific and fully reversible response to stress and thus have no biological consequence. Technically, this signifies the required complexity of DNA microarray experiments. Dose-response studies are a powerful tool to differentiate adaptive versus toxic responses and to establish ecotoxicogenomic thresholds that need to be exceeded prior to the initiation of the cascade of molecular responses leading to adverse effects.

Microarray analysis

DNA microarrays are an organized arrangement of multiple DNA probes fixed on an immobilized surface. Several thousand DNA spots, usually ranging from 60 to 100 μm in diameter, can be spotted onto a single slide. Each spot represents a gene or DNA sequence which is either unique or spotted in multiples. DNA microarrays can be categorized into two general types: oligonucleotide arrays and PCR-amplicon arrays. Oligonucleotide arrays comprise short immobilized DNA sequences (approximately 25–80 DNA bases) designed as complementary probes to specific genes or genome sequences. The advantage over PCR-amplicon arrays is their increased specificity and greater discrimination between closely related gene sequences. However, issues of signal intensity and cost are the limiting factors of oligonucleotide arrays. In consequence, it is important that microarrays are tailored towards the need of the specific experimental design. For example, to facilitate the identification of novel genes that are responsive to a chemical stressor requires arrays that ideally incorporate the maximum number of spots, including those of unknown function. In contrast, a focused approach may be useful when determining the potential of a pollutant to induce a particular toxic response or pathway. In this case it may be more appropriate to use an array comprising a well-defined, small number of carefully selected genes of known function. A typical DNA microarray experiment is shown in Fig. 11.2.

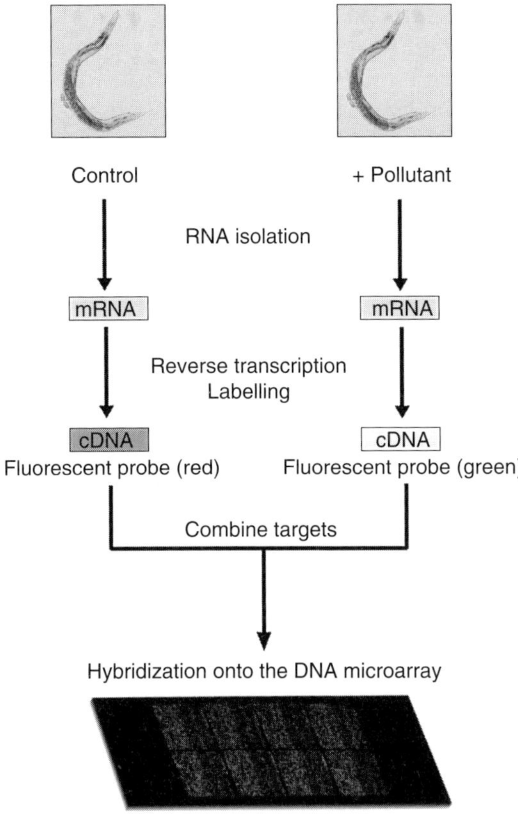

Fig. 11.2. Diagram of a typical dual-colour DNA microarray experiment. RNA is isolated from control and pollutant exposed organisms. The respective mRNAs are reversed transcribed to cDNA and labelled with two different fluorophores. Both samples are combined and allowed to hybridize to the DNA on a microarray slide. The probes spotted onto the microarray surface can be oligonucleotides, cDNA or small fragments of PCR products that correspond to mRNAs. Finally, the microarray is scanned in a laser scanner and the relative intensities of each fluorophore are quantified to identify up- and down-regulated genes.

Normalization

Stoichiometry dictates that the relative quantity of nucleic acid bound to a probe is a function of concentration. In other words, high signal intensity is caused by an increase in hybridization which, in turn, implies elevated expression levels. However, intrinsic variations in DNA microarrays may stem from deviations across replicated slides, inconsistent mRNA levels, selective incorporation of fluorescent dyes, hybridization conditions, differences in scanning parameters and technical inaccuracies. These non-biological variations represent significant challenges and highlight the importance and necessity of data normalization, a downstream application that balances array-to-array variability prior to statistical analysis. Numerous normalization procedures have been developed and their strengths and weaknesses have been exquisitely reviewed by Altman (2005). Normalization by centring on a small number of spots may reduce array effects, but tends to introduce considerable variation in the results. Focusing on the bulk of spots on the array is less variable but requires that the expression of the majority of transcripts remain static. An alternative approach is based on the use of external controls, which are also known as spikes or spike-in controls. These spikes are RNA molecules that are synthetically produced by *in vitro* transcription and added in defined amounts to the biological samples. Each spike is designed to hybridize only to the external control probe on the target array and form an ideal benchmarking system for microarray technology.

Finally, dye-swap designs using biological replicates (rather than technical replicates) with alternate fluorescent dyes have been used to improve efficiency of DNA array analysis, and in some cases even un-normalized data sets have been useful where only the very basic filtering methods (e.g. flag setting 'present/absent') have been utilized.

Statistical analysis

Analysis of DNA microarrays poses a large number of statistical challenges. Because of the importance of statistical analysis (Nadon and Shoemaker, 2002; Allison *et al.*, 2006), at least two crucial points need to be emphasized. First, although a large change in gene expression may reflect a response of biological importance, a small change does not imply the opposite. A simple numerical listing of up- and down-regulating is therefore not a sufficient evaluation of DNA array data. Second, the large number of genes that are typically analysed in concert highlights a further problem. Even if the statistical *P*-value, assigned to a given change in gene expression, indicates that it is extremely unlikely to have occurred at random, the large number of genes simultaneously analysed must, by definition, include false positives (type I errors). Multiple test correction methods are considered to be conservative, and as a consequence result in a high false negative (type II) error rate. Several approaches are available to address these problems, the most common being False Discovery Rate (FDR) analysis, introduced by Benjamini and Hochberg (1995).

Clustering

Even following rigorous statistical analysis, it is often difficult to identify important expression patterns. Numerous mathematical procedures have been introduced to aid in the analysis of pattern and cluster analysis (Aldenderfer and Blashfield, 1984). Clustering is a so-called unsupervised method, which does not use any kind of sample classification as input but asks the simple question: 'Are there unexpected, but biologically meaningful patterns that exist in the data?' Clustering can identify similar expression patterns within a cohort of genes and may suggest a similar responses cascade or common regulatory circuit. Two of the most widely used unsupervised approaches are hierarchical clustering (Weinstein *et al.*, 1997; Eisen *et al.*, 1998) and *k*-means partitional clustering (Soukas *et al.*, 2000). Hierarchical algorithms find successive clusters using previously established clusters, whereas partitional algorithms determine all clusters at once. Distilled to its simplest form, both algorithms use certain features of the data and a rule for determining relationships to a group of genes (or samples) that share similar patterns of expression. An analysis usually requires two steps. The first step measures relationships (e.g. distance or similarity) of gene expression in a pairwise fashion. The second step clusters the genes based on the measures derived in the first step. In most cases, results are displayed as a scatter plot, where good clustering will result in non-overlapping clusters.

Functional annotation

Gene ontology (GO) annotation represents a controlled vocabulary that describes species overlapping attributes of gene products. The use of functional annotation offers the possibility to study coordinated changes that control specific pathways or processes. Currently three GO sets are available: the 'Molecular Function' of a gene product; its role in multi-step 'Biological Processes'; and its localization to 'Cellular Components'. Each GO term is assigned to one of the three subdivisions and consists of a unique alphanumerical identifier, a common name, and a definition. GO term analysis facilitates uniform queries across DNA microarray experiments and will highlight trends that have been previously characterized. Several programs have been developed to assist with the functional GO annotation. The Database for Annotation, Visualization and Integrated Discovery, DAVID, from the National Institute of Allergy and Infectious Diseases (NIAID) is free of charge, and easy to use. It functionally groups transcripts using predefined levels of stringency. Further suitable tools are listed in Table 11.1.

A weakness of functional annotation and pathway mapping is that it fails to identify novel relationships and coordinated processes, or gene regulatory networks. A simple network consists of one or more input signalling pathways, regulatory proteins that integrate the input signals, several target genes and the RNA and proteins produced from those target genes. Such networks often include dynamic feedback loops that provide for further regulation of network architecture and output. New gene regulator network models may help to gain better insight into this important topic (Álvarez-Buylla *et al.*, 2007).

Validation

Even if the well designed and accurately performed microarray experiment leads to robust and reproducible results, it is common practice to confirm a subset of the data. Typically, Northern blot or quantitative RT-PCR analyses are performed, and in some cases examined and confirmed at the protein level. A comprehensive validation is not feasible and would defeat the purpose of a DNA microarray study. It should be added that this kind of data confirmation produces an important but only technical validation and repeating expression measurements with a different quantitative technique does not provide additional information considering the original goal.

Table 11.1. DNA microarray databases with gene expression information and software tools for analysing data sets.

Name	Website (http://…)	Description
Databases:		
Gene Expression Omnibus (GEO)	www.ncbi.nlm.nih.gov/projects/geo	Searchable database of microarray experiments submitted to NCBI
ArrayExpress	www.ebi.ac.uk/arrayexpress	Searchable database of microarray experiments submitted to EBI
Tox-MIAMExpress	www.ebi.ac.uk/tox-miamexpress/	Allows to link biological parameters (clinical observation, clinical pathology, pathology and histopathology) with gene expression data
ArrayTrack	www.fda.gov/nctr/science/centers/toxicoinformatics/ArrayTrack/	Provides an integrated solution for managing, analysing, and interpreting microarray gene expression data associated with pharmaco- or toxico- genomics studies
Chemical Effects in Biological Systems (CEBS)	cebs.niehs.nih.gov/	House data from multiple gene expression platforms, protein expression, and changes in low molecular weight metabolite levels aligned by their detailed toxicological context
Comparative Toxicogenomics Database (CTD)	ctd.mdibl.org/	Interactions between chemicals and genes/proteins in diverse organisms are curated from literature and integrated with diverse data (human diseases, references, sequences, vertebrate and invertebrate organisms, and the gene ontology) to facilitate environmental health research
Annotation/Pathway tools:		
SOURCE	source.stanford.edu	Database unification tool designed to facilitate analysis of genomic data from various organisms, developed at Stanford University
Database for Annotation, Visualization and Integrated Discovery (DAVID)	david.abcc.ncifcrf.gov/	A comprehensive set of functional annotation tools for investigators to understand biological meaning behind large lists of genes, developed by NIAID
AffyMiner	bioinfo-srv1.awh.unomaha.edu/affyminer/ /	For associating genes of interest of Affymetrix GeneChip microarray replicate data with gene annotation and gene ontology information

GoMiner	discover.nci.nih.gov/gominer/	To identify the biological processes, functions and components represented in microarray gene lists
GenMAPP	www.GenMAPP.org	Permits mapping of gene expression data to known and user-defined biological pathways and functional groupings
Selected data analysis software:		
Significance Analysis of Microarrays (SAM)	www-stat.stanford.edu/~tibs/SAM	Excel-based tool for statistical analysis. Balances type I and type II error rate with FDR
Gene Set Enrichment Analysis (GSEA)	www.broad.mit.edu/tools/software.html	A computational method that determines which in a given set of genes show statistically significant differences between two biological states
Bioconductor	www.bioconductor.org/	Open source and open development software project for the analysis and comprehension of genomic data
Cluster/TreeView	rana.lbl.gov/EisenSoftware.htm	Clustering and visualization tools from the Eisen lab at UC-Berkeley
GenePattern	www.broad.mit.edu/tools/software.html	For the analysis of molecular profiles, e.g. genome-wide microarray expression signatures by clustering
rVista	rvista.dcode.org/	Compares gene sequences and identifies conserved response elements
Ingeneue	rusty.fhl.washington.edu/ingeneue/	Constructs and analyses models of genetic networks

Data mining of existing data sets and standards

An important extension to performing DNA microarray experiments is to (re-) analyse experiments that are available from public databases or the literature. Indeed, data mining is a fruitful concept as it not only prevents superfluous replication, but allows detailed comparisons with related or even apparently unrelated research projects and may well substantiate knowledge and the mechanistic understanding of gene function. Several useful databases for data mining are presented in Table 11.1. The largest and most comprehensive of these repositories is the Gene Expression Omnibus (GEO) at NCBI.

In the past, the quality of data sets has been variable. Some experiments provided replicates while others did not, some entries included raw array data and others only processed data. Keeping the complexity of DNA microarray assays in mind, the Microarray and Gene Expression Data (MGED) society has developed guidelines and standards for the user community. Now widely accepted, MIAME[1] (*Minimum Information About a Microarray Experiment*) defines the minimum quantity and quality of information that is required to interpret and verify results (Brazma *et al.*, 2001). Indeed, many journals now require that MIAME compliant microarray data is uploaded to public databases as a prerequisite for publication in an effort to ensure unhindered public access to primary, good quality, data (Ball *et al.*, 2004).

Ecotoxicogenomics Using DNA Microarrays in *Caenorhabditis elegans*

The long life cycle of mammalian model systems limits the identification and evaluation of chronic or delayed effects. Even short-lived lower vertebrates (fish and amphibians) have long life cycles compared to many invertebrates. The nematode *Caenorhabditis elegans* is not only known for the availability of exquisite molecular genetic tools, but also satisfies all typical criteria necessary to function as an ecotoxicological test organism (see Höss and Williams, Chapter 9, this volume and Lagido, Chapter 10, this volume). Moreover, due to its completely sequenced genome, *C. elegans* is ideally suited for ecotoxicogenomics. For example, several studies have demonstrated that specific genes can affect lifespan via the modulation of signal transduction, transcription, and mitochondrial function (Lund *et al.*, 2002; Wang and Kim, 2003; Golden and Melov, 2004). This chapter focuses on approaches addressing issues in toxico- and ecotoxicogenomics using *C. elegans* as the test organism. The presentation of the examples follows the chronology of the publications rather than a classification of the chemical stressors.

Response to steroid hormones

In order to assist in the identification of possible endocrine disrupting chemicals (EDC), Custodia *et al.* (2001) studied the global gene expression patterns of *C. elegans* to three natural steroid hormones, testosterone, 17β-estradiol, and progesterone. The nematodes were exposed for 3–4 days to the hormones and all experiments were performed in triplicate using Stanford Microarray facilities and the Stanford Microarray Database (SMD).

The study showed that the *C. elegans* genome is widely responsive to the three vertebrate steroid hormones, but the overall pattern of gene expression differed among the three treatments. Whilst a serial dilution (10 μmol, 100 nmol, and 1 nmol) of progesterone did not result in concentration-dependent responses, 17β-estradiol (10 μmol) and progesterone (100 nmol) induced most dramatic effects. 17β-estradiol and progesterone induced expression of 1496 and 1077 genes, respectively. Moreover, 349 genes were found to be down-regulated by 17β-estradiol, and 939 by progesterone (the cut-off was set to 2.6 fold up/down regulation). In particular, genes involved in xenobiotic metabolism and responses (e.g. glutathione *S*-transferases, cytochromes P450, metallothionein and several HSP70 proteins) were shown to be upregulated. Transcription of several vitellogenin genes was increased by 17β-estradiol but decreased by progesterone. Although the expression patterns could be partially confirmed by Western blot experiments, the overall results remained preliminary, but at the very least substantiated that *C. elegans* is a useful model for screening EDC.

Response to ethanol

Three years later it was shown that *C. elegans* responds to acute ethanol exposure in a similar way to humans, mice, and fruit flies (Kwon *et al.*, 2004). This toxicogenomic work provides an excellent example of the evolution of DNA microarray quality. Seven independent experiments were performed in which nematodes were exposed to ethanol for different time periods, namely four sets of 6 h exposure, two sets of 15 min exposure, and one set of 30 min exposure. As before, the DNA microarray experiment was performed by SMD and the data sets were uploaded to their database. Statistical analysis was performed as follows: first genes that showed at least a twofold average change in transcript level were identified, and then subjected to both Bayes's posterior odds test (Tusher *et al.*, 2001) and t-test (slightly modified for small sample comparison) criteria by running SAM (Significance Analysis of Microarrays, see also Table 11.1).

The results yielded 230 genes, which were deemed to be significantly affected by ethanol. Fifty randomly selected genes were confirmed by Northern analysis. While the ethanol response of some of these genes was significantly changed at the early time points (indicative of genes that may be involved in the ethanol stress response), the majority of changes were observed at 6 hours, representing the physiological consequence of the ethanol exposure. The early response genes included many heat shock protein genes, confirming that the concentration used (7% ethanol) acts as a strong stressor to the nematodes. By analysing the promoter regions of the early response genes, a regulatory element (TCTGCGTCTCT) was identified that was necessary for the expression of a subset of ethanol response genes. All ethanol responsive genes were plotted onto the gene expression map (Kim *et al.*, 2001) which highlighted that many down-regulated genes were found to be clustered on mountain 8, corresponding to an enrichment of genes expressed in the intestine. The descriptions of the corresponding genes imply that intestinal proteins play a role in the response to ethanol stress and may probably reflect a global lowering of metabolism in the nematodes.

Following the identification of over 200 ethanol response genes, the precise molecular genetic basis that explains why animals become uncoordinated by ethanol exposure remained elusive. Nevertheless, many of the genes identified have human homologs and their expression profile may be used for an index of the extent of ethanol exposure or the state of alcohol-related diseases of a given individual.

Response to humic substances

Besides investigating the impact of xenobiotics on organisms it is also of substantial interest to determine the influence of naturally occurring substances that may chemically stress organisms. Humic materials are complex organic molecules constituting the most abundant source of natural organic matter (NOM) in freshwater and soil ecosystems (Steinberg and Münster, 1985; Thurman, 1985). Over the past years they have been shown to interfere with biological systems, e.g. via the induction of biotransformation enzymes, the inhibition of photosynthetic oxygen release (in freshwater plants), the production of internal oxidative stress or through the feminization of fish and amphibians (Steinberg, 2003; Steinberg *et al.*, 2006). In *C. elegans*, the majority of humic materials tested were found to stimulate reproduction (Höss *et al.*, 2001). Moreover, it has been shown that humic substances act as powerful chemoattractants in *C. elegans* (Menzel *et al.*, 2005b). The corresponding DNA microarray experiments (Menzel *et al.*, 2005b) were based on the Array-Ready Oligo Set™ for the *C. elegans* genome, version 1.1 (Qiagen), which were spotted on glass slides. Five independent experiments were performed in which nematodes were exposed to 6.4 mg/l NOM (a spring 2002 reverse osmosis isolate from Lake Fuchskuhle, Germany, which contains humic substances and, as minor components, polycarbohydrates and peptides) and 7.6 mg/l of the synthetic humic-like substance HS1500 (an auto-oxidation product of polyphenols with alkyl bridges and dominant aromatic and quinoide structures, sourced from Sopar Pharma GmbH, Mannheim, Germany), respectively. In both experiments synchronized L1 nematodes were cultured in liquid medium and treated with humic material until young adult stage (typically after 48 h). To identify significantly regulated genes, the data were analysed with the GeneSight (BioDiscovery Inc., USA) software using a threshold of ≥2.0-fold transcript abundance difference between the reference and experimental samples (student t-test; $P<0.05$) and followed by hierarchal clustering methods (Eisen *et al.*, 1998) using the same software. A validation of the global DNA microarray approach was confirmed by semi-quantitative RT-PCR of six randomly selected genes from different gene families.

The humic-like substance HS1500 had the strongest effect, marked by an up-regulation of 554 genes, compared to 240 in NOM Fuchskuhle. Moreover, 685 genes were found to be down-regulated by HS1500, and 350 by NOM Fuchskuhle. Because the main intention of this work was to focus on similarities between both humic materials, a combined analysis was preferred, which resulted in 84 up- and 210 down-regulated genes. Taking into account that both humic materials sources were able to attract the nematode, it is remarkable that 18 genes coding for putative chemoreceptors were found to be differentially regulated. Furthermore, a limited number of enzymes involved in biotransformation were found to be differentially expressed, namely cytochromes P450, glutathione *S*-transferases, UDP-glucuronosyltransferases as well as acyl- and glycosyltransferases. Humic materials are likely to produce oxidative stress caused by H_2O_2 and other reactive oxygen species (Pflugmacher *et al.*, 2006), which in turn may generate oxidative stress in the nematode. Interestingly, one of the most responsive genes induced was F32A5.2, a putative peroxidase. Similarly, the nuclear hormone receptor SDC-1 was found to be significantly over-expressed in response to both humic materials; *sdc-1* encodes a zinc-finger protein of the C_2H_2-type that affects egg laying and vulval development. Although the mode of action for all these effects still remains hazy, humic materials seemingly act as a messenger that impact on essential bodily functions and may have the potential to act as an environmental signal. This study on humic material as natural chemical stressors and simultaneous attractants to *C. elegans* led to a subsequent paper which tried to provide an answer to the following question: 'Why does an organism actively seek a stressful environment?' Although the exposed organisms have to spend energy to rid the exposed chemicals they also have the potential to expand their lifespan (Steinberg *et al.*, 2007) – the ecological consequence remains obscure.

Response to a polychlorinated biphenyl (PCB52)

Polychlorinated biphenyls (PCBs) are ubiquitous organic chemicals that pose a global environmental health problem. PCB52, an *ortho*-substituted 2,2',5,5'-tetrachlorobiphenyl is a non-coplanar congener; the mechanisms whereby it acts appear very complex and continue to be a matter of discussion. Menzel *et al.* (2007) performed whole genome DNA microarray experiments using synchronized *C. elegans* populations exposed to 5 mg/l PCB52 for 48 h, and harvested at the young adult stage. Independent replicates (4 controls and 4 PCB52 exposures) were prepared and analysed using two technical repeats utilizing the Array-Ready Oligo Set™ for the *C. elegans* genome, version 1.1 (Qiagen). The MIAME compliant data have been deposited in NCBI's Gene Expression Omnibus (GEO) and are accessible through the accession number GSE7125. Besides using the GeneSight software (BioDiscovery Inc., USA) to identify significantly up- and down-regulated genes (threshold of ≥2.0-fold transcript abundance difference; student t-test, $P < 0.05$), this work integrated gene enrichment analysis using the web-accessible program DAVID to obtain a ranking of functional categories based on co-occurrence with sets of genes in the gene lists ($P<0.05$). The DNA microarray approach was confirmed by semi-quantitative RT-PCRs of 34 up-regulated genes selected from three gene families of special interest, the expression status of 30 genes could be validated.

Overall, the DNA microarray experiment identified 1158 up-regulated and 560 down-regulated genes, which significantly responded to PCB52 treatment. The results from the gene ontology screen were analysed further by aligning all identified gene classes with each other to reveal over-represented and overlapping functional gene classes. This additional approach resulted in clearly arranged Venn diagrams (Fig. 11.3). The up-regulated genes encompassed in particular the processes of oxidative catabolism, in addition transcriptional regulation was found to include numerous members, followed by the esterase/lipase/thioesterase class, lipid binding,

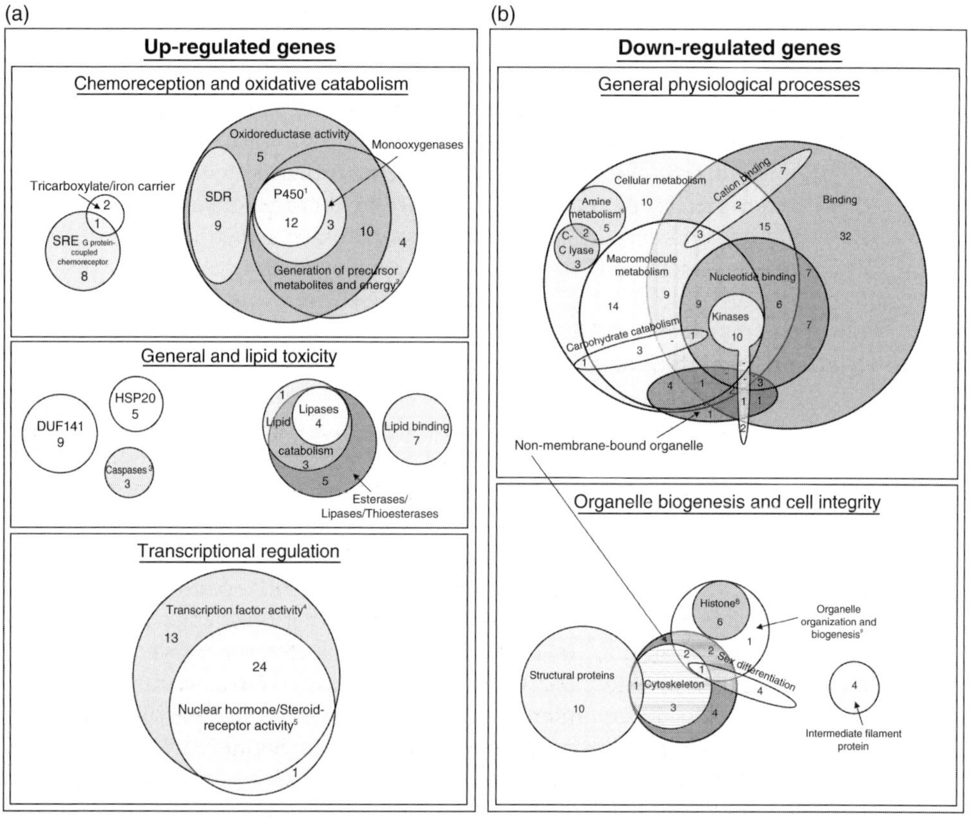

Fig. 11.3. Venn diagrams illustrating the overlap between over-represented gene classes of significantly PCB52 induced (a) and repressed (b) genes, respectively. Included are only gene class members displaying changes in expression of at least twofold ($P<0.05$). Reproduced from Menzel *et al.* (2007) with permission from Elsevier.

and chemoreceptors of the serpentine receptor class epsilon (SRE). Furthermore, the presence of general toxicity was evidenced by the induction of several small *hsp-20* genes and caspases. The increased oxidoreductase activity suggests that the nematode has indeed the ability to metabolize PCB52. The considerable up-regulation of lipid modifying enzymes firmly implies that PCB52 disrupts *C. elegans'* lipid homeostasis. The vast majority of down-regulated genes were found as members of general physiological processes such as cellular metabolism, binding, organelle biogenesis, and cell integrity.

Beyond the DNA microarray experiments the knockdown of 30 genes by RNA interference (RNAi) was used to determine the consequences in reproductive fitness of the presence and absence of PCB52. The RNAi screen resulted in PCB52 hypersensitivity (*dhs-2* and *dhs-23*) as well as hyposensitivity (*cyp-14A3 and cyp-35A1-5*). Fig. 11.4 summarizes these results, but presents also all corresponding expression results for three gene classes and the validation by RT-PCR. Localization experiments by using promoter-GFP fusions indicated that all genes investigated are predominantly expressed in the intestine of *C. elegans*. Four of the five selected *cyp-35A1-5* cytochromes P450 were shown to be involved in fat storage, with PCB52 exposure increasing the fat content in N2 wild type as indicated by Nile Red staining. CYP-14A3 may be considered as an enzyme that can hydroxylate PCB52. In summary, this paper provides strong evidence of the molecular mechanisms that underlie the toxicity of non-coplanar PCBs, and verify that the activation of lipid metabolism and increase in lipid storage are likely to be major factors that drive the toxic effect of PCB52. Moreover, it is a positive example of how experiments can succeed in revealing toxic modes of action on the basis of DNA microarray data.

Response to a phthalate (DEHP)

Di(2-ethylhexyl)phthalate (DEHP) is added as a softener to polyvinyl chloride (PVC), an abundant component of cables, floor tiles, garden hoses, containers, footwear, and clothing. Although DEHP has been shown to be present in the environment, little is known about its ecotoxicological properties. Roh *et al.* (2007) investigated DEHP toxicities to *C. elegans* using multiple toxic parameters, such as mortality, growth, reproduction, and stress-related gene expression. Based on the results of the acute toxicity test (24 h), a concentration corresponding to 1/100 of the LC50 was selected for subsequent experiments. Three replicates, for both exposure and control conditions, were conducted. The DNA microarray experiments were performed using GeneChip® *C. elegans* Genome Arrays (Affymetrix, Santa Clara, USA), which contains 22,500 probe sets against 22,150 unique *C. elegans* transcripts and analysed via the GeneChip Operating Software (GCOS, Affymetrix).

Although true statistical evaluation of the data set could have improved the message of the paper, the authors were able to demonstrate that exposure to DEHP induces a strong and differential induction/repression of specific genes. Using a cut-off setting of a twofold change, 20 genes were found to be

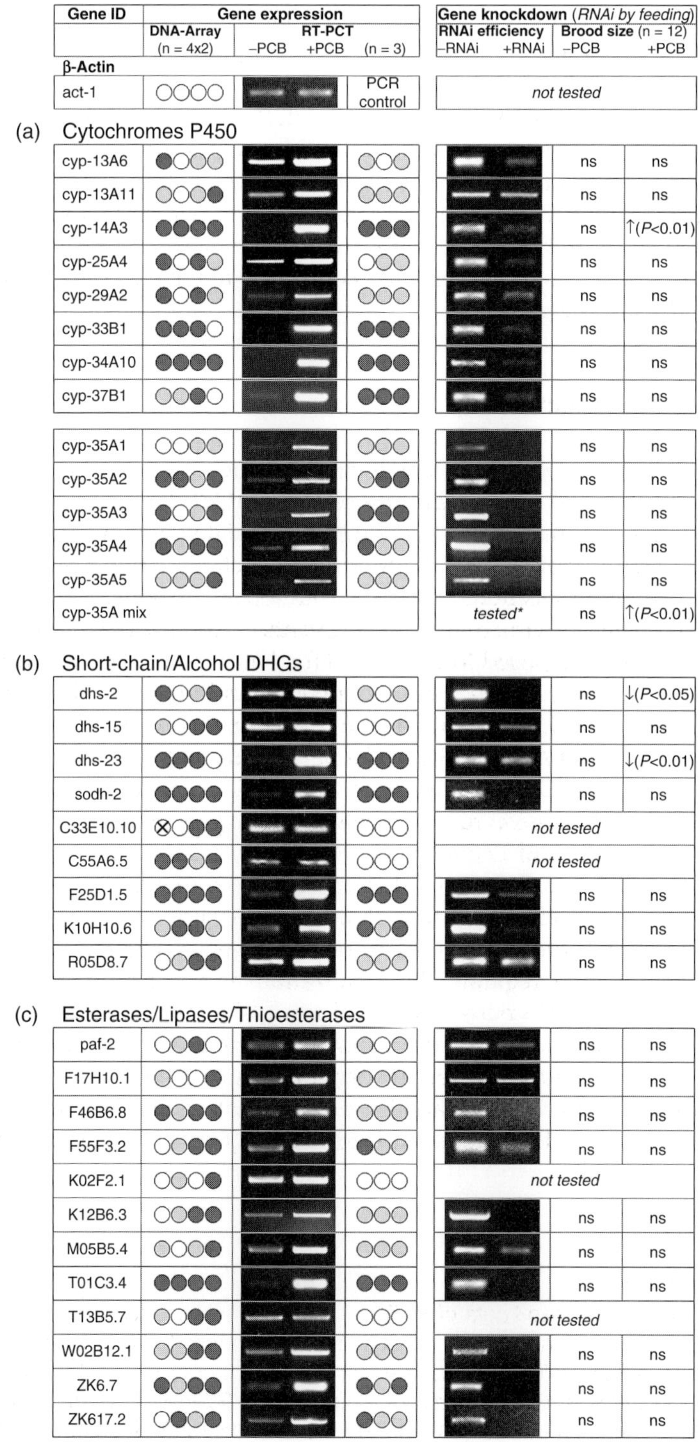
Gene ID
Gene expression
DNA-Array (n = 4x2)
RT-PCT
–PCB +PCB (n = 3)
Gene knockdown (RNAi by feeding)
RNAi efficiency
–RNAi +RNAi
Brood size (n = 12)
–PCB +PCB
β-Actin
act-1
PCR control
not tested
(a) Cytochromes P450
cyp-13A6 ns ns
cyp-13A11 ns ns
cyp-14A3 ns ↑(P<0.01)
cyp-25A4 ns ns
cyp-29A2 ns ns
cyp-33B1 ns ns
cyp-34A10 ns ns
cyp-37B1 ns ns
cyp-35A1 ns ns
cyp-35A2 ns ns
cyp-35A3 ns ns
cyp-35A4 ns ns
cyp-35A5 ns ns
cyp-35A mix tested* ns ↑(P<0.01)
(b) Short-chain/Alcohol DHGs
dhs-2 ns ↓(P<0.05)
dhs-15 ns ns
dhs-23 ns ↓(P<0.01)
sodh-2 ns ns
C33E10.10 not tested
C55A6.5 not tested
F25D1.5 ns ns
K10H10.6 ns ns
R05D8.7 ns ns
(c) Esterases/Lipases/Thioesterases
paf-2 ns ns
F17H10.1 ns ns
F46B6.8 ns ns
F55F3.2 ns ns
K02F2.1 not tested
K12B6.3 ns ns
M05B5.4 ns ns
T01C3.4 ns ns
T13B5.7 not tested
W02B12.1 ns ns
ZK6.7 ns ns
ZK617.2 ns ns

up-regulated and 20 down-regulated. The expression of heat shock proteins *hsp-16.1* and *hsp-16.2* was decreased by DEHP exposure, a trend that could not be confirmed in transgenic *hsp-16::GFP* strains following the exposure to DEHP.

Cytochrome P450 *cyp-35A2*, the UDP-glucuronosyltransferase *ugt-21*, the ABC transporters *pgp-1*, and the vitellogenin *vit-1* were all strongly induced. This occurred concomitantly with the deterioration of the physiological state, which suggests an increase in expression of those genes is likely to be a reaction to toxicity, rather than a homeostatic response. The data obtained from this study certainly contribute to the toxicological knowledge of DEHP in *C. elegans*.

Response to a heavy metal (Cd)

The final study mentioned here focused on global transcriptional changes in the nematode following cadmium exposure (Cui *et al.*, 2007). Cadmium is a persistent environmental toxicant that is associated with a variety of diseases. At low levels of exposure, the toxicological effects can be prevented by the activation of intracellular defence and repair systems, namely the stress response. The DNA microarray experiments of Cui *et al.* (2007) were performed using *C. elegans* exposed to 100 μM Cd for 4 and 24 h. For each condition, samples were prepared from three independent cultures, and the RNA was isolated in triplicate. Labelled probes were hybridized to *C. elegans* oligonucleotide microarrays from Agilent (Santa Clara, USA) using dye-flips, statistical analysis was performed using Agilent's GeneSpring and the data are accessible through GEO accession number GSE7535. The gene ontology for significantly changed genes (fold change ≥1.5, $P < 0.001$) was assigned using GoMiner which identified 237 up-regulated and 53 down-regulated genes that significantly changed following either 4 h or 24 h exposure to cadmium. These genes were clustered into early and late response genes. The former encompasses pathways, which regulate the localization and transportation of different chemical species (in particular metal ions) (Fig. 11.5). This suggests that the first response to cadmium intoxication is a transcriptional adjustment to maintain ion homeostasis and readjustment of perturbed energy supply. During the 24 h exposure period, metabolic and localization pathways were enriched within up- and down-regulated gene lists (Fig. 11.5). Cadmium exposure

Fig. 11.4. Summarized data of PCB52-induced gene expression, RNAi-coupled reproduction assay and RNAi efficiency. (a) CYP genes; (b) SDR genes; (c) Esterase/Lipase/Thioesterase genes; ● > 4-fold increase; ◎ 2–4-fold increase; ○ < 2-fold increase; ⊗ no analysable result. RNAi efficiency was determined only in PCB52 treated nematodes. (ns = not significantly different). Reproduced from Menzel *et al.* (2007) with permission from Elsevier.

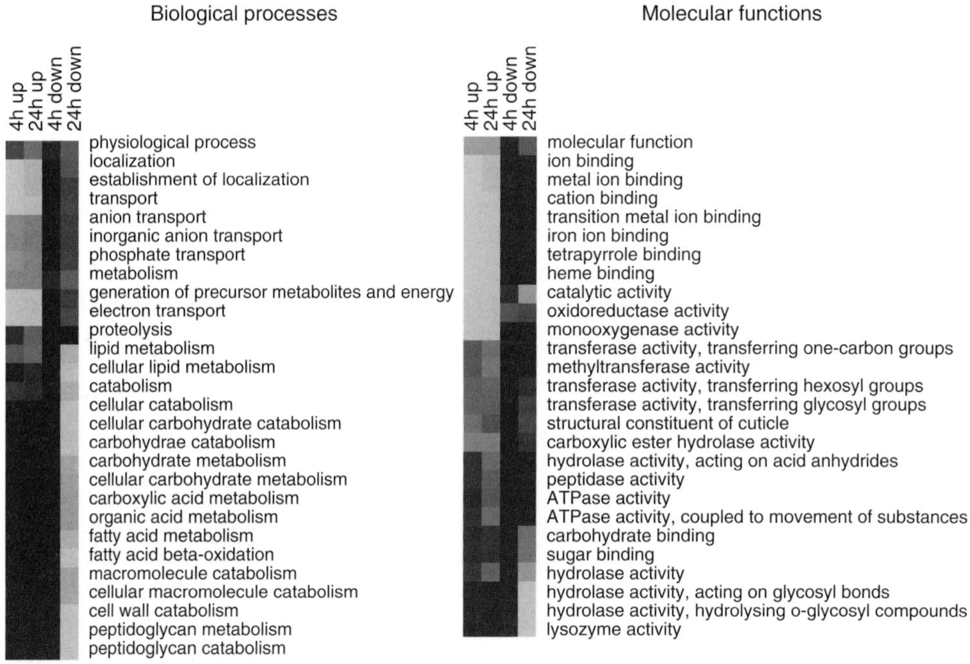

Fig. 11.5. Biological processes and molecular functions enriched with cadmium-responsive genes. Only GO terms with $P < 0.05$, and ≥4-fold changed genes in at least one of four conditions (up- or down-regulated after 4 or 24 h cadmium exposures) are displayed. Brightness (pale grey) translates into significance and enrichment of the pathway. This figure was taken from Cui *et al.* (2007).

resulted in the over-expression of fourteen cytochrome P450 genes, six glutathione *S*-transferase (GST) genes and five UDP-glucuronosyltransferase (UGT) genes. In addition to these biotransformation enzymes, the expression of four ABC transporters, *pgp-1*, *pgp-8*, *pgp-9* and *mrp-3*, was induced by cadmium, and *pmp-5* was repressed. Moreover, proteolysis was significantly enriched, possibly leading to an accumulation of damaged proteins. Pathways that were overpopulated with down-regulated genes included fatty acid metabolism, cellular lipid metabolism and cell wall catabolism, indicating that multiple cellular functions may be disrupted by cadmium toxicity.

Of the 53 down-regulated genes, several have documented RNAi phenotypes (http://www.wormbase.org), including embryonic lethality, slow growth, larval growth arrest, and sterility. To study the biological role of up-regulated genes, their expression was inhibited by RNAi in an *mtl-2* null (Cd-hypersensitive) background. In the presence of cadmium, 50 of the 92 genes tested resulted in reduced growth. Although several of the genes have previously been reported to be involved in cadmium detoxification, the majority are novel targets involved in heavy metal toxicity. The discovery of novel genes involved in the resistance to cadmium toxicity provided valuable information in understanding the biological function of the transcriptional changes caused by cadmium.

Conclusions and Outlook

The application of DNA microarrays to ecotoxicogenomics is still at an early stage, many uncertainties and complex challenges remain. Nevertheless, the case studies confirm the value of a DNA microarray approach, as it facilitates the identification and classification of gene responses to drugs and environmental pollutants, and potentially may contribute toward aiding in risk assessment and biomarker research. However, it must be emphasized that most studies lack appropriate bioinformatic and statistical support and thus fail to exploit the true power of global transcript screening. Moreover, a continuing struggle is to link molecular genetic data to observed phenotypic effects of ecotoxicological importance. Indeed, once reliable differential expression profiles have been determined the following question arises: 'What do these differences mean in an ecotoxicological context?' Ideally, the integration of selected gene (and also protein and metabolite) data with traditional ecotoxicological parameters will identify mechanistically based agglomerative biomarkers and elucidate mechanistic networks that could be used to develop predictive quantitative models. Then this approach could be used to establish thresholds of toxicity and predict exposure levels of a contaminant or complex mixture required to elicit a particular biomarker or adverse response (Boverhof and Zacharewski, 2006). Adverse effects, however, can rarely be attributed to an individual event. Most responses to a pollutant will involve complex interactions between genes, proteins, and metabolites. At present, only ecotoxicogenomics provide the key to determine (simultaneously) the broad molecular status of an organism experiencing toxicological effects. Ecotoxicogenomics, in its present form, typically identifies genes that are differentially expressed in a largely descriptive manner and devoid of hypotheses and assumptions. Another promising approach is the inter-species comparison of gene expression profiles, which will reveal evolutionary conserved molecular events in response to toxicosis. It has been suggested that cross-species DNA microarray hybridization is feasible at least in mammals (Hittel and Storey, 2001) as well as aquatic species (Renn *et al.*, 2004; von Schalburg *et al.*, 2005). Ecotoxicogenomics is therefore not limited to identifying novel candidate genes as biomarkers but also aims to pinpoint stressor specific expression signatures and reveal stress modulated molecular pathways, finally transforming DNA microarrays from a laboratory bench exercise to the future environmental safety assessments – undoubtedly the holy grail of ecotoxicogenomics.

Note

[1] Please refer to http://www.mged.org

References

Aldenderfer, M.S. and Blashfield, R.K. (1984) *Cluster Analysis*. Sage, Newbury Park, California.

Allison, D.B., Cui, X., Page, G.P., Sabripour, M. (2006) Microarray data analysis: from disarray to consolidation and consensus. *Nature Reviews Genetics* 7, 55–65.

Altman, N. (2005) Replication, variation and normalisation in microarray experiments. *Applied Bioinformatics* 4, 33–44.

Álvarez-Buylla, E.R., Benítez, M., Davila, E.B., Chaos, A., Espinosa-Soto, C. and Padilla-Longoria, P. (2007) Gene regulatory network models for plant development. *Current Opinion in Plant Biology* 10, 83–91.

Ball, C.A., Brazma, A., Causton, H., Chervitz, S., Edgar, R., Hingamp, P., Matese, J.C., Parkinson, H., Quackenbush, J., Ringwald, M., Sansone, S.A., Sherlock, G., Spellman, P., Stoeckert, C., Tateno, Y., Taylor, R., White, J. and Winegarden, N. (2004) Submission of microarray data to public repositories. *PLoS Biology* 2, E317.

Benjamini, Y. and Hochberg, Y. (1995) Controlling the false discovery rate: a practical and powerful approach to multiple testing. *Journal of the Royal Statistical Society: Series B (Statistical Methodology)* 57, 289–300.

Boverhof, D.R. and Zacharewski, T.R. (2006) Toxicogenomics in risk assessment: applications and needs. *Toxicological Sciences* 89, 352–360.

Brazma, A., Hingamp, P., Quackenbush, J., Sherlock, G., Spellman, P., Stoeckert, C., Aach, J., Ansorge, W., Ball, C.A., Causton, H.C., Gaasterland, T., Glenisson, P., Holstege, F.C., Kim, I.F., Markowitz, V., Matese, J.C., Parkinson, H., Robinson, A., Sarkans, U., Schulze-Kremer, S., Stewart, J., Taylor, R., Vilo, J. and Vingron, M. (2001) Minimum information about a microarray experiment (MIAME) – towards standards for microarray data. *Nature Genetics* 29, 365–371.

Burczynski, M.E., McMillian, M., Ciervo, J., Li, L., Parker, J.B., Dunn, R.T. II, Hicken, S., Farr, S. and Johnson, M.D. (2000) Toxicogenomics-based discrimination of toxic mechanism in HepG2 human hepatoma cells. *Toxicological Sciences* 58, 399–415.

Chapman, R.W. (2001) EcoGenomics – a consilience for comparative immunology? *Developmental and Comparative Immunology* 25, 549–551.

Cui, Y., McBride, S.J., Boyd, W.A., Alper, S. and Freedman, J.H. (2007) Toxicogenomic analysis of *Caenorhabditis elegans* reveals novel genes and pathways involved in the resistance to cadmium toxicity. *Genome Biology* 8, R122.

Custodia, N., Won, S.J., Novillo, A., Wieland, M., Li, C. and Callard, I.P. (2001) *Caenorhabditis elegans* as an environmental monitor using DNA microarray analysis. *Annals of the New York Academy of Sciences* 948, 32–42.

Eisen, M.B., Spellman, P.T., Brown, P.O. and Botstein, D. (1998) Cluster analysis and display of genome-wide expression patterns. *Proceedings of the National Academy of Sciences of the USA* 95, 14863–4868.

Golden, T.R. and Melov, S. (2004) Microarray analysis of gene expression with age in individual nematodes. *Aging Cell* 3, 111–124.

Hamadeh, H.K., Bushel, P., Paules, R. and Afshari, C.A. (2001) Discovery in toxicology: mediation by gene expression array technology. *Journal of Biochemical and Molecular Toxicology* 15, 231–242.

Hittel, D. and Storey, K.B. (2001) Differential expression of adipose- and heart-type fatty acid binding proteins in hibernating ground squirrels. *Biochimica et Biophysica Acta* 1522, 238–243.

Höss, S., Bergtold, M., Haitzer, M., Traunspurger, W. and Steinberg, C.E.W. (2001) Refractory dissolved organic matter can influence the reproduction of *Caenorhabditis elegans* (Nematoda). *Freshwater Biology* 46, 1–10.

Kim, S.K., Lund, J., Kiraly, M., Duke, K., Jiang, M., Stuart, J.M., Eizinger, A., Wylie, B.N. and Davidson, G.S. (2001) A gene expression map for *Caenorhabditis elegans. Science* 293, 2087–2092.

Kwon, J.Y., Hong, M., Choi, M.S., Kang, S., Duke, K., Kim, S., Lee, S. and Lee, J. (2004) Ethanol-response genes and their regulation analyzed by a microarray and comparative genomic approach in the nematode *Caenorhabditis elegans. Genomics* 83, 600–614.

Lund, J., Tedesco, P., Duke, K., Wang, J., Kim, S.K. and Johnson, T.E. (2002) Transcriptional profile of aging in *C. elegans. Current Biology* 12, 1566–1573.

Menzel, R, Rödel, M. Kulas, J. and Steinberg, C.E.W. (2005a) CYP35: Xenobiotically induced gene expression in the nematode *Caenorhabditis elegans. Archives of Biochemistry and Biophysics* 438, 93–102.

Menzel, R., Stürzenbaum, S., Bärenwaldt, A., Kulas, J. and Steinberg, C.E.W. (2005b) Humic material induces behavioral and global transcriptional responses in the nematode *Caenorhabditis elegans. Environmental Science and Technology* 39, 8324–8332.

Menzel, R., Yeo, H.L., Rienau S., Li, S., Steinberg, C.E.W. and Stürzenbaum, S. (2007) Cytochrome P450s and short-chain dehydrogenases mediate the toxicogenomic response of PCB52 in the nematode *Caenorhabditis elegans. Journal of Molecular Biology* 370, 1–13.

Nadon, R. and Shoemaker, J. (2002) Statistical issues with microarrays: processing and analysis. *Trends in Genetics* 18, 265–271.

Oberemm, A., Onyon, L. and Gundert-Remy, U. (2005) How can toxicogenomics inform risk assessment? *Toxicology and Applied Pharmacology* 207, S592–S598.

Pennie, W.D., Tugwood, J.D., Oliver, G.J. and Kimber, I. (2000) The principles and practice of toxicogenomics: applications and opportunities. *Toxicological Sciences* 54, 277–283.

Pflugmacher, S., Pietsch, C., Rieger, W. and Steinberg, C.E.W. (2006) Influence of structural moieties of dissolved organic matter on the photosynthetic oxygen production of aquatic plants. *Science of the Total Environment* 357, 169–175.

Renn, S.C., Aubin-Horth, N. and Hofmann, H.A. (2004) Biologically meaningful expression profiling across species using heterologous hybridization to a cDNA microarray. *BMC Genomics* 5, 42.

Roh, J.Y., Jung, I.H., Lee, J.Y. and Choi, J. (2007) Toxic effects of di(2-ethylhexyl)phthalate on mortality, growth, reproduction and stress-related gene expression in the soil nematode *Caenorhabditis elegans. Toxicology* 237, 126–133.

Snape, J.R., Maund, S.J., Pickford, D.B. and Hutchinson, T.H. (2004) Ecotoxicogenomics: the challenge of integrating genomics into aquatic and terrestrial ecotoxicology. *Aquatic Toxicology* 67, 143–154.

Soukas, A., Cohen, P., Socci, N.D. and Friedman, J.M. (2000) Leptin-specific patterns of gene expression in white adipose tissue. *Genes and Development* 14, 963–980.

Steinberg, C.E.W. (2003) *Ecology of Humic Substances in Freshwaters – Determinants from Geochemistry to Ecological Niches*, Springer Verlag, Heidelberg.

Steinberg, C.E.W. and Münster, U. (1985) Geochemistry and ecological role of humic substances in lakewater. In: Aiken, G.R., McKnight, D.M., Wershaw, R.L. and MacCarthy, P. (eds) *Humic Substances in Soil, Sediment and Water. Geochemistry, Isolation and Characterization*, John Wiley and Sons, New York, pp. 105–146.

Steinberg, C.E.W., Kamara, S., Peuthert, A., Prokhotskaya, V.Yu., Grigutytė, R., Manusadžianas, L., Karasyova, T., Timofeyev, M.A., Zhang, J., Paul, A., Meinelt, T., Farjalla, V.F. and Menzel, R. (2006) From individuals to ecosystems: dissolved humic substances as ecological driving forces in unproductive freshwater lakes. *Freshwater Biology* 51, 1189–1210.

Steinberg, C.E.W., Saul, N., Pietsch, K., Meinelt, T., Rienau, S. and Menzel, R. (2007) Dissolved humic substances facilitate fish life in extreme aquatic environments and have the potential to extend lifespan of *Caenorhabditis elegans. Annals of Environmental Science* 1, 81–90.

Thurman, E.M. (1985) *Aquatic Humic Substances. Organic Geochemistry of Natural Waters*, Nijhoff-Junk, Dordrecht, The Netherlands.

Tusher, V.G., Tibshirani, R. and Chu, G. (2001) Significance analysis of microarrays applied to the ionizing radiation response. *Proceedings of the National Academy of Sciences of the USA* 98, 5116–5121.

von Schalburg, K.R., Rise, M.L., Cooper, G.A., Brown, G.D., Gibbs, A.R., Nelson, C.C., Davidson, W.S. and Koop, B.F. (2005) Fish and chips: various methodologies demonstrate utility of a 16,006-gene salmonid microarray. *BMC Genomics* 6, 126.

Wang, J. and Kim, S. (2003) Global analysis of dauer gene expression in *Caenorhabditis elegans. Development* 130, 1621–1634.

Waters, M., Boorman, G., Bushel, P., Cunningham, M., Irwin, R., Merrick, A., Olden, K., Paules, R., Selkirk, J., Stasiewicz, S., Weis, B., Van Houten, B., Walker, N. and Tennant, R. (2003) Systems toxicology and the Chemical Effects in Biological Systems (CEBS) knowledge base. *Environmental Health Perspectives Toxicogenomics* 111, 15–28.

Weinstein, J.N., Myers, T.G., O'Connor, P.M., Friend, S.H., Fornace, A.J. Jr., Kohn, K.W., Fojo, T., Bates, S.E., Rubinstein, L.V., Anderson, N.L., Buolamwini, J.K., van Osdol, W.W., Monks, A.P., Scudiero, D.A., Sausville, E.A., Zaharevitz, D.W., Bunow, B., Viswanadhan, V.N., Johnson, G.S., Wittes, R.E. and Paull, K.D. (1997) An information-intensive approach to the molecular pharmacology of cancer. *Science* 275, 343–349.

Yue, H., Eastman, P.S., Wang, B.B., Minor, J., Doctolero, M.H., Nuttall, R.L., Stack, R., Becker, J.W., Montgomery, J.R., Vainer, M. and Johnston, R. (2001) An evaluation of the performance of cDNA microarrays for detecting changes in global mRNA expression. *Nucleic Acids Research* 29, E41.

12 Commercial Aspects of the Use of Nematodes as Bioindicators

Marcus W. Trett,[1,2] Beatriz Calvo Urbano,[1,2] Simon J. Forster[1] and Simon P. Trett[1]

[1]*Physalia Limited, Consultant and Forensic Ecologists, Harpenden, UK;*
[2]*Nebalia S.L. Consultoría Medioambiental, Madrid, Spain*

Introduction

In its opening chapter, the UK Government's Biodiversity Action Plan presents a meiofaunal monitoring survey of domestic and industrial effluent discharges to illustrate the value of species-rich bioindicators as a means of assessing environmental stresses and, in turn, helping to protect and conserve biodiversity (Anon., 1994). The key meiofaunal group in the analyses were free-living nematodes which are most commonly the major contributors to the biodiversity of a given habitat (see Atrill *et al.*, 1996). This chapter examines the background to the use of nematodes as bioindicators in commercial situations. It highlights some of the key benefits that they can offer regulatory bodies, industry and other sectors in the assessment of prevailing environmental conditions, the problems that can arise when undertaking work in this non-academic realm, the methods and protocols that are adopted for the nematode analyses and the standard and possibly less standard analyses that are undertaken to examine the resulting data. In a final section, case histories are also presented to illustrate some of the different uses to which nematodes have been put. Looking back, it would appear that the research and development phase for this particular monitoring and assessment tool has been progressing for about 100 years and so this chapter may be a timely summary of present practice as well as an opportune introduction as to how nematodes can help to meet some of the needs of a modern commercial society.

Background to the use of nematodes in ecological assessment

To understand how nematodes gained a commercial role as bioindicators, it is useful to appreciate some of the legal and regulatory factors that shaped the marketplace for ecological monitoring and assessment tools. Less than 30 years ago it was rare to find an environmental impact assessment (EIA) for a

 Nematodes as Environmental Indicators (eds M.J. Wilson and T. Kakouli-Duarte)

development that included reference to the natural environment in anything other than aesthetic terms. In Europe this began to change following the implementation of a European Communities Directive (Anon., 1985), which resulted in legal requirements in each of the member states for the inclusion of ecological components in environmental impact studies. The development of a legal framework paved the way for the evolution of *ecological* impact assessment (EcIA) as a field of study in its own right and, by raising the profile of potential hazards to the natural environment, led to an expansion in related areas such as ecological risk assessment (EcRA). Within a few years of the implementation of the Directive, demand for the services of fully-trained terrestrial and aquatic ecologists had increased significantly. It also brought about the emergence of professional bodies such as the Institute of Ecology and Environmental Management (IEEM) based in the UK, that set standards and regulated the emerging profession of the 'consultant ecologist'.

In addition to baseline surveys, studies were needed to determine ecological effects of ongoing operations, and these investigations demanded the adoption of quantitative approaches. Fully-quantitative, field-based surveys remain the only practical means of assessing the spatial and temporal changes that underpin the detection and description of actual ecological effects and these approaches are now used routinely at sites throughout the world. However, in some sectors of industry, particularly the wastewater treatment and chemical manufacturing industries, fully quantitative ecological surveying had been standard practice for a considerable period of time (at least since the mid-1950s). In countries worldwide, protection of aquatic receiving environments had been identified as a high priority and many effluent dischargers were already required by statutory regulatory bodies to monitor and assess the effects of altered environmental conditions associated with their waste streams. It was against this background that many of the present authors became involved in ecological assessments.

At this stage in the history of impact studies, macrofaunal monitoring surveys (i.e. surveys based on assessment of the communities of larger aquatic invertebrates) were the norm. To a large extent they still are and in several instances, quantitative macrofaunal surveys not only continue to form part of the operational requirements for manufacturing facilities as prescribed by the authorities, but their results also provide the basis for granting of discharge consents. These, in turn, help to determine the quantities of waste materials that can be released by the operators of discharge pipelines. Ultimately, it was the industries that were already obliged to monitor their ecological effects using a macrofaunal approach that created the demand for a move towards the use of integrated faunal monitoring systems – programmes that combined assessments of macro- and meiofauna (or in a few cases purely meiofauna), in parallel with physico-chemical surveys. Indirectly, this led to the formal appearance of Nematoda as a functional component of ecological monitoring systems.

What caused this change? The shift in emphasis arose from concerns over the natural patchiness and seasonal variability inherent in the communities of the larger invertebrate species (macrofauna; see section on Benefits of Using Nematodes as Bioindicators below). The natural variation in the assemblages of these invertebrates generated disquiet about potential false allegations of environmental harm that could arise in relation to point source discharges as well as to diffuse inputs of materials entering aquatic habitats, such as leachates from landfills or terrestrial run-off from spill and leak incidents. In the light of this, a change to the higher resolution monitoring ostensibly offered by meiofaunal assessments (see Ferris and Ferris, 1979; Platt and Warwick, 1980) became not only a logical step forwards but also a commercially viable one. This was assisted by the fact that fewer and smaller volume replicate samples appeared to be necessary to provide adequate descriptions of meiofaunal variance than for the macrofauna.

Materialization of meiofauna in the ecological monitoring tool box for use in aquatic environments occurred gradually between the late 1970s and mid-1980s; early patrons included pigment manufacturers, explosives factories, halogenated chemicals companies and pharmaceutical producers operating at sites around the world. In these early days, meiofaunal samples were often collected during the macrofaunal monitoring survey and then preserved for possible future analyses. The rationale behind this 'voucher sampling' approach was that it provided a second tier of assessment that could be resorted to if the macrofaunal survey indicated the existence of acute or unexplained stresses or if the macrofaunal data proved to be too poor to provide a serviceable interpretation of conditions within a survey area. Several of our staff members still refer to this as the *meiofaunal nematode court of appeal* in recognition of the fact that it was the nematode component that carried the bulk of the information (see below).

The storage of the samples had the perceived advantage that if an incident did occur at any point in time, a set of samples was available that could provide, in effect, dynamic baseline data enabling the extent and degree of any impacts resulting from the incident to be determined. This was clearly worth a king's ransom in terms of evaluating liability as well as determining the extent of any environmental harm that might have occurred. Both are important criteria in the establishment of fines and in the settlement of insurance claims. However, the voucher sample system eventually disappeared, largely because of the logistic problems of curating the samples and ensuring their adequate long-term preservation. Also extremely large numbers of samples were being accumulated and this created logistical problems. Eventually, many of the industrial customers that needed this type of survey adopted hybrid macrofaunal and meiofaunal surveys or, in a very few cases, exclusively meiofaunal surveys, in which all samples collected were analysed.

Published field and laboratory studies as well as experience gained from surveying and monitoring in a wide range of situations showed that within the meiofauna it was the nematodes and the patterns of variance in the structures of their assemblages that provided the most serviceable information on prevailing environmental conditions (see for example Somerfield *et al.*, 1995; Muthumbi *et al.*, 2004; Fraschetti *et al.*, 2006; Gyedu-Adabio and Baird, 2006). With suitable biometric and statistical expertise, this information is relatively easily retrieved through the profusion of modern multivariate analytical techniques (see

Bongers and Ferris, 1999 and Neher and Darby, Chapter 4, this volume). This is perhaps the key to nematodes' success not only as bioindicators but also as biosentinels and their use in forensic studies (see section on Assessment of Nematode Assemblage Data below).

Another driver for the use of nematodes as bioindicators in applied ecological studies came from the ecological risk assessment domain. Single-species toxicity tests, including nematode-based screening of materials (see review by Höss *et al.*, 2006 and Höss and Williams, Chapter 9, this volume), have been in use for a long time. However, there was, and still is, a tendency to extrapolate from laboratory-based toxicity testing on single species, or artificially maintained assemblages, to predict effects of toxic materials in the environment. Arguments against this practice have been rehearsed by Chapman (2002a, b; Chapman *et al.*, 2003), who described toxicity testing as 'scientifically-valid but ecologically irrelevant' on several grounds. He suggested that measurement of actual ecological effects is a far more valid, but by no means simple, approach to determining and setting appropriate and precautionary soil, sediment and water guideline concentration values for contaminants. He can also be credited with raising the issue of statistical significance versus ecological significance (Chapman *et al.*, 2002) – an area that will be returned to later in connection with the commercial use of nematodes as bioindicators (see Level III analyses and Case history 4 below).

Specific Needs Addressed by Nematodes as Bioindicators

Since the early 1980s ecological assessments utilizing meiofauna and meiofaunal nematodes have been employed in commercial contexts in a wide variety of situations and at sites worldwide. These have included baseline and annual pre- and post monsoon marine monitoring surveys in South East Asia, deep sea dumping ground surveys off the southern coast of Spain, annual surveys to monitor effects of industrial discharges into the St Lawrence River, Canada, assessments of impacts of landfill and contaminated land leachates entering rivers and estuaries in Scotland, Northern Ireland and north-west England, monitoring of oil production facilities using meiofaunal colonization units (MCUs) in the North Caspian Sea and quarterly meiofaunal monitoring assessments for the UK Environment Agency of sampling stations located along the entire 110 km tidal reach of the Thames estuary.

In each of the cases listed, the studies addressed the highly specific needs of the customers, which included industry, regulators and government bodies. By way of example, Table 12.1 presents a list of the most common, generic requirements of each of the customer sectors that we have worked with in the UK. Many, but not all, of these needs can and have been addressed in some way by meiofaunal nematode assessments and examples are touched on below while a few are described in more detail in the case histories later in this chapter.

The industrial sectors remain some of the principal customers for ecological monitoring surveys and the benefits of nematodes as monitoring tools, described below, are readily appreciated by industrial environmental

Table 12.1. A list of the principal areas in which different customer sectors in the UK require inputs from ecological surveys and assessments that can and in many cases have been met by the use of nematodes as bioindicators. While the industrial sectors have historically represented the largest group of nematode-based assessment consumers, other sectors continue to grow steadily.

Customer Sector	Needs	Comments
Industry (Chemical/petro-chemical, explosives, pharmaceutical, power generation, construction, transport, mineral and aggregate extraction, paper, food processing and drinks manufacturing including some brewing companies...)	Detection, description and delineation of *actual* ecological effects of operations on surrounding habitats	Part of planning permissions granted for discharges to controlled waters and discharge consent compliance monitoring; for larger industries this forms part of Environmental Management/Stewardship Systems
	Environmental performance testing	Assessment of efficacy of treatment systems
	Incident monitoring	Ecological assessments of effects following and arising from incidents (e.g. spills, leaks and 'loss of containment')
	Early warning and *in situ* monitoring systems	Biosentinel monitoring where risks perceived to be high
Water Authorities	Abstraction licensing	Baseline assessments of habitats that will be exposed to potential stressors (contaminants/ disturbance)
	Monitoring of wastewater treatment works	Periodic assessments (annual) of ecological effects
Regulatory Bodies	Baseline assessments	Often delegated to developers forming part of planning applications for developments
	Post-development assessments and monitoring	Ostensibly as follow-up to above; *rarely enforced with developers*
	Compliance monitoring and certification	Part of their statutory role; work often sub-contracted
Insurance Industry	Assessments for ascription of liability	Varied ecological appraisal work – overlap with legal sector

Continued

Table 12.1. Continued

Customer Sector	Needs	Comments
Legal Profession	Evidence of any detectable impacts arising from incidents and their characterization	Description and 'quantification' of harm caused – used to assess the scale of fines for damages
	Source identification for materials leading to detectable harm	Confirmation/rebuttal of alleged cause(s) of harm; includes elements of forensic ecology
Local Authorities (+ Industry and Developers)	Status of contaminated land sites and prioritizing contaminated sites for clean-up and the targeting and design of remediation programmes	Now specific UK requirement to assess and rank contaminated land and to identify potential ecological receptors under Environmental Protection Act 1990 Part IIA
Remediation Industry	Delineation of areas within sites requiring remediation	Landfills, contaminated land and former factory/brownfield sites; principal use of terrestrial meiofauna surveys
	Assessment of success of the measures adopted	Certification of works undertaken for site closure/'sign-off'

managers. In most cases there is a well-defined need to monitor effects of industrial operations on the environment which is usually set down in the planning permissions granted to individual factories and in the regularly reviewed licences to operate. While we are unaware of any operational specifications that stipulate the use of nematodes as sole monitoring tools, the direct value to industry of incorporating these in surveys has included:

- Early warning of adverse changes in effects associated with effluents.
- Accurate tracking of variations in natural conditions in the receiving environment (e.g. changes in sediment granulometry and total organic carbon contents of sediments resulting from coastal storms). This now includes failures of monsoon rains and other changes associated with increased variation in climate regimes.
- Identification of contaminant or other stress effects emanating from neighbouring sources (e.g. other industrial operations or variable river and drain discharges).
- Appraisal of net effects of altered effluent treatment processes, changes in effluent quality due to new manufacturing procedures and modified discharge patterns or effluent dispersal.

- Detection and delineation of terrestrial and aquatic effects arising from industrial landfills and historically contaminated sites.
- Detailed descriptions of the extents of effects arising from spill and leak incidents and assessment and documentation of any ecological repercussions associated with clean-up operations.

Regulators clearly share an interest in all of the points listed above. However, apart from incident monitoring, contracts using nematodes as indicators for these authorities revolve around the assessment of baseline conditions, particularly in urbanized environments, and in connection with changes in communities associated with reduced surface water flows (e.g. during drought periods and/or when increased freshwater abstraction occurs). As regulatory bodies are closely involved in identifying potential impacts of proposed developments, and as they are able to specify the ecological information that needs to be gathered, they also have an indirect but potent effect on the requirements of developers. Assessment of altered sediment conditions through changed patterns of scouring in tidal habitats, leading to re-suspension and mobilization of historically contaminated sediments and monitoring of shading impacts are examples where nematode assessments are 'best practice'. This is particularly true where the macrofaunal assemblages are depauperate, for example in docklands and along industrialized reaches of rivers and canals.

Where their cases relate to environmental issues, the requirements of insurers and the legal profession are usually highly specific. These most commonly centre on an incident that has resulted in environmental harm. Here the brief can be to provide evidence indicating who was responsible, to identify when the incident occurred, whether it was an acute event or if not, a chronic effect arising resulting from something that was overlooked, to determine how large the affected area is and to establish if there is evidence that recovery is already occurring or if measures taken at the time may have worsened the impacts. The distinct advantages of nematode assessments in the detection, description and delineation of effects (see below) are not lost on lawyers who know that the sizes of areas affected determine the scale of fines. In the case of insurance companies, their primary concern is usually whether their client really was responsible in the first place followed by the scale and magnitude of the damages.

In Europe, the remediation industry is growing rapidly as brownfield sites and contaminated land come under increasing pressures for redevelopment. In the UK there is now a requirement to undertake ecological assessments and to identify potentially affected ecological receptors (see Table 12.1). In the majority of cases these studies have not been undertaken as yet, which is perhaps no surprise given that a single borough in a major UK city has over 1000 sites on its contaminated land register. For the remediation companies, nematode assessments are already providing a flexible and sensitive means of identifying migration of contaminant materials from sites (see Trett and Thurgood, 2008). This is especially true where the sites are situated beside watercourses or on coastal land, which applies to many former industrial plants and landfills. Once the remediation works have been undertaken, the same surveys are now furnishing information that is helping in the documentation of recovery and the assessment of the success of the measures that have been adopted (e.g. soil washing, the use of reactive barriers, *in situ* chemical treatment and removal and treatment of contaminated materials).

Benefits of Using Nematodes as Bioindicators

As organisms for ecological monitoring and for the evaluation of prevailing environmental conditions, nematodes stand out head-and-tail in front of all other contenders. The group offers the unequalled ability to undertake studies and assessments in nearly every natural habitat examined so far as well as in many man-made environments. Their community structures can detect stresses ranging from the subtle disturbance caused by stingrays feeding (Sherman *et al.*, 1983), through the chronic effects of different agricultural soil management regimes (Sánchez-Moreno *et al.*, 2006), and the impacts of whale corpses on deep sea sediments (Debenham *et al.*, 2004) to the acute effects associated with a range of industrial and domestic discharges to surface waters (see Moore and Bett, 1989; Fraschetti *et al.*, 2006). Evidence continues to grow to show that their community structures can help to track global warming (Bakonyi and Nagy, 2000; Doran *et al.*, 2002) and, from a retrospective point of view, their global distributions may even reflect the processes of continental drift (Gerlach, 1977) and their biochemistry, the history of the Earth's chemosphere (Boaden, 1975, 1977; Jensen, 1986).

From the earliest days of detailed nematode sampling, stretching back to the late 1800s and the studies of Bastian, Butschli and Cobb, it was clear that nematodes were more abundant and more diverse than any of the other metazoan groups that occupied the same habitats. These, among many other features, were first listed by Ferris and Ferris (1979) and Platt and Warwick (1980) as key advantages of free-living nematodes over other faunal groups for their use in ecological monitoring and assessment. From a commercial and forensic perspective these are equally valid and can be summarized as follows.

Abundance

Feature

Nematodes exhibit higher total abundances than other phyla in a given particulate habitat (see Yeates *et al.*, Chapter 1, this volume) and our own surveys support this; densities of up to 40 million nematodes/m^2 were observed at the Millennium Dome Terraces, Greenwich, UK, where geotextile matting had been used to stabilize newly created, artificial saltmarsh terraces along the Thames shores (survey undertaken for the Environment Agency, UK).

Benefit

Their high abundances mean that statistically valid sampling can be achieved more easily for nematodes than for other groups such as the macro-invertebrates. In certain stressed and/or coarse sediment habitats, to achieve macrofaunal sampling sufficiency would mean that the survey itself would exert an unacceptable impact. Smaller, more-easily processed samples are adequate for meiofaunal nematode assessments. In muddy marine sediments, single 10 cm diameter core samples can yield over 20,000 nematodes (Forster, 1998).

High species richness

Feature

Nematode assemblages exhibit higher species richness in a given particulate habitat than other phyla. Although total numbers of species for all macrofaunal phyla may be greater than the numbers of nematode species, nematodes usually exceed the species richness of any single macrofaunal phylum and often by an order of magnitude. We have noted exceptions in certain fouling and epigrowth communities (e.g. on hulls of ships and the stanchions of offshore oil platforms where recruitment is from the water column).

Benefit

The high species richness and the resulting complexity of the nematode assemblages, means that in each habitat there is a broad spectrum of biochemistries, physiologies, feeding and micro-habitat preferences, life-history strategies, behavioural groups, reproductive biologies and sensitivities to stresses. This forms the ecological engine room for the integration of environmental effects and it ensures a more balanced assessment of prevailing environmental conditions than is possible with macrofaunal communities alone.

Tolerance-sensitivity range

Feature

Nematodes possess an intrinsic broad spectrum of responses and sensitivities to differing physico-chemical conditions, ranging from intolerance and elevated sensitivity to resistance and tolerance. It is frequently noted that they have been found in nearly every extreme environment examined so far, including hot volcanic springs, anoxic sediments, seeps and hydrothermal vents and have been recovered from a wide variety of heavily industrially contaminated substrata.

Benefit

Their broad spectrum of tolerances means that nematodes include species that are the last to surrender as conditions deteriorate and, equally, some of the first to re-appear as conditions improve. At the other end of the spectrum, the more sensitive species may be readily displaced by comparatively mild physical or chemical stresses. From a commercial point of view, this means that changes can be detected, described, delineated and, therefore, monitored across almost the entire gamut of environmental conditions. This is a significant point in favour of nematodes as bioindicators in all habitats. Industry and regulators need reliable, *in situ* methods of assessing changes and this cannot be delivered across the same range of conditions using other 'indicator' groups.

Low mobility

Feature

The mobility of nematodes within the particulate environments is comparatively restricted. Whilst bulk movements of substrata, such as sediment transport in active aquatic systems or erosion of terrestrial soils during storm incidents may lead to major natural translocations of nematodes and other interstitial species assemblages (see for example Sherman and Coull, 1980; Eskin and Palmer, 1985; Nkem *et al.*, 2006), nematodes, that measure mostly between 0.5 and 3 mm in length, are only capable of limited self-propelled excursions which may in many cases be restricted to periodic vertical migrations (see Hunt *et al.*, 2001).

Benefit

A confounding factor in ecological assessments based on macro-invertebrate communities in terrestrial, wetland and aquatic habitats is the fact that active, mobile species, frequently predators and scavengers but also including some grazers, migrate away from localized stresses. These animals may then return as and when the environmental constraint diminishes. This can result in:

- Reduced spatial resolution of the survey
- Impaired ability to detect the effects of a former incident (i.e. reduced temporal resolution)
- Interference with community analyses that examine the patterns of variance in community structures (multivariate approaches) by reducing 'signal to noise' ratios, and
- Masking of spatial dependency patterns (see section on Level III analyses).

In contrast, the high resolution of surveys based on analyses of nematode communities is a direct consequence of their low mobility. This, in turn, leads to high spatial fidelity and the ability to delineate effects accurately as the assemblages are continuously subjected to the constraints of any noxious materials that enter their environment (see below).

Life-cycle times

Feature

Nematode life-cycles cover short to intermediate periods of time. Life-cycles reported in the literature span from between 2.5–4 days (*Caenorhabditis elegans* (see Jager *et al.*, 2005); *Rhabditis marina* (Vranken and Heip, 1983);

Pristionchus pacificus (Sommer, 2006)) to between 1 year and 18 months (*Theristus* (*Penzancia*) *anoxybioticus* (Jensen, 1995) and *Oncholaimus oxyuris* (Smol *et al.*, 1980), respectively).

Benefit

The range of lengths of nematode life-cycle times enables the effects of short- and long-term stresses, acute and/or chronic, to be integrated rapidly and reflected in the structures of their communities. This increases the temporal resolution of effects and facilitates the description of temporal changes which is essential in monitoring studies. Evidence from incident surveys shows that communities of larger invertebrates are slower to respond to intermediate and chronic stresses and take longer to either recover or to reach a new stable state.

Interstitial mode of life

Feature

With some exceptions, free-living nematodes exploit the interstitial architecture of soils and sediments for the major part, if not all, of their lives.

Benefit

Having evolved to live and move in the moisture present in the pore spaces between soil and sediment particles, free-living nematodes are in intimate and near continuous contact with the interstitial pore water system. Accordingly, their exposure to any contaminants entering this realm is more likely to be a significant factor in determining the survival of the nematode species than that of a larger macroinvertebrate species. In the case of soils, this lies at the core of proposals that equilibrium partition models for nematodes and soil pore water contaminants might provide a superior basis for the establishment of soil quality guidelines to safeguard food webs (Trett *et al.*, 2000).

Conservative reproductive strategies

Feature

Meiofaunal nematode species have evolved conservative reproductive strategies and under given conditions populations of individual species are intrinsically stable (Ferris and Ferris, 1979; Platt *et al.*, 1984).

Benefit

In general, whilst processes in community ecology drive changes at the population level, the effects of these processes are most easily detected where they lead to net alterations in patterns in communities – hence the value community structure analyses. As nematode populations are inherently stable, observed changes in community structure can be more readily related to the effects of changes in environmental conditions. This is reflected in the results of the multivariate correlation analyses used in the commercial assessments to identify and rank ecologically significant environmental factors (see Level III mathematical analyses below). Nematode community structures regularly exhibit stronger relationships with measured soil and sediment physico-chemical parameters than macroinvertebrate groups.

Commercial Nematode Surveys

Constraints imposed on commercial nematode surveys

While in an ideal world there are many things that we would wish to do in relation to studying, say, the impacts of an industrial effluent, we have to accept that in a practical, commercial situation these are not always possible to achieve. Indeed, in several cases there are some very real and serious constraints and to deny these would be deceitful. For present purposes, we need to be aware of these constraints and how they affect ecological surveying and the use of nematodes as bioindicators. Experience shows that the problems fall into three categories. In each case it transpires that nematode assessments offer distinct benefits over using other faunal groups that can help to minimize the effects of these limitations.

Historical constraints

Although this is changing slowly, it is rare to find established operations such as factories, refineries, power plants and even domestic wastewater treatment works that have pre-construction, baseline ecological data relating to their surroundings and receiving environments for any of their waste discharges. If such data are available, it is even rarer to find that they encompass assemblages of soil or sediment nematodes. Consequently, an ecological assessment is forced to examine spatial patterns in the structures of communities and to make comparisons with assemblages present at what are assumed to be appropriate control sites. In this situation, nematode assessments have a clear advantage over those that employ other groups. The complexity of their communities makes statistically rigorous assessment of relationships between physico-chemical parameters and community structure possible. This approach can and usually does incorporate bioreactive materials specific to the relevant industrial operation, which, in the absence of the baseline data, can establish the likely extent of the effects that have been caused.

Financial constraints

Restricted funding of regulatory bodies and competing budgetary demands in industry, can preclude collection and analyses of large numbers of samples, which, in turn, can affect the size of survey areas and/or sampling sufficiency and the numbers of control sites that can be examined. This leads to direct and obvious consequences and overall will compromise the ability to discriminate between anthropogenic effects and natural variation in communities. Again nematode-based assessments can partially offset this problem. As noted in the above section Benefits of Using Nematodes as Bioindicators, assemblages of nematodes encapsulate higher spatial and temporal information than communities of larger invertebrates and, at the same time, fewer, smaller samples may be required to achieve sampling sufficiency.

Time constraints

For surveys to be of value to customers there are several situations where results are needed as soon after the fieldwork as possible. This is particularly true where loss of containment incidents (spills and leaks) are being assessed and where the surveys form part of compliance monitoring and are needed for the issue of operating consents. A year or six months later is not necessarily what insurers, lawyers, regulators and factory environmental managers want to hear. While this places demands on laboratory and taxonomic staff, it is usually the biometrics and statistical studies that suffer. Not until all faunal and physicochemical data sets are in place and have been checked and signed off, can the mathematical analyses commence. Beyond the standard indices (Level I analyses) and community structure (Level II) analyses, the pressure of limited final reporting time can curtail the range of exploratory analyses that can be undertaken (Level III analyses; see descriptions of analyses given above and protocol in Fig. 12.1). 'If we had had more time we would have…' usually pre-empts a wish list in the notes returned with the outputs of the analysis. Again, nematode surveys help in this area. Investment of time and resources in analysing species-rich nematode data sets is rewarded more rapidly with mathematically coherent results. This can be seen in the agreements between multivariate classification and ordination analyses and in the results of indicator species analyses (ISAs), which use tiers of Monte Carlo permutation tests to assess statistical significances of the relationships between individual species and mathematical clusters of communities (Dufrêne and Legendre, 1997).[1] Reasons for this are many but primarily derive from the high species richness of nematode communities which retain valuable ecological information in the presence and patterns of abundances of large numbers of locally 'rare' taxa (see Cao *et al.*, 1998, 2001). It is also a reflection of their greater spatial fidelity, their shorter generation times and of the intimacy of their contact with the soil/sediment pore water which mediates the effects of bioavailable soil and sediment contaminants.

On a professional front, all limitations that may have affected the results of the survey and/or affected their interpretation have to be reported to the customer, (see Institute of Ecology and Environmental Management code of professional conduct (IEEM, 2005), and their survey guidance documentation (IEEM, 2006)). Equally, these have to be spelled out in any evidence prepared on the basis of the results of the survey data.

Standard commercial nematode survey protocols

The incorporation of standard, published soil and aquatic nematode surveying techniques into the existing commercial survey programmes is straightforward (see surveying methods outlined in McIntyre and Warwick, 1984; Been

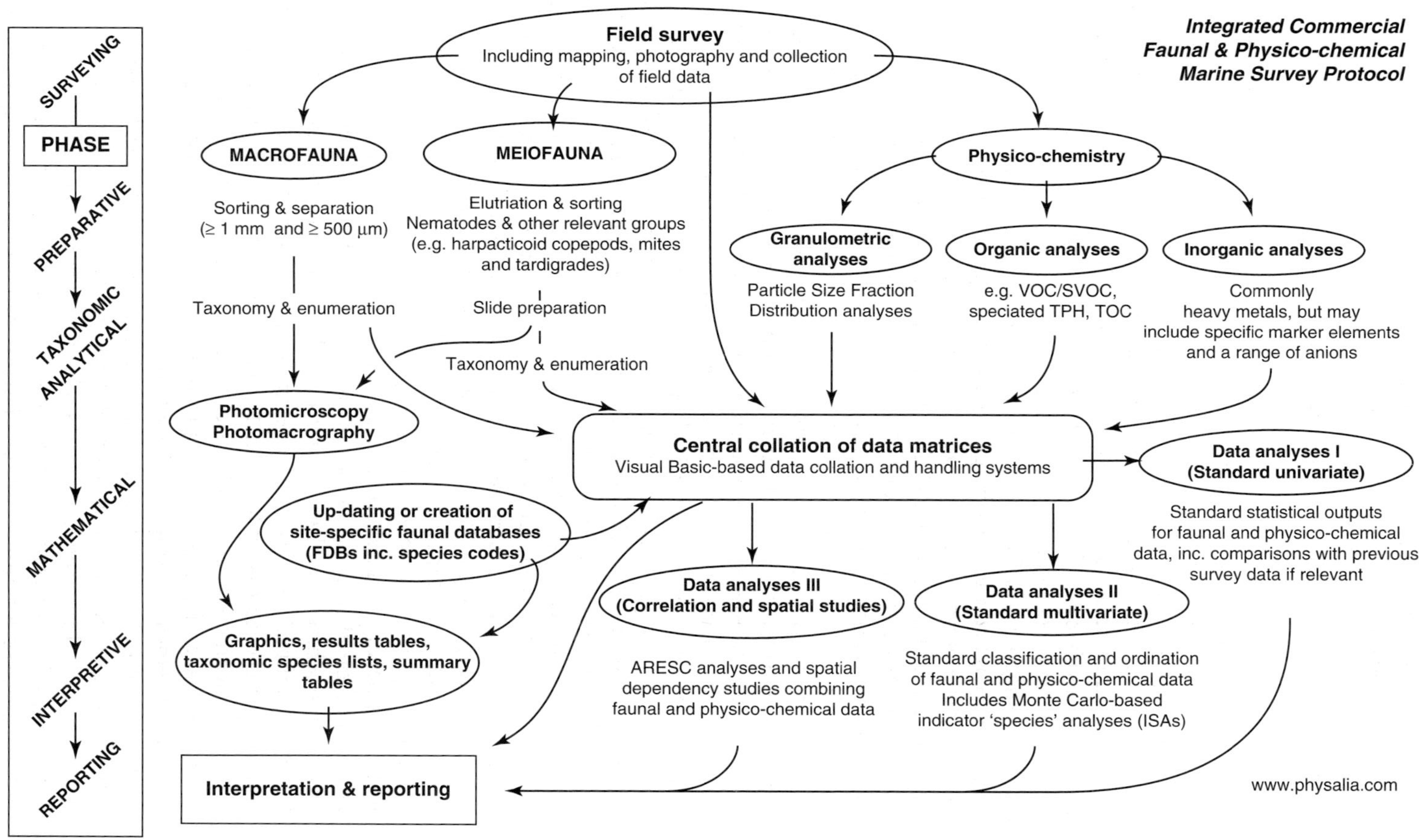

Fig. 12.1. The key operational stages in an integrated macrofaunal, meiofaunal and physico-chemical marine monitoring survey undertaken by Physalia Limited and Nebalia S.L. for industry and regulators.

and Schomaker, 2000 and 2006; Bert *et al.*, 2007). As an example, Fig. 12.1 shows the main stages in an offshore marine invertebrate monitoring survey programme which integrates meiofaunal analyses, and hence nematodes, with analyses of groups of larger invertebrates. Meiofaunal core samples, which, in this particular case for quantitative assessments of nematodes and harpacticoid copepods, are collected at the same sites as samples for two size classes of macrofauna (≥ 1 mm and ≥ 500 µm). In addition, samples are collected at each site for sediment physico-chemical analyses. Granulometric samples provide data on the proportions of different particle size fractions which, in aquatic (and terrestrial) habitats prove to be important natural determinants of nematode community structures (see the section Case Histories presented towards the end of this chapter). The samples for chemical analyses can include organic and inorganic contaminants, depending on the nature of the putative stress being investigated and the presence of any confounding influences (e.g. riverine inputs, neighbouring discharges and offshore moorings). In many cases, specific elements are also assayed for use in inter-element ratio analyses where they function as markers for materials of terrestrial and/or aquatic origin.

Marine survey areas comprise fixed grids of stations in which the sampling sites are usually distributed in a 'herringbone' pattern across a feature of interest such as a point of discharge, a dredging area or a dumping ground. In some marine studies the survey grids may become modified over successive surveys due to observed patterns of net sediment transport and the associated movements of deposited materials on sea and estuary beds or where natural features such as reefs or even wrecks are encountered and preclude collection of grab and core samples. Terrestrial nematode surveys utilize similar grids and are in many ways considerably easier to organize and execute than aquatic surveys. However, a crucial difference in the case of land-based studies is the natural three-dimensional patchiness in the distributions of nematodes in terrestrial and wetland soils (Whitehead, 1977; Powers *et al.*, 1994; Liang *et al.*, 2005, Been and Schomaker, 2006; Jiang *et al.*, 2007). If time permits, pilot studies can enable assessment of a range of site-specific variables. These include distribution of soil types within the study area, effects of existing vegetation on resident nematode populations, depth to which sampling will be undertaken (given soil type and nature of stress being investigated) and the likely numbers of samples needed to gain an adequate description of natural variation (see Been and Schomaker, 2006). In the commercial world, this may not be possible, for example where a spill incident occurs. In these circumstances, the 'if in doubt, do nothing' strategy cannot be applied. As time may be of the essence, over-sampling is necessary followed by selective analyses of the samples collected from within and beyond the areas presumed to have been affected.

A summary of the laboratory protocol used for the processing aquatic meiofaunal nematode samples is shown in Fig. 12.2. With some modifications, this basic protocol has been used in aquatic studies since the early 1980s and has given rise to several separate procedures, including replacement methods for the Whithead tray separators (Whitehead and Hemming, 1965) that were formerly used in terrestrial studies. Special modifications of the illustrated protocol are used for different types of quantitative nematode samples. These

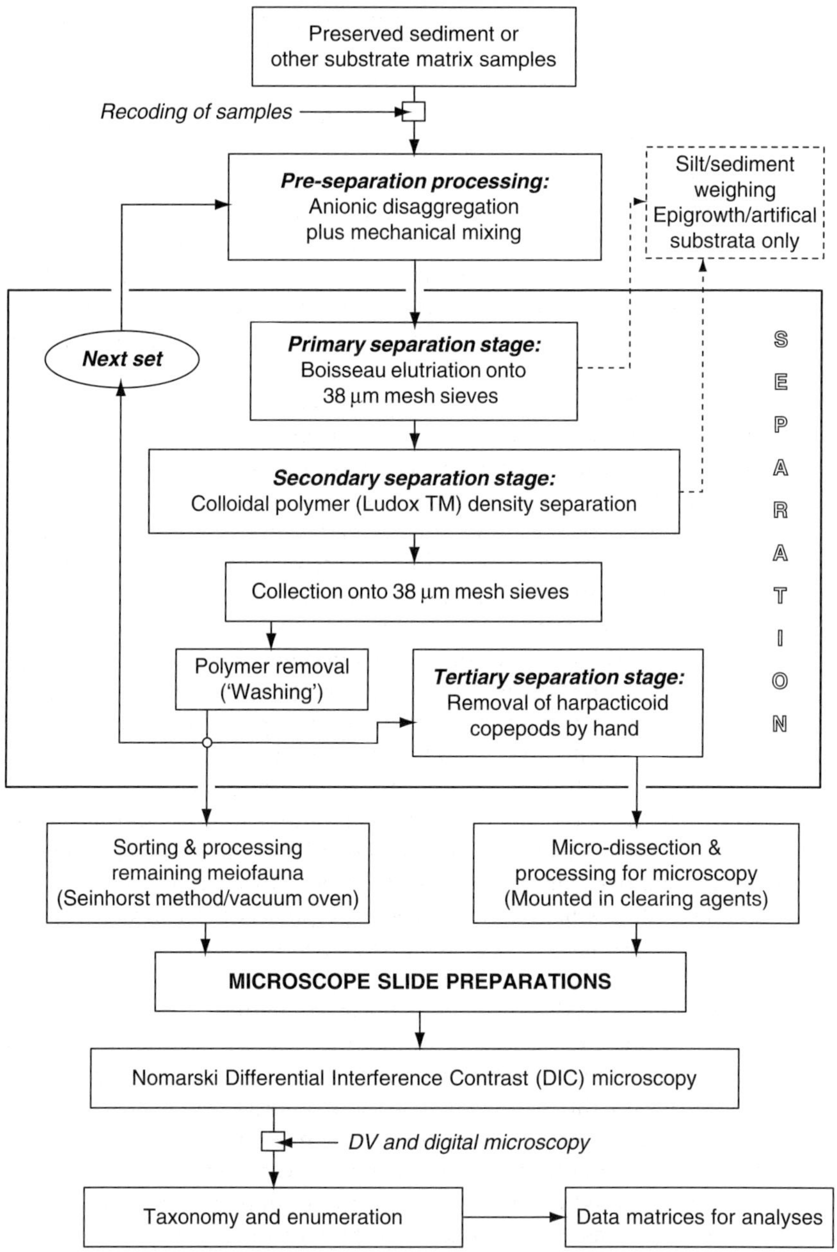

Fig. 12.2. The basic meiofauna processing protocol used and modified by staff at Physalia over many years. This hybrid, three-stage system is further modified to accommodate non-standard meiofaunal samples (e.g. epigrowth samples, artificial substrata collected from nematode colonization units and nondescript materials for forensic examination such as sediment and soils from shoes or the tread of car tyres).

include samples of intertidal algal epigrowth communities present on pier stanchions and river frontages and the artificial substrata from nematode colonization units (NCUs) used in a range of natural and man-made habitats (see below and Selected Case Histories section). Note the use of 38 µm mesh sieves. These were adopted from the earliest surveys and prevent significant losses of small 'straight' adult and juvenile nematodes as opposed to coiled worms (David J. Hooper, Harpenden, 1980, personal communication; see also Hooper, 1986).

All nematode taxonomic work is carried out using Nomarski differential interference contrast (DIC) microscopes with slides made using the Seinhorst rapid glycerol technique (Seinhorst, 1959). This is used in preference to lactophenol, which, although a more rapid technique, is avoided on health grounds. Use of lactophenol (or a modified in-house formulation) is presently restricted to meiofaunal harpacticoid copepod preparations where the clearing of the cuticular structures is important for the copepod setotaxy. In standard integrated macrofaunal-meiofaunal surveys the first 100 adult nematode specimens encountered in a random sample will be identified to species level (or to an operational taxonomic unit within a genus). Total counts are then made of all other nematodes in the sample. In preliminary studies for specific surveys where, for example, post-development assessments will be undertaken or a future monitoring programme is planned, and in cases where a new procedure is being developed, such as the use of nematode colonization units in a new location, the first 150 or even 200 nematodes specimens in a sample may be identified to provide a better understanding of the taxa present and their relative abundances.

Results from the taxonomic analyses include total abundances of nematode feeding-type groups which, for marine/outer estuarine species, are still based on Wieser's original classification (Wieser, 1953) rather than on more recent and potentially more valuable systems (namely Moens and Vincx, 1997). For freshwater and oligohaline habitats, feeding type classifications have corresponded with those described by Traunspurger (1997). These data, along with the final site-by-site results, are entered into standardized spreadsheets. With up to nearly 400 nematode taxa occurring in single 50 station monitoring surveys (e.g. Bass Strait, Tasmania), and cumulative totals of over 1000 taxa in the nematode faunal databases for some survey sites (e.g. Malaysia and South Africa), data entry and the cross-checking with microscope record sheets are time-consuming and arduous procedures. However, quality control and attention to detail is an extremely important part of the work and, just as chain-of-custody forms are required for handover of samples, faunal data sets are not accepted until signed-off as checked and correct by the taxonomists.

Data handling and analyses and the presentation and illustration of the results of the nematode analyses for commercial customers are considered later in this chapter.

Less conventional nematode sampling and assessment methods

In a limited number of circumstances, where the accurate assessment of conditions is required and where nematodes are the monitoring tool of choice, owing to either the paucity or even absence of macrofauna or to the nature of the stress under investigation, standard surveying protocols may be neither appropriate nor possible. Examples include:

- Assessment of the impacts on marginal habitats of increased shading caused by developments located beside watercourses; now a standard requirement of the Environment Agency in the UK for developments along built-up canals and estuaries (e.g. the Thames).
- Evaluation of the extent and degree of effects arising from marine and terrestrial spills that impinge on rocky shores at low or on falling tides or assessment of the stability of patterns of ecological effects in the same habitats arising from domestic and industrial effluents that disperse along tidal rocky shores.
- Monitoring of prevailing water quality using submerged surfaces of a variety of structures such as barrages, pier stanchions, marine wind farm turbines and mooring or channel-marking buoys.

In these situations, nematode communities can be readily sampled using quantitative epigrowth samples. Samples, comprising the algal and/or moss matrices that develop on submerged or on wetted and water-retentive substrates in intertidal habitats, can be collected from a known area, ranging from 25 × 25 cm down to 10 × 10 cm depending on the depth (thickness) of the epigrowth mats. Aspect, depth of immersion and, for intertidal habitats, degree of exposure to wave action and position within the littoral zone are clearly important factors and must be recorded accurately. These data can then be incorporated into multivariate correlation analyses (see section below on Assessment of Nematode Assemblage Data for Commercial Studies).

The laboratory separation protocols outlined above can be used with these samples and the sediment present in the epigrowth matrices is usually collected from the elutriation and Ludox separation stages. These are washed, if necessary, and are then dried slowly to constant weight, and used as a measure of the equilibrium between accretion and erosion rates that existed at the site sampled.

One step removed from the use of a natural epigrowth matrix is the deployment of artificial matrices that become colonized by modified species assemblages. These pads have several benefits. Foremost among them is that the matrix, a coarse, inert fibre pad, can be standardized, reducing the effects of variation in microhabitat. These can also be deployed in situations where natural, photosynthetic matrices (plant and algal epigrowth) could not occur (e.g. within cooling water systems and water abstraction culverts). This approach has been used in a range of projects where controlled sampling would otherwise have been difficult and where the role of nematodes as bioindicators was more important than characterization of effects on natural assemblages (see section on Selected Case Histories below). For sample processing, the standard protocol is modified to include two or three 'washing cycles' in which the colonist fauna is flushed from the pad using running water.

Assessment of Nematode Assemblage Data for Commercial Studies

Ecological data are noisy and in this respect nematode community data are no exception. However, some data are noisier than others and nematode data, thankfully, do not fall into this category, especially when their communities are constrained by adverse environmental conditions. The reasons for this have already been presented in and are the key to understanding the value of nematodes as bioindicators. In spite of this, there is no single community statistic or set of results from a mathematical analysis that gives access to a definitive appraisal of the effects of prevailing environmental conditions. To extract this information, we need to resort to a variety of techniques. In a commercial situation, this data mining process needs to be efficient and well focused but should not prevent exploratory investigation or in any way blinker the final interpretation. This, coupled with judicious use of Occam's razor,[2] leads to the most parsimonious and defensible analysis of nematode survey results and provides the basis for a robust ecological interpretation of prevailing conditions.

The analyses used in most Physalia commercial studies can be divided into two categories; Level I analyses, which encompass standard uni- and bivariate approaches, and Level II analyses that use multivariate community techniques. A third category (Level III analyses) is employed in more advanced studies where there may be a need to identify and rank measured environmental parameters, natural and anthropogenic, in terms of their relationships with nematode community structures or specialized, bespoke analyses for a wide variety of non-standard survey situations. Examples of these three levels of analyses are presented in the sections below. As will be seen, graphics summarizing the results of the analyses form an important part of the reporting process and strong emphasis is placed on relating the analytical outputs back to field situations (see, for example, Case history 3, below). This helps in the process of communicating the findings of surveys to audiences that are usually experts in areas other than ecology.

Level I analyses: standard univariate studies

Standard community description parameters are generated automatically when the data sets enter the system. For each site, these include standard parameters such as species richness and densities recorded as numbers/litre soil or sediment or as numbers/m^2 substrate, depending on the type of sample analysed. Diversity indices are also calculated, usually reciprocal infinite and finite Simpson's dominance indices (1-D; Simpson, 1949; see also Hill, 1973), preferred as this is less influenced by the occurrence of rarer nematode taxa and therefore small variations in sample size (Whittaker, 1972). Finally, simple dominance indices (percentage abundance of the most abundant or the two most abundant species) are generated along with densities of individual nematode feeding types and ratios of Weiser's type 1B to type 2A ('non-selective'[3] deposit feeders and detritivores to epistrate feeders) and their equivalents in freshwater systems (see Moens *et al.*, 2006). At present the maturity index (MI; Bongers, 1990) is not part of the suite of standard parameters examined in the analyses. This may change if the MI can be modified to give higher resolution than family level and/or if the index and the ascription of colonizer-persister (c-p) values can accommodate the revisions that are now occurring rapidly in the taxonomy of the Nematoda following the most recently published genomic studies (see, for example, De Ley *et al.*, 2006 and Meldal *et al.*, 2007).

As well as being reported directly in the final survey documents, the standard nematode community parameters fuel other analytical lines. Overlaying plots of proportionally-scaled symbols for the parameters onto maps of the survey areas provide a rapid visual means of assessing the distribution of the data with respect to features of interest (e.g. factory discharges; Fig. 12.3). When they are combined with appropriate parametric and non-parametric statistical tests, these are particularly valuable in detecting change in time series data derived from monitoring surveys. Standard *k*-dominance curves are generated from the ranked dominance values and allow communities to be compared *en mass*. These are valuable Level I tools for coarse screening of the nematode data sets and also enable the validity of comparisons between their diversity indices to be determined (see Platt *et al.*, 1984).

In many cases data are available from the same sites as the nematodes for two size classes of macrofauna (≥ 1 mm ≥ 500 µm size classes). The Level I analyses then generate *Nebalia plots*[4] which highlight differential responses in the communities by rescaling the species richness data for each group at each site (Fig. 12.4). The species richness index (S_i') is derived simply as:

$$S_i' = ((S_i \;/\max(S_i ,\dots,S_n))^2$$

where S_i is the species richness recorded for one of the faunal groups at a given site and max ($S_i ,\dots,S_n$) is the maximum species richness for that faunal group recorded at all sites examined. Values for the index S_i' range from 0.20 to 1.00.

Level II analyses: multivariate studies

Avoiding models that assume linear responses of species to environmental gradients (e.g. principal components analysis and discriminant analysis), patterns in the structures of the nematode assemblages are examined using transformed data matrices and a suite of both classification and ordination methods (see Neher and Darby, Chapter 4, this volume). This strategy has distinct advantages. As the two groups of multivariate techniques differ fundamentally in their underlying rationales, and hence the algorithms that they employ, the coherence of the clusters of assemblages indicated in the classification studies can be examined in the light of the results of the ordination analyses. Any deviations between the implied relationships for the

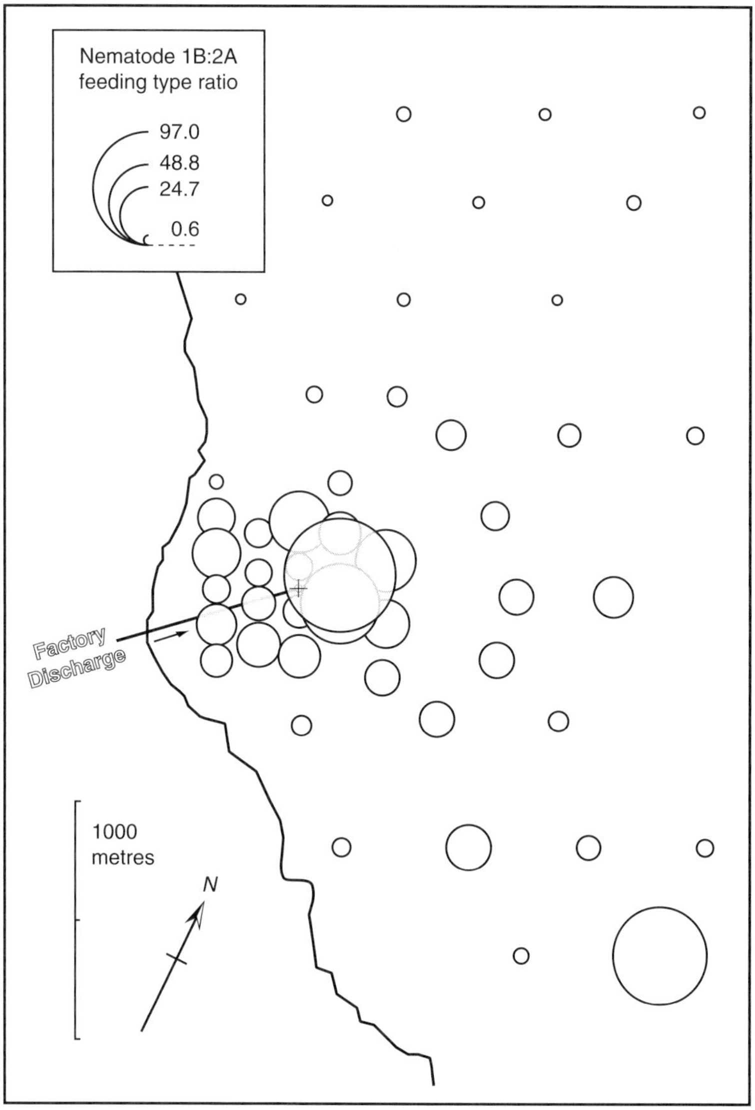

Fig. 12.3. Example of a scaled-symbol plot used in reports to relate ecological and physico-chemical data back to features within survey areas. This marine example shows the nematode 1B:2A feeding type ratio (i.e. the ratio between the densities of deposit-feeding species and epistrate feeders at each site) which can highlight differential effects on the two feeding groups. Here the high ratio seen at the end of the industrial outfall reflects the elevated abundances of xyalid species as well as reduced numbers (and species richnesses) of members of the families Chromadoridae, Cyatholaimidae and Microlaimidae. The offshore high value at the southerly end of the grid reflects a deeper water site with higher proportions of fine sediment fractions (< 63 μm particles) that naturally select for deposit feeders (Type 1B species).

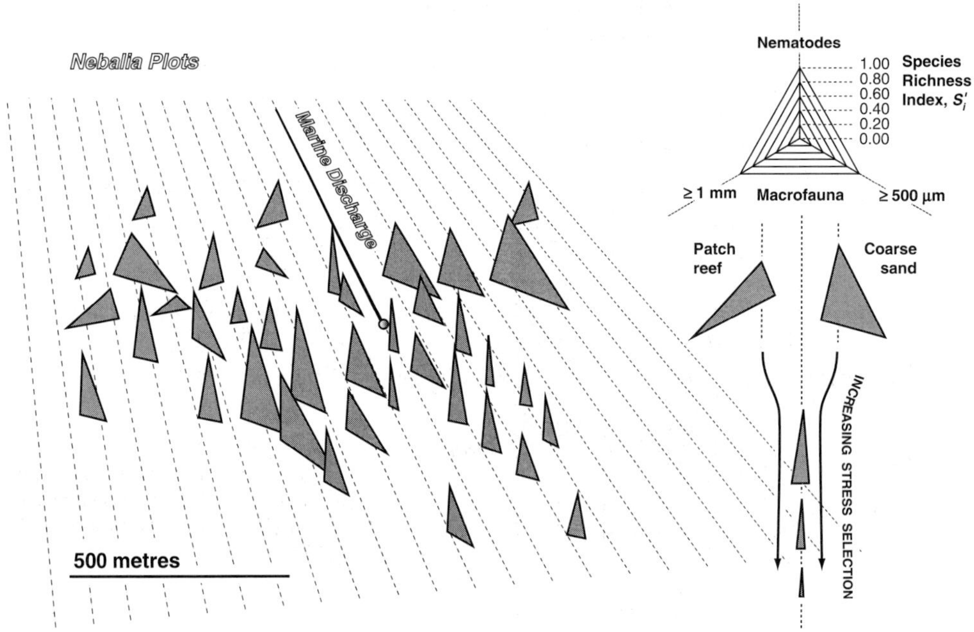

Fig. 12.4. Nebalia plots examining the responses to an industrial effluent of species richness indices of two size classes of macrofauna (≥ 1 mm and ≥ 500 µm) and of meiofaunal nematodes. The detail, taken from a central section of a survey area, encompasses patch reef sediment as well as coarse, mobile sand communities. Note the change in the community structures reflected in the shape and size of the triangular symbols. These effects did not relate to sediment granulometry and evidence from correlation studies indicated that effluent products present in the sediments were implicated in the shaping of the assemblages.

assemblages are then investigated more closely. It should come as no surprise that fewer inter-analysis anomalies occur with nematode data sets than with those derived from the macrofaunal groups.

An example of the results of this process of cluster coherence mapping (CCM) is presented in Fig. 12.5. The plot includes the identities of the mathematical indicator species wherever these showed statistically significant relationships with the clusters[1]. These species are identified using indicator species analyses (ISAs) in which maximum indicator values (IV_{max}) for each species in each cluster are tested against those generated by randomization procedures based on Monte Carlo permutation tests (see Dufrêne and Legendre, 1997).

Level III analyses: multivariate and other studies

No formal protocols exist for the Level III analyses and the biostatisticians among us appear to enjoy the limited freedom that this gives them to explore different possibilities (as opposed to probabilities?). Multivariate correlation analyses are commonly included at this stage to probe the combined nema-

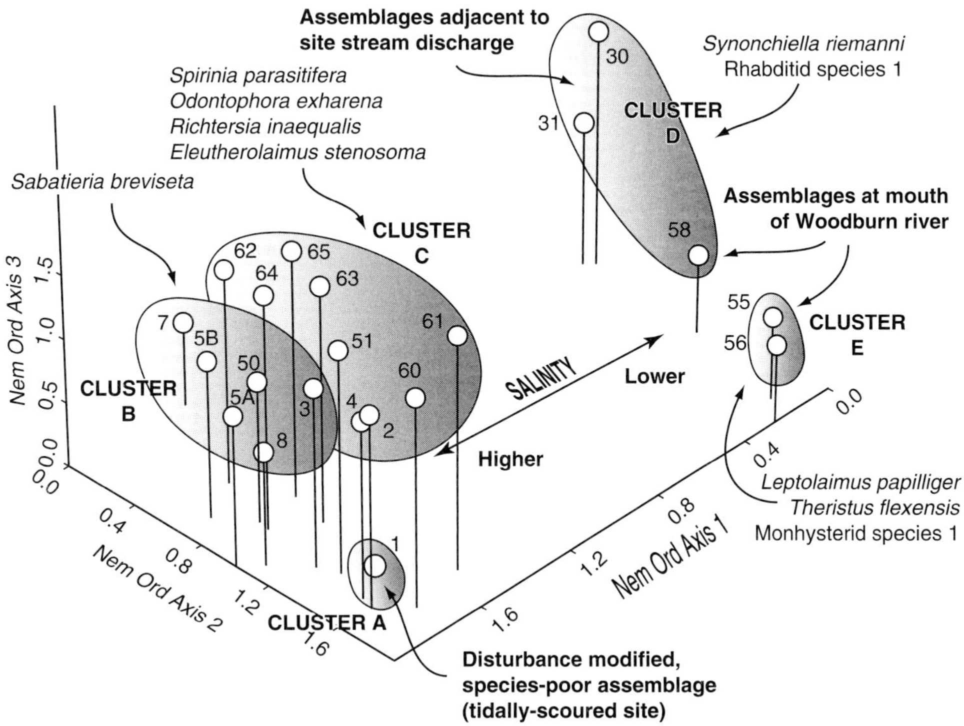

Fig. 12.5. A summary ordination plot taken from a Physalia nematode monitoring survey of a contaminated land site bordering Belfast Lough, Northern Ireland, following remediation works. This multi-purpose plot was used to show the coherence of the clusters identified in the multivariate classification analyses (CCM), to display the results of the indicator species analyses (ISAs; $P = 0.05$) and to highlight the natural effects of a salinity gradient that existed in the intertidal area adjacent to the site that became the principal determinant of nematode community structures following the remediation works.

tode community and soil/sediment physico-chemical data matrices (see examples shown in Case history 4, below). While these approaches tend to focus on *statistically* significant relationships between community structures and the measured environmental factors, assessment and ranking of *ecologically* significant contaminants (ARESC) studies acknowledge the fact that lower order relationships when multiplied up over many generations can lead to marked changes in community structures. This is particularly relevant when using nematodes as bioindicators where generation times (egg-to-egg) can be as low as three days (Vranken and Heip, 1983; see also above section on Benefits of Using Nematodes as Bioindicators).

In recent years (2000 onwards) our commercial monitoring surveys have focused on detecting the effects of these more subtle environmental factors on meiofaunal nematode assemblages. This is linked to industrial and regulatory interest in gaining early warning of potential changes associated with chronic as opposed to acute stresses. A spin-off of this work has been the use of techniques that examine spatial dependency (see Haining, 1990). An example of this work is presented in Fig. 12.6. In this particular case, evidence was

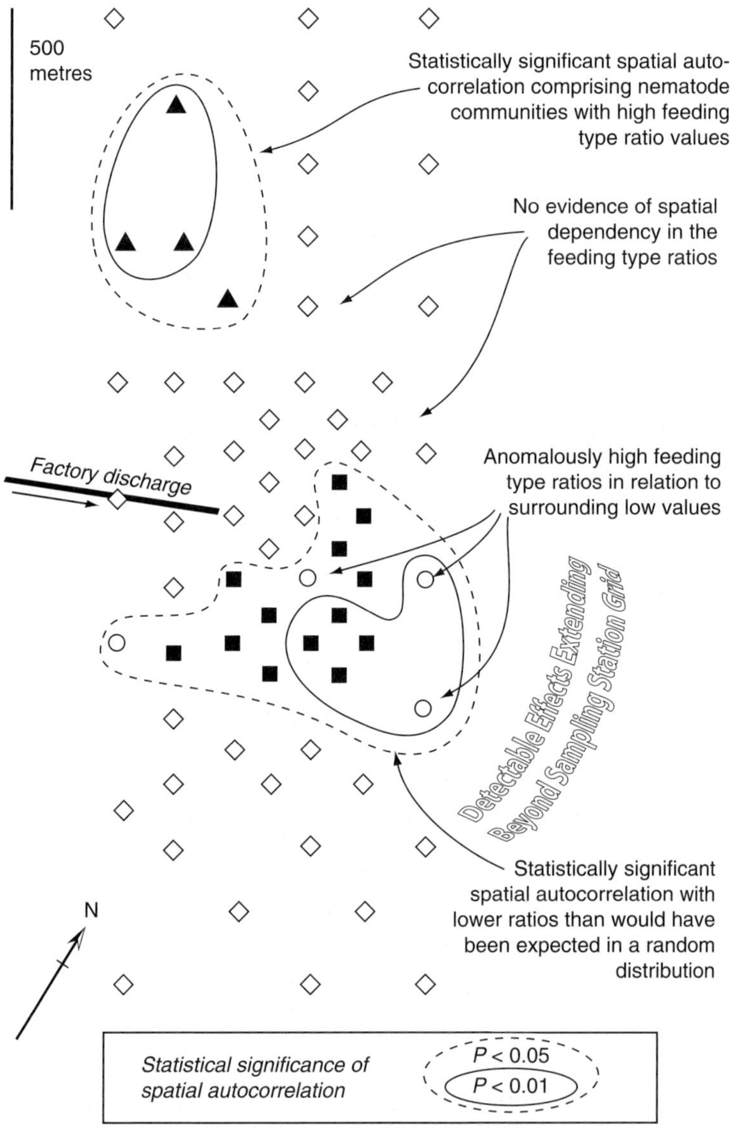

Fig. 12.6. Spatial dependency studies are used by Physalia and Nebalia in industrial and regulatory surveys to examine the extent to which ecological parameters at any site may be associated with those at surrounding sites. This is accomplished by a combination of spatial autocorrelation and local indicators of spatial association (LISA) analyses. In the present example, nematode 1B:2A feeding type ratios were not found to show 'global' autocorrelation effects but statistically significant local effects were detected. The high values (top left) correlated spatially with a zone of finer sediments. The low values present in a zone adjacent to the end of the outfall were all lower than would have been expected based on a random distribution of values. The study indicated that the survey area was not sufficiently large to describe this 'subtle' nematode community feature adequately. The following year, additional sites were established to rectify this.

found that the survey area was insufficiently large to assess the full extent of these more subtle effects and to gain a full description of the patterns generated in the nematode communities.

Where Level III studies are undertaken in connection with long-term monitoring survey programmes they may also include the development of site-specific (customer-specific) indices that act as a measure of 'ecological performance'. These frequently include nematode community data and, although most commonly used for internal reporting purposes, they are occasionally reported to regulatory authorities.

Selected Case Histories

Case history 1, fouling studies: early warning of biocide failure

Customers: An international service-chemical producer and several industries (UK, Ireland, Belgium and Spain)

'Once-through' cooling water systems are used for a variety of industrial plants including power stations, refineries and chemical manufacturing facilities. Owing to the constraints placed on temperatures of waters returned to inland waterways designed to prevent the adverse ecological effects that poorly dissipated hot waters can exert, most of these cooling systems are used for outer estuarine and coastal installations. However, this potentially exposes the internal architecture of the cooling systems to colonization by a range of marine and brackish water fouling species (e.g. hard-fouling species, such as the bivalves *Mytilus edulis* and *Dreissena polymorpha*, serpulid polychaetes and a variety of barnacle species, as well as soft-fouling species, such as hydroids, Porifera, Bryozoa and Ascidiacea). Traditional prevention includes continuous chlorination of the incoming cooling water which kills the settlement stages (larvae/propagules) at the point of intake.

With the general move away from the use of chlorine, not just on health and safety grounds, but also due to the quantities of total residual oxidant (TRO) that can remain in discharged cooling water and the potential for chloramine formation, many operators, including power stations, have changed to pulsed biocide systems where 6 or more, 15 to 30 minute treatments are applied in a 24-hour period. These use various biocide formulations but the most effective appear to be based on hypobromous acid combined with proprietary additives as these rapidly attack membranes of fouling species as well as the byssal holdfasts of mussels. An important feature associated with pulsed biocide systems is that the majority of the time, the target organisms have already settled on the walls and internal surfaces of the cooling systems. Therefore any failure in the treatment system resulting from poor mixing and circulation of the biocide or periodic increases in particulate organic matter in the intake waters that reduce the biocidal effect, (e.g. on some spring tides or during autumn months) can result in a gradual build-up of fouling communities within the cooling system. The cost of this in terms of facilitated anoxic corrosion (mediated by sulphate-reducing bacteria), increased hydraulic resistance to pumping, reduced heat transfer efficiencies (and therefore, in power plants, reduced generating capacities) and in some rare cases, physical damage to condenser systems where hard-fouling species detach and shred condenser pipes, is immense.

Fouling studies undertaken by Calvo Urbano (1995) had already shown that following microbial biofilm formation, aquatic nematodes were often among the first metazoans to colonize the new substrata. This led to the development of a biocide calibration and monitoring system in which settlement units, created from inert plastic matrices held within open-ended, high density polypropylene tubes, were introduced into and secured at different locations within a given cooling water system. Establishment of meiofaunal nematodes within the settlement tubes was used to indicate that target concentrations of the pulsed biocide were not being achieved in a given section of the cooling water system and this formed the basis for calibration of dosages. On an academic note, in UK outer estuarine situations it was noted that Leptolaimidae (e.g. *Leptolaimus* species) and some Monhysteridae (e.g. *Diplolaimella ocellata*) were frequently amongst the earliest colonists, even though these were either poorly represented or in many cases not recorded at all in samples of sediments collected in the immediate catchment of the intakes. The reason for this remains uncertain.

Case history 2, source identification: paper mill complex

Customer: A mill producing paper for newsprint in southern England

In 1990, a papermaking machine that had been producing newsprint paper almost non-stop for over 25 years, save for annual maintenance outages, suffered a series of failures over a period of two weeks that led to a costly close-down of the production unit. Examination of paper that was ejected from the machine on each occasion showed that its tensile strength had been compromised by the presence of pinkish, gelatinous materials which led to tearing at the point at which the paper was drawn off at high speed from the felt belt used for de-watering the paper pulp. When the gelatinous materials were examined by the mill's technicians under low power stereo microscopes their attention was attracted by the presence of live nematodes. We identified these as juvenile rhabditids belonging to several species (Rhabditida; Rhabditidae (incl. *Diploscapter* species) and Neodiplogastridae (incl. *Mononchoides* species cf. *striatus*).

The source of these nematodes was a mystery as was that of the gelatinous materials which transpired to be rapidly growing colonies of slime-forming bacteria. The pulp that fed this machine was formed at high temperature using chemically sterilized waters drawn from lower salinity reaches of an adjacent estuary. The feed stock was a mixture of 'virgin stock', made from Scandinavian pine chips, and a percentage (less than 20%) of de-inked, recycled paper. This was cooled and thinned using process water supplied by the site operators to all mills in the complex from one of several deep boreholes in the surrounding chalk.

Checks failed to reveal any construction works being undertaken on the estuary that might have disturbed sediments and caused elevated quantities of organic matter to be drawn into the plant. This could potentially have affected the chemical sterilization procedures and might have enabled live nematodes to enter via this route. Equally, the nematode species, being primarily freshwater species, were not recorded in sediments near to the estuary intake. It was also clear from this part of the investigation that the sterilization procedures at that time were highly efficient.

Working back from the pulp production facility and collecting samples from the cooling water feed line, live nematodes were discovered in several samples. The search eventually led to three of the deep boreholes situated beyond the site boundary. Samples taken from two of these yielded live nematodes and the remains of several crustaceans, including Cladocera (*Bosmina* and Chydoridae species), as well as Ostracoda (possibly *Eucypris* species). Geophysical surveys undertaken as a result of these findings indicated that the exceptionally dry summers of the late 1980s had probably led to fracturing of the upper chalk substrata and a dried out surface watercourse was identified that was most probably responsible for the contamination of the waters drawn up from the two boreholes. The nematodes that survived the subduction process and enabled the tracing to be carried out successfully were subsequently identified in samples collected from a tributary of the former stream. While this untreated deep borehole would have explained the source of both the bacteria and the nematodes, it did not explain why other mills which also used water from these boreholes had not experienced similar problems.

The answer to this came from the pumping records for the mill's own estuarine abstraction. To meet increased production demand, water had started to be drawn at the turn of the tide instead of on the full ebb tide. This resulted in the entrainment of the effluent plume from the sewage works downstream of the paper mill complex. It is highly likely that this change provided the nutrients in the chemically sterilized waters that fuelled the growth of the slime-forming bacteria in the water used to cool the pulp feed stock. The warm conditions would undoubtedly have promoted the proliferation of the bacteria and the saprophagous/microbivorous nematodes.

Apart from its annual maintenance outages, this papermaking machine has been running continuously since we examined the plant in July 1990.

Case history 3, contaminated land assessment: explosives factory

Customer: ICI Nobel Explosives, Scotland

Aware that the history of usage of an explosives burning ground may have left a legacy of contamination at their factory site on the west coast of Scotland, the factory's technical staff investigated options for assessing the nature and extent of any problems that might have existed. The site, located on sand dunes, had ceased to be used for 'deactivating sensitive chemical materials' some eight or more years beforehand. Areas that had formerly been used for the burning could no longer be identified as they had grown over with marram grass, *Ammophila arenaria*. Chemical analyses of the sandy soils failed to detect the residues of any organic compounds associated with the explosives that were burned on the land. However, significant concentrations of several heavy metals were identified in some areas. These metals derived from batches of 'electrochemical devices' (detonators) that were destroyed on the site from time to time.

As part of their company's site assessment procedures, they wanted to identify and describe areas showing *actual ecological effects* and to define and prioritize any zones that might require remediation. The macroinvertebrates present on the site were primarily surface-dwelling, mobile arthropods (chiefly spiders (e.g. Lycosidae) but also beetle species (Carabidae and Staphilinidae)). Clearly, the distribution and densities of these would not provide an accurate reflection of any stresses operating within the soils. Below ground, larger soil-dwelling invertebrates were sparse and patchily distributed and were characterized by low densities of dipteran larvae (e.g. Tipulidae) and some campodeiform beetle larvae (species indet.), while annelid species were almost completely absent from the sandy soil (observations restricted to low numbers of Enchytraeidae).

At the same time, meiofaunal analyses were being used elsewhere within the company to identify potential ecological effects associated with former landfill sites. This approach appeared to offer an ideal solution for the burning ground investigation. The surveys that were undertaken examined communities of soil meiofauna in a grid of 30 fixed sampling stations. At each site, samples for soil physico-chemistry were collected along with the meiofauna. In later studies, samples were also collected for soil microbial analyses (see Ellis *et al.*, 2002).[5] Of the 11 transient and permanent meiofaunal groups recorded in the soils, Nematoda were, as always, the most diverse and abundant group with up to 77 species, including representatives of over 21 families, being recorded in a single survey. Nematode densities were comparatively modest, ranging from just 25 nematodes/litre soil to more than 2200/litre (equivalent to over 660,000 nematodes/m^2). Whilst this clearly reflected the sandy soils of the dunes, evidence from the dominance-diversity structures of the communities indicated that stress-selection was occurring at several sites. Mapping of nematode co-dominance values (percentage abundance of the two most abundant species) indicated that these effects were concentrated in the central section of the survey area (Fig. 12.7). As can be seen, these high values coincided with the clusters of structurally-distinct nematode assemblages, identified in multivariate classification analyses and found to be coherent in ordination analyses. Equally noteworthy was that these clusters exhibited contiguous distributions within the survey area.

Distributions of total metal burden indices (TMBIs – site-specific measures of the net loading of soils with metals; see Trett *et al.*, 2000), provided

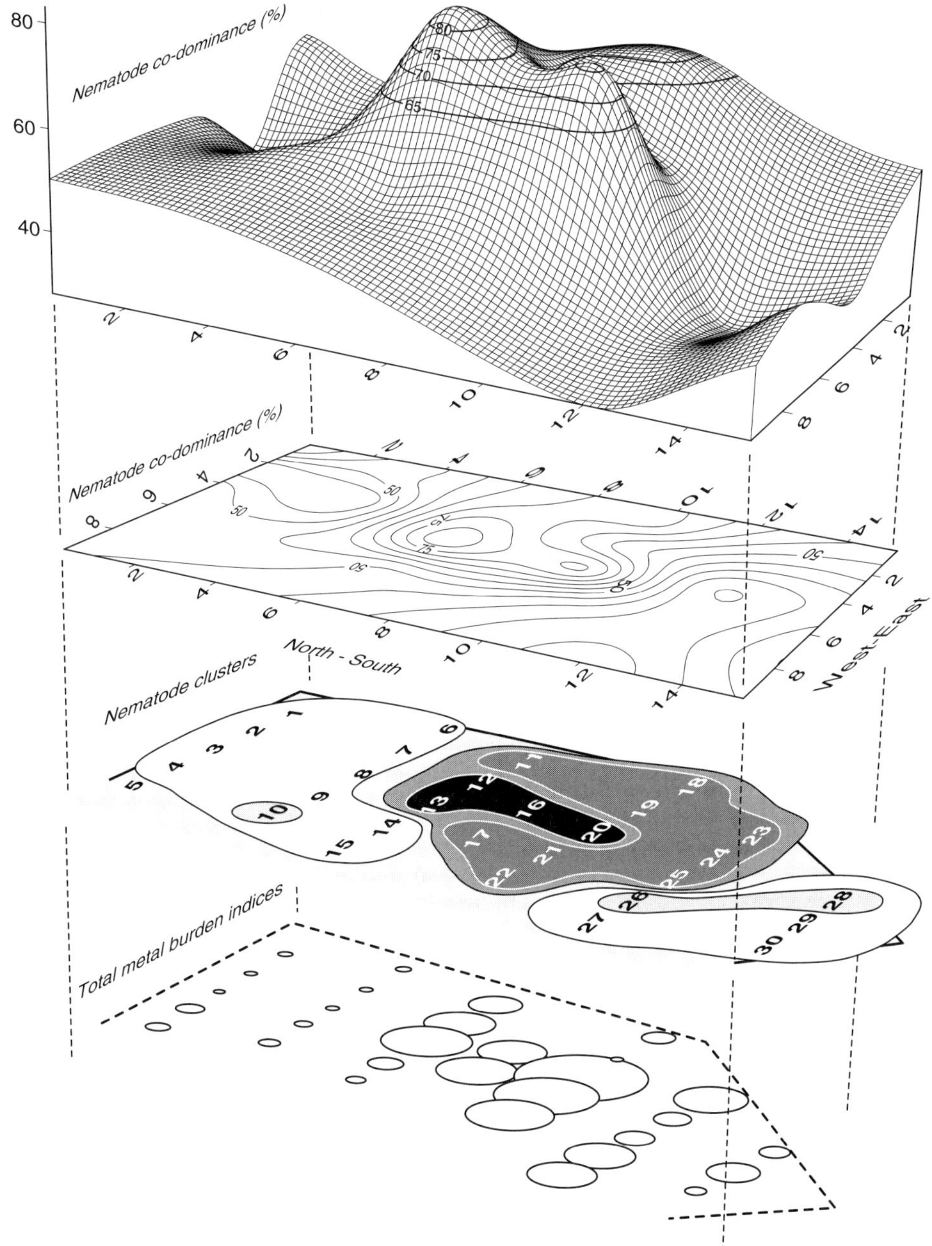

Fig. 12.7. A composite diagram showing the results of one of the terrestrial nematode surveys undertaken on the former burning ground at ICI Nobel Explosives in Scotland (see Trett *et al.*, 2000). The poorer nematode assemblages exhibiting the highest dominance (% co-dominance) values were contained in two, spatially contiguous clusters identified in the multivariate community analyses (see darker shaded clusters on the cluster map). These correlated spatially with elevated total metal burden indices (measures of the net metal loading of the soils) and were shown to be statistically significantly correlated with soil concentrations of cadmium, copper, vanadium and aluminium (see text).

strong indication that these were associated with poorer, more modified nematode assemblages (Fig. 12.7). This was confirmed using the ARESC correlation approach (see section above on Level III mathematical analyses), in which the soil concentrations of four metals were shown to have statistically significant correlations with the structures of the nematode communities, namely cadmium, vanadium, copper and aluminium. Of these, the soil aluminium was a surprise as the observed concentrations (range: 2290–34,800 ppm) were below or well within the 'typical soil ranges' (10,000 – 300,000 ppm; Levinson, 1974; Bowen, 1979). However, closer examination of records of site use showed that the detonators included aluminium-cased devices and this indicated that aluminium was a coincidental correlate. In other words, the patterns of variance in the aluminium distribution would have closely matched those of the metal sensitizers used in the ignition charges.

These studies, reported in part in Trett *et al.* (2000), enabled the areas requiring remediation to be identified and mapped accurately. Given the history of the site usage, the size of area identified as supporting modified assemblages was smaller than might otherwise have been expected. The benefits of this were that it enabled resources to be targeted on sites requiring remediation while minimizing disturbance to surrounding land. This, in turn, permitted a larger reservoir of natural plant and animal communities to be conserved, aiding recolonization of land disturbed during subsequent site works. The site itself has now been sold on and has been developed by the new owner.

Case history 4, performance monitoring: liquid wastes discharges to coastal waters

Customers: Operators on a coastal industrial complex in South Africa

The coastal industrial complex supports over 20 manufacturing plants which include paint and dye factories, a pigment manufacturer and several specialist chemical producers. After treatment, liquid wastes from these factories are pumped to sea via two long outfalls, which discharge at a distance of between 1.5 and 2 km offshore in approximately 30 m depth of water. The seabed of the highly active coastal zone system is characterized by medium to coarse sands and patch sandstone reef. Principal effluent components of ecological concern were heavy metals and one acidic effluent stream, which was particularly rich in iron, manganese and zinc, contained lower concentrations of chromium, vanadium and arsenic. Since the early 1980s, annual sea surveys had been undertaken examining macrofaunal communities in a 6 × 2 km survey area located across the ends of the outfalls and parallel to the shoreline. This comprised a grid of up to 130 sampling stations.

Natural variability of the macrofaunal communities collected on 1 mm mesh sieves had proved to be too high to reliably identify and monitor effects year-on-year. As a move towards increasing the sensitivity of the monitoring programme, additional sediment samples were elutriated onto 500 μm mesh sieves to yield what were termed 'small invertebrate' community fractions. These were dominated by interstitial polychaete worms, amphipod shrimps and groups such as Turbellaria, Nemertea, Sipuncula, Cumacea, Echiura and bivalve Mollusca. More useful than the macrofauna, these provided evidence of affected communities extending up to 1.5 km across the ends of the outfalls. This encompassed a zone of strongly iron-stained sediments and this was taken to indicate that causes were associated with the metal discharges.[6]

From the late 1980s onwards, meiofaunal analyses (quantitative meiofaunal nematode and harpacticoid copepod studies) were added to the survey. From the earliest days, over 300 nematode taxa were routinely recorded in the studies and the distributions of community parameters such as species richness, diversity indices and community dominance/co-dominance values indicated the existence of extensive effects on the seabed that did not necessarily coincide with the sites of iron-stained sediments. Standard multivariate techniques (classification analyses combined with ordination studies) also enabled zones of nematode communities with modified structures to be mapped.

With a higher resolution monitoring system in place, attention was focused on the causes of the observed impacts and steps were taken to reduce these. By 1997, modifications to effluent densities and improved effluent mixing at sea had reduced the extent of the affected areas on the seabed (see 1997 marine map in Fig. 12.8). At this stage, correlation studies using data from more comprehensive sediment physico-chemical analyses were being incorporated into the surveys specifically for use in the meiofaunal studies. Figure 12.8 shows the 1997 statistically significant correlations between the sediment physico-chemistry and nematode community structures. Note that concentrations of copper, lead and nickel are naturally high in these coastal waters and are marker elements for materials of terrigenous (lithogenic) origin.

Over the following seven years, significant capital expenditure was made by the owners of the factory complex to address the effects at sea. The year-on-year results of the meiofaunal nematode and physico-chemistry correlation analyses were used to track and document changes at sea. Over this period the measures implemented included:

- Reductions in metal loads (mass) discharged to sea
- Reductions in net volumes of effluent
- Reclamation of materials from waste streams that could be sold on as co-products
- Closure of older effluent holding facilities
- Modifications to the designs of diffuser sections on the discharge pipelines to work more efficiently at reduced pressures
- Partitioning of the two effluent mixing zones by modifications to the pipeline geometry
- Pre-dilution within the pipeline of one discharge stream by mixing with seawater drawn from beneath the beach.

The last measure listed above was highly innovative, requiring the construction of a state-of-the-art pumping and effluent mixing facility that both neutralized

1997

☆ AXIS 1 Determinand	Pearson's *r*	AXIS 2 Determinand	Pearson's *r*
■ Fe	–0.482 **	125–250 µm	0.798 **
> 2000 µm	–0.477 **	500–1000 µm	–0.706 **
Ni	–0.473 **	1000–2000 µm	–0.632 **
Al	–0.460 **	63–125 µm	0.472 **
Cu	–0.457 **	250–500 µm	0.463 **
TMBI	–0.452 **	□ Mn	–0.434 **
TOC%	–0.448 **	Ni	–0.343 **
■ Pb	–0.444 **	■ Fe	–0.338 **
□ Cr	–0.415 **	Cu	–0.299 *
□ V	–0.357 **	Al	–0.288 *
□ Zn	–0.303 *	< 63 µm	0.255 *
500–1000 µm	0.234 *	□ Zn	–0.236 *
□ Mn	–0.233 *		

2004

☆ AXIS 1 Determinand	Pearson's *r*	AXIS 2 Determinand	Pearson's *r*
125–250 µm	0.891 **	K	–0.625 **
500–1000 µm	–0.705 **	63–125 µm	–0.617 **
1000–2000 µm	–0.693 **	Al	–0.542 **
K	0.597 **	Mg	–0.531 **
> 2000 µm	–0.595 **	Ni	–0.524 **
63–125 µm	0.581 **	Cu	–0.521 **
Al	0.578 **	Na	–0.519 **
Ni	0.543 **	TOC%	–0.475 **
■ Fe	0.439 **	■ Fe	–0.469 **
Mg	0.421 **	■ Pb	–0.418 **
Cu	0.360 **	500–1000 µm	0.405 **
TMBI	0.297 *	Ca	–0.301 *
□ Mn	0.295 *	Sr	–0.285 *
□ Zn	0.295 *		

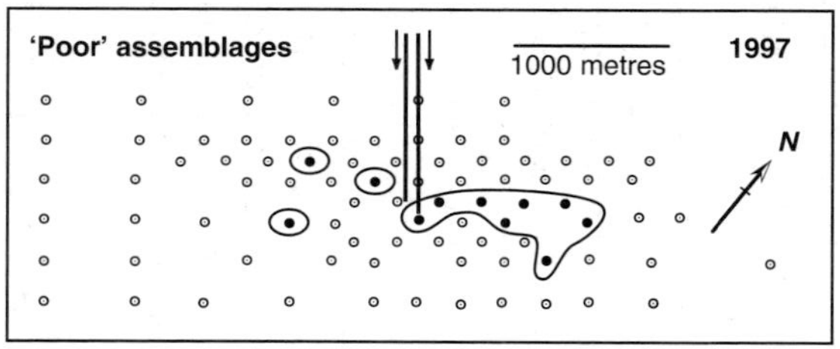

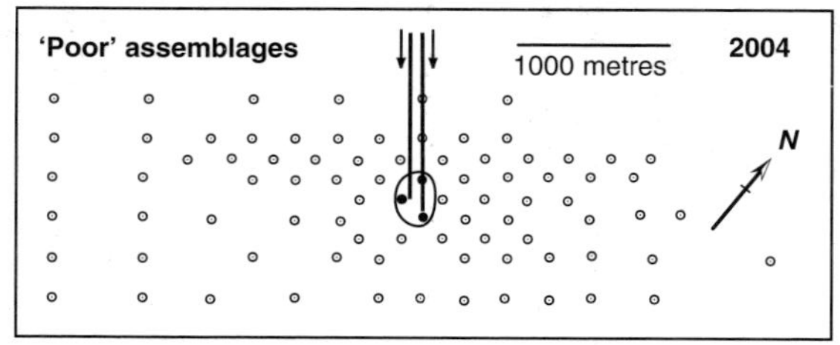

☆ Axes accounting for the highest proportions of the observed variance in the meiofaunal nematode community structures

□ Elements closely associated with the industrial complex discharges ■ Elements associated with the industrial discharges and natural coastal riverine inputs

Statistical significance: * $P < 0.05$; ** $P < 0.01$

Fig. 12.8. Comparison of the nematode multivariate correlation tables for monitoring surveys undertaken in 1997 and 2004 showing the ranked, statistically significant environmental determinands and their correlation coefficients in relation to the nematode community structure axes. The maps shown illustrate the sites at which 'poor' nematode assemblages were recorded in the years shown in the tables above and are based on the same lower quartile criteria (namely depressed species richness, depressed diversity indices, elevated 1B:2A feeding type ratios in medium-coarse sand sediments and elevated dominance values). In both studies Axis 1 accounts for the highest proportion of the variance observed in the nematode community structures. Note that between 1997 and 2004 the significance of effluent associated elements declined and that natural factors (primarily sediment particle size fractions) increased in their significance, migrating from Axis 2 in 1997 to Axis 1 in 2004. These changes were tracked in annual monitoring surveys and paralleled the changes made by the factory complex in their effluent treatment systems (see Case history 4).

effluent acidity in-line and reduced final concentrations of effluent components entering the marine environment. This meant that even if effluent impinged on the seabed, for example during thermocline periods in the summer months, benthic communities were not exposed to acute effects of the wastes.

By 2004, the results of the correlation analyses showed that natural factors (primarily the proportions of different sediment particle size fractions as well as marker elements for materials of terrestrial origin) had become primary correlates with the structures of the nematode communities. Using the same criteria as those employed in 1997, the zone of residual impacts in 2004 is illustrated in Fig. 12.8 (see 2004 marine map). These lie within the mixing ('boil') zone for the discharges (computed as a function of volume of effluent emitted and mean sea depth at the point of discharge) and this is now recognized by the regulatory authorities (Department of Water Affairs and Forestry) as an 'acceptable footprint'.

Case history 5, bespoke *in situ* monitoring system: Caspian Sea

Customer: Oil industry and service company

The shallow north Caspian Sea is part of an extensive and internationally important Ramsar-designated site[7] and supports numerous endemic species, some of which are listed as critically endangered. Accordingly, the relatively recent discovery of exploitable oil reserves beneath the sea (late 1990s), set against significant changes in world politics in relation to traditional Middle East oil supplies and subsequent increases in oil demand, has placed considerable pressures on these habitats and their species, as surrounding countries permit the construction of offshore production facilities. In recognition of the ecological importance of the shallow water habitats, monitoring and assessment programmes were agreed between international oil companies and several of the governments concerned. Clearly these programmes needed to be sensitive enough to detect any subtle, long-term degradation relating to fugitive materials arising from the construction and operation of the production platforms, as well as the acute effects of loss of containment incidents and the impacts of clean-up (see Green and Trett, 1989).

In some quarters, the agreements to monitor were based on a belief that traditional benthic macroinvertebrate surveys would meet all requirements. The commencement of baseline sampling soon dispelled this notion. Offshore, the sediments in the oligohaline waters (salinity ca. 1.2%) support a comparatively depauperate macroinvertebrate fauna in which larvae of salt-tolerant chironomids (Diptera: Nematocera) are seasonally the dominant faunal group along with the small polychaete worm, *Manayunkia caspica* (see Green, 1968). This poverty of the macrofauna may reflect the fact that the shallow northern sea (minimum depth ca. 2.5 metres) freezes over in December and may not thaw until late March by which time ice drift can also have caused significant physical impacts on the seabed.

From earlier studies that we had undertaken (1997/8), it was clear that the North Caspian meiofauna was considerably richer than the macrofauna and in excess of 100 nematode taxa have now been recorded from this region. To overcome the problems of ice and those of restrictions placed on the collection of samples from the nature reserve, a monitoring system based on nematode colonization units (NCUs) was used. These comprise inert, 9 × 11 cm, coarse fibre pads, which are held in large mesh, heavy-duty Netlon® bags, similar to those used in marine oyster farms. Between 2003 and 2005, the colonization units were suspended at three different depths around the shores of artificial islands that were created to support land-based types of drilling rigs. After a period of just eight weeks' immersion, densities of up to 20,000 nematodes per pad were recorded, equivalent to over 2 million individuals/m^2 of the artificial substrate. Although up to 47 taxa were recorded in a sampling round, at most representatives of 12 nematode species were present in any one single pad sample. Representatives of all the major nematode trophic groups were present in the sample assemblages. However, 'non-selective' deposit feeders (Weiser type 1B species, e.g. Axonolaimidae and Xyalidae) predominated in the lower water column samples while selective epistrate feeders and diatomivorous species (Weiser type 2A species, e.g. Chromadoridae and Microlaimidae) were dominant in the upper water column samples.

To date, chemical stresses have not been detected by these 'biosentinel' units and physical disturbance caused by wind and wave action in the shallow waters appear to be the major determinants of the structures of the assemblages that colonize the pads.

Future Commercial Development of Nematodes

If asked 30 years ago to predict a future role for nematodes as bioindicators for industry and regulators, we might have hazarded a guess that use in laboratory-based toxicity assessments would have expanded, either in single species tests or as part of the then emerging field of mesocosm studies (see Steele, 1979). In the late 1970s, ecological assessment using nematodes was truly in its infancy. The few studies that had been published showed that nematode communities offered some interesting and potentially valuable responses to different stresses (see Ferris and Ferris, 1979; Platt and Warwick, 1980). However, most of us lacked the practical experience that was later to be gained from studies undertaken in a wide variety of situations. More importantly, we did not have routine access to the computing power required to extract information on community structures from what were then seen as almost unmanageably large data sets. At that time it would clearly have been unthinkable that international chemical manufacturers would one day spend several millions of pounds on modifying and installing new effluent systems based substantially on the results of nematode analyses-derived annual monitoring surveys (see Case history 4, above).

For the present, we can identify some areas in existing commercial use of nematodes that have the potential to expand. These would include the use of artificial substrata in aquatic habitats to assess the integrated effects of conditions in a range of situations that are not easily examined using other faunal groups. Standardized nematode colonization units can only be used where there is sufficient interaction between sediments and the overlying waters (e.g. in shallow water bodies) or in pumped water systems. In other circumstances, methods using pre-colonized artificial substrata could be developed. The ease of deployment and collection of these units and the reduction of variance due to microhabitat-type effects are attractive features. A simple applied example of an area that would benefit from this approach would be biocide treatments of ballast water in ships. This ensures that alien species are not seeded into waters at sites around the world. However, the effectiveness of the sterilization procedures are not easily determined owing to the complex hull structures of ships. As nematodes include many highly tolerant species that are among the last to disappear in chemically-stressed environments, the *in situ* technique could form an important part of the certification process.

Targeted environmental assessments using nematodes as a means of examining and monitoring potential foci of contamination may also become routine in the near future. Staff at Physalia currently undertake sediment transport analyses (STAs) and mapping of sediment transport pathways in a range of coastal situations. These types of study identify what are believed to be stable subtidal sites of net sediment accretion which represent 'sinks' for sediment-bound contaminants. Focused monitoring in these areas using nematodes as bioindicators should provide invaluable information on the integrated effects of industrial, urban and domestic inputs and could form the basis of a network for the assessment of coastal water quality.

Höss *et al.* (2006) point out that environmental assessments based on freshwater nematodes are scarce. The same is true of terrestrial situations, although this is beginning to change (Trett *et al.*, 2000; Van der Wurff *et al.*, 2004, 2007; Trett and Thurgood, 2008; and Danovaro *et al.*, Chapter 6, this volume). Both of these areas need more work to overcome technical problems, many of which relate to habitat heterogeneity (spatial and temporal) and achieving sampling sufficiency under these variable conditions. However, this should not preclude longer term monitoring studies of single sites with persistent stress problems such as contaminated land sites and rivers receiving effluents. Refinement of maturity indices along the lines already noted in the above section on Level I analyses, to take into account the most recent changes in our understanding of nematode taxonomy and improved sub-family level c-p scores, will benefit both fields of study.

The distant future for nematodes as bioindicators is less clear. In the short- to medium term, it seems unlikely that molecular approaches will take on a significant role in stream-lining nematode surveys although this field is moving rapidly (see Neilson *et al.*, Chapter 8, this volume). However, it is not beyond the realms of belief that specific markers in nematode genomes will be recognized as reflecting tolerances to given stresses in the environment. Clearly this would be of value in identifying sites at which a contaminant or a closely-related group of contaminants was exerting selection pressure on the nematode assemblages and this approach could be adopted for targeting appropriate clean-up measures, or for the identification of rational monitoring sites. Equally, there is the promise that molecular libraries will permit the rapid identification of the nematode species present in a single, small sample of soil or sediment. If translated into a practical tool, this would circumvent the issues of time delays between field sampling and the interpretation and reporting which is important when using nematodes as bioindicators in spill- and leak incident surveys (see above section Constraints imposed on commercial nematode surveys). However, the present authors note that with high proportions of morphologically-distinctive nematodes defying attribution to described taxa in tropical marine surveys and having to be assigned operational taxonomic unit (OTU) status, the value of this approach may be restricted to sites with well described nematode faunas.

What is clear is that the use of nematodes as bioindicators in a commercial context has come a long way in a comparatively short period of time. As can be seen from the present chapter and others in this book, the success of nematodes in this field is reflected in their almost unique ability to fulfil all the key criteria for monitoring systems as set out by UNESCO in its definition of *SMART environmental indicators* (i.e. *S*pecific, *M*easurable, *A*ttainable, *R*elevant and *T*rackable; UNESCO, 2005). For the future, with increasing pressures on finite global resources and uncertainty surrounding the impacts of methods used to exploit these, particularly set against changes in climate patterns, it would appear that nematode bioindicators will remain the preferred tool for the accurate assessment and monitoring of actual ecological effects.

Notes

[1] In this context, 'indicator species' relates to the mathematical fidelity of a species and the degree to which its abundances are concentrated in a particular group (cluster) of communities as recognized in the multivariate analyses. It should not be confused with the concept of a single species that may (or may not!) indicate a particular set of prevailing environmental conditions.

[2] '*Lex parsimoniae*' – summarized by the 14th century English Franciscan friar, William of Occam. This sets out a logistic approach for deciding between multiple competing theories that are equal in all other respects and suggests that the theory that introduces the fewest assumptions and that postulates the fewest hypothetical entities is probably the correct one.

[3] Behavioural observations indicate that non-selective is probably an inappropriate term (see Moens and Vincx, 1997 and discussions in Moens *et al.*, 2006).

[4] Named after the Madrid-based company, Nebalia S. L. Consultoría Medioambiental, who developed the indices and triangular plots for use in situations where stresses may exert differential effects on nematode communities and on other faunal elements (e.g. meiofaunal harpacticoid copepods and halacarid mites or different size classes of macrofauna).

[5] This was the first published study to demonstrate that soil nematode communities and microbial assemblages provide compatible, coherent evidence

in relation to the ecological impacts of heavy metal contaminants (Professor John C. Fry, University of Cardiff, 2001, personal communication).

[6]It is not clear that iron exerts a direct ecological effect in this situation and it may represent an accidental correlate (cf. aluminium in Case history 3). There is even some circumstantial evidence from marine chemists that the flocculated iron hydroxide at sea may chelate some of the effluent metals reducing their bioavailability in the sediments.

[7]The Ramsar Convention is an international treaty for the protection and conservation of wetlands, their species and their communities. Developed at a meeting held at Ramsar, Iran (coincidentally situated on the shores of the Caspian), in February 1971, this includes the shallow water sites present in the North Caspian Sea.

References

Anon. (1985) Council directive on the assessment of the effects of certain private and public projects (Directive 85/337/EEC). *Official Journal of the European Communities* LR175/40, 5 July 1985.

Anon. (1994) *Biodiversity; The United Kingdom's Action Plan*. Her Majesty's Stationery Office, London, 188 pp.

Attrill, M., Ramsay, P., Thomas, R. and Trett, M.W. (1996) An estuarine biodiversity hot-spot. *Journal of the Marine Biological Association of the United Kingdom* 76, 161–171.

Bakonyi, G and Nagy, P. (2000) Temperature- and moisture-induced changes in the structure of the nematode fauna of a semiarid grassland; patterns and mechanisms. *Global Change Biology* 6, 697–707.

Been, T.H. and Schomaker, C.H. (2000) Development and evaluation of sampling methods for fields with infestation foci of potato cyst nematode (*Globodera rostochiensis* and *G. pallida*). *Phytopathology* 90, 647–656.

Been, T.H. and Schomaker, C.H. (2006) Distribution patterns and sampling. In: Perry, R.N. and Moens, M. (eds) *Plant Nematology*. CABI Publishing, Wallingford, pp. 302–326.

Bert, W., Messiaen, M., Hendrickx, F., Manhout, J., De Bie, T. and Borgonie, G. (2007) Nematodes of small farmland ponds. *Hydrobiologia* 583, 91–105.

Boaden, P.J.S. (1975) Anaerobiosis: meiofauna and early metazoan evolution. *Zoologica Scripta* 4, 21–24.

Boaden, P.J.S. (1977) Thiobiotic facts and fancies; aspects of the distribution and evolution of anaerobic meiofauna. *Mikrofauna des Meeresbodens* 61, 45–63.

Bongers, T. (1990) The maturity index; an ecological measure of environmental disturbance based on nematode species composition. *Oecologia* 83, 14–19.

Bongers, T. and Ferris, H. (1999) Nematode community structure as a bioindicator in environmental monitoring. *Trends in Ecology and Evolution* 14, 224–228.

Bowen, H.J.M. (1979) *Environmental Chemistry of the Elements*. Academic Press, London.

Calvo Urbano, B. (1995) The structure and dynamics of a marine fouling community at Felixstowe, (Suffolk). Ph.D. thesis, Centre for Research in Aquatic Biology, University of London. 286 pp.

Cao, Y., Williams, D.D. and Williams, N.E. (1998) How important are rare species in aquatic community ecology and bioassessment? *Limnology and Oceanography*, 43, 1403–1409.

Cao, Y., Larson, D.P. and Thorne, R. St. J. (2001) Rare species in multivariate analyses for bioassessment: some considerations. *Journal of the North American Bentological Society* 20, 144–153.

Chapman, P.M. (2002a) Integrating toxicology and benthic ecology: putting the 'eco' back into ecotoxicology. *Marine Pollution Bulletin* 44, 7–15.

Chapman, P.M. (2002b) Ecological risk assessment (EcRA) and hormesis. *Science of the Total Environment* 288, 131–140.

Chapman, P.M., Ho, K., Muns, W., Solomon, K. and Weinstein, M.P. (2002) Issues in sediment toxicity and ecological risk assessment. *Marine Pollution Bulletin* 44, 271–278.

Chapman, P.M., Wang, F., Janssen, C., Goulet, R.R. and Kamunde, C.N. (2003) Conducting ecological risk assessments of inorganic metals and metalloids – current status. *Human and Ecological Risk Assessment* 9, 641–697.

De Ley, P., Decraemer, W. and Eyualem-Abebe (2006) Introduction: summary of present knowledge and research addressing the ecology and taxonomy of freshwater nematodes. In: Eyualem-Abebe, Andrássy, I. and Traunsperger, W. (eds) *Freshwater Nematodes: Ecology and Taxonomy*. CAB International, Wallingford, UK pp. 3–30.

Debenham, N.J., Lambshead, P.J.D., Ferrero, T.J. and Smith, C.R. (2004) The impact of whale falls on the abundance of nematode in the deep sea. *Deep-Sea Research* 51, 701–706.

Doran, P.T., Priscu, J.C., Lyons, W.B., Walsh, J.E., Fountain, A.G., McKnight, D.M., Moorhead, D.L., Virginia, R.A., Wall, D.H., Clow, G.D., Fritsen, C.H., McKay, C.P. and Parsons, C.P. (2002) Antarctic climate cooling and terrestrial ecosystem response. *Nature* 415, 517–520.

Dufrêne, M. and Legendre, P. (1997) Species assemblages and indicator species: the need for a flexible, asymmetrical approach. *Ecological Monographs* 67, 345–366.

Ellis, R.J., Best, J.G., Fry, J.C., Morgan, P., Neish, B., Trett, M.W. and Weightman, A.J. (2002) Similarity of microbial and meiofaunal community analyses for mapping ecological effects of heavy-metal contamination in soil. *FEMS Microbiology Ecology* 40, 113–122.

Eskin, R.A. and Palmer, M.A. (1985) Suspension of marine nematodes in a turbulent tidal creek: species patterns. *Biological Bulletin* 169, 615–623.

Ferris, V.R. and Ferris, J.M. (1979) Threadworms (Nematoda). In: Hart, C.W. and Fuller, S.L.H. (eds) *Pollution Ecology of Estuarine Invertebrates*. Academic Press, London. pp. 1–33.

Forster, S.J. (1998) Aspects of the ecology of free-living nematodes assemblages in a temperate intertidal mudflat. Ph.D. thesis, Centre for Research in Aquatic Biology, University of London.

Fraschetti, S., Gambi, C., Giangrande, A., Musco, L., Terlizzi, A. and Danovaro, R. (2006) Structural and functional response of meiofauna rocky assemblages to sewage pollution. *Marine Pollution Bulletin* 52, 540–548.

Gerlach, S.A. (1977) Means of meiofaunal dispersal. *Mikrofauna des Meeresbodens* 61, 89–103.

Geyedo-Adabio, T.K. and Baird, D. (2006) Response of meiofauna and nematode communities to increased levels of contaminants in a laboratory microcosm experiment. *Ecotoxicology and Environmental Safety* 63, 443–450.

Green, J. (1968) *The Biology of Estuarine Animals*. Sidgwick and Jackson, London.

Green, J. and Trett, M.W. (1989) *The Fate and Effects of Oils in Freshwater*. BP International contribution to the European Year of the Environment. Elsevier Applied Science, London. 338 pp.

Haining, R. (1990) *Spatial Data Analysis in the Social and Environmental Sciences*. Cambridge University Press, Cambridge, 409 pp.

Hill, M.O. (1973) Diversity and evenness, a unifying notion and its consequences. *Ecology* 54, 427–432.

Hooper, D.J. (1986) Extraction of free-living stages from soils. In: Southey, J.F. (ed.) *Laboratory Methods for Work with Soil and Plant-parasitic Nematodes*. Her Majesty's Stationery Office, London, pp. 5–30.

Höss, S., Traunsperger, W. and Zullini, A. (2006) Freshwater nematodes in environmental science. In: Eyualem-Abebe, Andrássy, I. and Traunsperger, W. (eds) *Freshwater Nematodes: Ecology and Taxonomy*. CAB International, Wallingford, UK pp. 144–159.

Hunt, H.W., Wall, D., Decrappeo, N.M. and Brenner, J.S. (2001) A model for nematode locomotion in soil. *Nematology* 3, 705–716.

IEEM (2005) *Code of Professional Conduct.* Institute of Ecology and Environmental Management, Winchester, 3 pp.

IEEM (2006) *Guidance on Survey Methodology.* Institute of Ecology and Environmental Management, Winchester, 46 pp.

Jager, T., Alda Álvarez, O., Kammenga, J.E. and Kooijman, S.A.L.M. (2005) Modelling nematode life-cycles using dynamic energy budgets. *Functional Ecology* 19, 136–144.

Jensen, P. (1986) Nematode fauna present in the sulphide-rich brine seep and adjacent bottoms of the East Flower Garden, NW Gulf of Mexico. IV. Ecological aspects. *Marine Biology* 92, 489–503.

Jensen, P. (1995) *Theristus* (*Penzancia*) *anoxybioticus* n. sp. (Nematoda: Xyalidae) from sublittoral methane seepages in the northern Kattegat, Denmark. *Journal of Nematology* 27, 231–236.

Jiang, D., Li, Q., Liu, F., Jiang, Y. and Liang, W. (2007) Vertical distribution of nematodes in an age sequence of *Caragana microphylla* plantations in the Horquin Sandy Land, Northeast China. *Ecological Research* 22, 49–56.

Levinson, A. (1974) *An Introduction to Exploration Geochemistry.* Applied Publishing, Calgary.

Liang, W., Zhang, X., Li, Q., Ou, W. and Neher, D.A. (2005) Vertical distribution of bactivorous nematodes under different land uses. *Journal of Nematology* 37, 254–258.

Meldal, B.H.M., Debenham, M.J., De Ley, P., Tandingan De Ley, I., Vanfleteren, J.R., Vierstraete, A.R., Bert, W., Borgonie, G., Moens, T., Tyler, P.A., Austen, M.C., Blaxter, M.L., Rogers, A.D. and Lambshead, P.J.D. (2007) An improved molecular phylogeny of the Nematoda with special emphasis on marine taxa. *Molecular Phylogenetics and Evolution* 42, 622–636.

McIntyre, A.D. and Warwick, R.M. (1984) Meiofauna Techniques. In: Holme, N.A. and McIntyre, A.D. (eds), *Methods for the Study of Marine Benthos*, IBP Handbook 16, Blackwell Scientific Publications, Oxford, pp. 217–244.

Moens, T. and Vincx, M. (1997) Observations of the feeding ecology of estuarine nematodes. *Journal of the Marine Biological Association of the United Kingdom* 72, 211–227.

Moens, T., Traunsperger, W. and Bergtold, M. (2006) Feeding ecology of free-living benthic nematodes. In: Eyualem-Abebe, Andrássy, I. and Traunsperger, W. (eds) *Freshwater Nematodes: Ecology and Taxonomy*. CABI Publishing, Wallingford, pp. 103–131

Moore, C.G. and Bett, B.J. (1989) The use of meiofauna in marine pollution impact assessment. Presentation at the Meiofauna Specialist Group of the Linnean Society of London. *Zoological Journal of the Linnean Society* 96, 263–280.

Muthumbi, A.W., Vanreusel, A., Duineveld, G., Soetaert, K. and Vincx, M. (2004) Nematode community structure along the continental slope off the Kenyan coast, western Indian Ocean. *International Review of Hydrobiology* 89, 188–205.

Nkem, J.N., Wall, D.H., Virginia, R.A., Barrett, J.E., Broos, E.J., Porazinska, D.L. and Adams, B. (2006) Wind dispersal of soil invertebrates in the McMurdo Dry Valleys, Antarctica. *Polar Biology* 29, 346–352.

Platt, H.M. and Warwick, R.M. (1980) The significance of free-living nematodes to the littoral ecostystem. In: Price, J.H., Irvine, D.E.G. and Farnham, W.F. (eds) *The Shore Environment. Volume 2: Ecosystems*. Academic Press, London, pp. 729–759.

Platt, H.M., Shaw. K.M. and Lambshead, P.J.D. (1984) Nematode species abundance patterns and their use in the detection of environmental perturbations. *Hydrobiologia* 118, 59–61.

Powers, L.E., Freckman, D.W. and Virginia, R.A. (1994) Depth distribution of soil nematodes in Taylor Valley, Antarctica. *Antarctic Journal of the United States* 29, 175–176.

Sánchez-Moreno, S., Minoshima, H., Ferris, H. and Jackson, L.E. (2006) Linking soil properties and nematode community composition: effects of soil management on soil food webs. *Nematology* 8, 703–715.

Seinhorst, J.W. (1959) A rapid method for the transfer of nematodes to anhydrous glycerine. *Nematologica* 4, 67–69.

Sherman, K.M. and Coull, B.C. (1980) The response of meiofauna to sediment disturbance. *Journal of Experimental Marine Bioliology and Ecology* 46, 59–71.

Sherman, K.M., Reidenauer, J.A., Thistle, D. and Meeter, D. (1983) Role of a natural disturbance in an assemblage of marine free-living nematodes. *Marine Ecology Progress Series* 11, 23–30.
Simpson, E.H. (1949) Measurement of diversity. *Nature* 163, 688.
Smol, N., Heip, C. and Govaert, M. (1980) The life-cycle of *Oncholaimus oxyuris* (Nematoda) in its habitat. *Annales de la Société Zoologique de Belgique* 110, 87–103.
Somerfield, P.J., Rees, H.L. and Warwick, R.M. (1995) Interrelationships in community structure between marine meiofauna and macrofauna in relation to dredgings disposals. *Marine Ecology Progress Series* 127, 103–112.
Sommer, R.J. (2006) *Pristionchus pacificus*. The *C. elegans* Research Community, *Wormbook*, doi/10.1895/wormbook.1.102.1, http://www.wormbook.org
Steele, J.H. (1979) The uses of experimental ecosystems. *Philosophical Transactions of the Royal Society of London Series B* 288, 583–593.
Steyaert, M. Garner, N., Van Gansbeke, D. and Vincx, M. (1999) Nematode communities from the North Sea: environmental controls on species diversity and vertical distribution within the sediment. *Journal of the Marine Biolological Association of the UK* 79, 253–264.
Traunspurger, W. (1997) Bathymetric, seasonal and vertical distribution of feeding-types of nematodes in an oligotrophic lake. *Vie Milieu* 47, 1–7.
Trett, M.W. and Thurgood, R. (2008) Assessment and monitoring of actual ecological effects; nematodes in the service of industry and regulators. In Proceedings of the 10th International UFZ-Deltares-TNO Conference on Soil-Water Systems (CONSOIL 2008). Milan, June 2008. pp. 82–91.
Trett, M.W., Calvo Urbano, B., Forster, S.J., Hutchinson, J.D., Feil., R.L., Trett, S.P. and Best, J.G. (2000) Terrestrial meiofauna and contaminated land assessment. *Environmental Science and Technology* 34, 1594–1602.s
UNESCO (2005) Final report of the expert meeting on Ecological Science and Biosphere Reserves Setting Priorities in Coastal and Marine Areas in Europe as part of the Man and the Biosphere Programme, 12–13 May, 2005. UNESCO Regional Bureau for Europe ROSTE, Venice in association with MARBEF. 56 pp.
Van der Wurff, A., Van Megen, H., Riksen, J., Boivin, M.-E., Kools, S. and Kammenga, J. (2004) Impact of an irregularly distributed mix of toxicants on nematode communities. *Stimulation Programme System-oriented Ecotoxicological Research* 8, 8–12.
Van der Wurff, A., Kools, S., Boivin, M.-E., Van den Brink, P., Van Megen, H., Riksen, J., Dorzosuk, A and Kammenga, J. (2007) Type of disturbance and ecological history determine structural stability. *Ecological Applications* 17, 190–202.
Vranken, G. and Heip, C. (1983) Calculation of the intrinsic rate of natural increase, r_m, with *Rhabditis marina* Bastian 1865 (Nematoda). *Nematologica* 29, 468–477.
Whitehead, A.G. (1977) Vertical distribution of potato, beet and pea cyst nematodes in some heavily infested soils. *Plant Pathology* 26, 85–90.
Whitehead, A.G. and Hemming, J.R. (1965) A comparison of some quantitative methods of extracting small vermiform nematodes from soil. *Annals of Applied Biology* 55, 25–38.
Whittaker, R.H. (1972) Evolution and measurement of species diversity. *Taxon* 21, 213–251.
Wieser, W. (1953) Die Beziehung zwischen Munholengestalt, Ernährungsweise und Vorkommen beifreilebenden marinen Nematoden. *Arkiv für Zoologie* 4, 439–484.

Index

Page numbers in **bold** refer to tables or figures.

α, β and γ scale diversity 8, 28, 49, 108
abundance 2, 6, 27, 45, 52
  and bioindication 282–283
  as biomass 140
  in dredged sediments 154
  factors controlling 20–23
  and fish farms 152
  global 2, 124
  on hard substrates 151
  and heavy metal pollution 97–99, 148, 149
  indices 48, 49, 109, 110, 112, 113, 115, 131–132, 293–294, 302
    see also dominance indices
  in neutral community model 115
  and organic chemical pollution 147
  and organic enrichment 150–151
  simulation models 99–102
aggregations 56, 99
agricultural systems 13, 17, 18–20, 28, 29–30
  copper experiment 97–99, 131
  and diversity 7–8, 60, 64–65, 65–66, 131
algae feeders 4, 22, 23–24 *see also* trophic groupings
aluminium 176, 304
  toxicity tests 213–214
  transgenic indicator strains 242
anatomy 2, **3,** 9–10, 155
animal feeders *see* parasites
anoxic environments 4, 9, 24, 283
anthelminthic drenches *see* drugs, nematicidal
anthropogenic pollution 146–164, 189, 208, 254
Aphelenchoides spp. 16–17
aquatic nematodes 17, 27, **62,** 124
  dispersal 68
  diversity 64, 125–126
  environmental modification 25, 26
  feeding effects **14,** 15–16
  populations, controlling factors 21–24
  see also marine nematodes
ARB software 193–194
ARESC (assessment and ranking of ecologically significant contaminants) 297, 304
arsenic 150, 156, 173–174, 176, 304
  transgenic indicator strains 242
arsenite
  transgenic indicator strains 237, **238,** 240, 246
artificial substrates **290,** 291, 292, 300, 308, 309
assays *see* ecological monitoring surveys, commercial; toxicity tests; transgenic indicator strains

assemblage analysis
case studies using 146–164, 155–164, 172–185
indices for 126–140, 287
molecular methods 192–202
assemblages 6
community models 114–117
effect of soil disturbance 127–128
in plant communities 1, 26
assessment and ranking of ecologically significant contaminants (ARESC) 297, 304
attractants 5, 10, 264, 265
autoecology 107, 108

β-galactosidase 226–227
transgenic strains using 230, 234–235, 237, 239, 240, 242
bacteria 10–12
as biosensors 190–191
chemoautotrophs 24
community composition 15–16, 26, 27
influence on diversity 71, 74
growth stimulation 10–11, 25
bacteria feeders 4, 10–12, 18, **19,** 24, 26, 27, 65, 68, 71, 125 *see also* trophic groupings
barium 176–177
basal indices 134, **135,** 136–137, **138**
beaches, cleaning 154–155
Belgian Benthic Index (BBI) 155–156
bioassays *see* biosensors; ecological monitoring surveys, commercial; toxicity tests; transgenic indicator strains
biocide effectiveness monitoring 299–300, 309
Biodiversity: The United Kingdom's Action Plan 275
biofouling fauna 299–300
bioindication 46, 94–95, 188–189
accuracy of 99–104, 150
advantages of nematodes 30, 31, 125–126, 150, 189, 191, 209–210, 225, 262, 275, 277–278, 282–286, 286–287
using arthropods **210,** 212–213, 216, 302, 305
biosensors, non-nematode 190–191
commercial 275–311
conclusions based on 95–104
using copepods 150, 151, 289, **290,** 291, 305
using fish 102
using macrofauna 102, 155–156, 276–277, 282–283, 284, 287–289, 294, **296,** 302, 304, 307–308
using molluscs 24, **210,** 305
using worms **210,** 212–213, 217, 305
biological control
of arthropods 15
of nematodes 17, 28, 29
Biological Monitoring Working Party (BMWP) 156
biomass, as indicator **139,** 140, 152, 172
biomonitoring, commercial 275–311
biosensors, non-nematode 190–191
biosentinals 278, 308
bioturbation 10, 25
BMWP (Biological Monitoring Working Party) 156
body size 152
Brillouin's Diversity Indices **110,** 113
Brillouin's Evenness Indices **110,** 113
Bt corn 217–218
burning grounds *see* explosives contamination monitoring
Bursaphelenchus xylophilus 12

c-p groups *see* colonizer-persister classification
CA (Correspondence Analysis) 120
cadmium 148–149, 160, **162,** 173, 174, 177
C. elegans response to 269–271
genes responsive to 234–235
pollution, assemblage analysis 182–183, 304
toxic effects 232, 269–271
toxicity tests 212, 213, 214–215
transgenic indicator strains 232–**234,** 237, 239, 242–243, 246
Caenorhabditis elegans 1–2, 5, 125, 196
cadmium, response to 269–271
DNA microarray experiments 262–271
ethanol, response to 263–264

heat-shock response, transgenic strains 230–232
heavy metal response 269–271
heavy metal response, transgenic strains 231, 232–**234,** 234–235, 236, 237, **238,** 239, 240–241, 242–243, 246
humic substances, response to 264–265
metabolic stress response, transgenic strains 236–237
metallothionein response, transgenic strains 232–**234**
novel stress promoter, transgenic strains 234–236
novel transgenic strains 244–247
organic chemical response, transgenic strains **231, 233,** 236, 237, **238,** 239–240, 243, 245, 246
PCBs, response to 226, **231,** 265–267, **268**
phthalate, response to 243, 267–269
reporter gene expression 226–228
steroid hormones, response to 262–263
toxicity tests using 209–211, 212–219
transformation techniques 226, 228–230
as transgenic biosensors 225–247
transgenic strains, response characterization 237–242
Canonical Correspondence Analysis (CCA) 120
capilliary sequence single strand conformation polymorphism (CE-SSCP) 197
carbon cycling 9, 11, 74
CCM (cluster coherence mapping) 296
Cephalobidae 70
channel indices 136–137
chemical oxygen demand (COD) 158–159, 163
chemical stressors 308
4-nonylphenol (NP) 160, **161,** 163–164
biocides 212–213, 213–214, **233,** 237, **238,** 239–240, 245, 299–300
ethanol 263–264
naturally occurring 254, 264–265
organic chemicals 146–148, 211–214, 215–216, 217, **231, 233,** 236, 246
phthlate **231,** 243, 267–269
polychlorinated organic chemicals 212–213, 215–216, 226, **233,** 236, 265–267, **268**
steroid hormones 246, 262–263
toxicity tests 211–214, 215, 217
transgenic indicator strains **231, 233,** 236, 237, **238,** 239–240, 243, 245, 246
chemical tests for pollutants 208
chemoautotrophic bacteria 24
chemotaxis 5, 10, 264, 265
chironomids 307
test using 216
chromium 149, 173–174, 177, 304
pollution, assemblage analysis 183–184
Cloacina 6, 24
cluster analyses 117–119, 294–296, **297,** 302, **303**
DNA microarrays 258–259
cluster coherence mapping (CCM) 296
coastal systems 20, **22,** 127, 154–155, 304–307, 309
cobalt
transgenic indicator strains 232–233
COD (chemical oxygen demand) 158–159, 163
Collembola 212–213
colonizer-persister (c-p) classification 49, 94, 129–130, 157, 160, 174, 294
accuracy 163–164
and functional guild indicators 134–140
and maturity indices 130–134
triangles 132–134, 137
commercial biomonitoring 275–311
commercial survey methods, ecological monitoring 287–292
communities
ecological 5–6
ecological succession models 114–115
indices for analysis *see* indices
neutral models 115–117
parameters for describing *see* indices

competition
among algae 23
in plant succession 13
concentration–effect relationships
239–240, 256
containment incidents **279,** 281, 287, 307
contamination *see* pollution
cooling systems, monitoring 299–300, 301
copepods 150, 151, **288,** 289, **290,** 291, 305
copper 148, **306**
field experiment 97–99, 131
pollution, assemblage analysis 149, 173–174, 174–175, 177, 178, 179–181, 304
tolerance of 149
toxicity tests 212, 213, 215
transgenic indicator strains 232–233, 236, 237–239, 240, 242, 246
copper sulphate 181
transgenic indicator strains **238**
Correspondence Analysis (CA) 120
Council directive on the assessment of the effects of certain private and public projects (EU Directive 85/337/EEC) 276
crop rotation 29, 97
Cry1Ab protein, toxicity tests 218
customer-specific indices 299
cyanide pollution, toxicity tests 217
cytochrome P450 (CYP) gene induction
transgenic strains using 235–236

Δ^+ (Taxonomic Distinctness Index) 157, **158**
Daptonema tenuispiculum 154
databases
DNA microarray **260–261,** 262, 263
ecotoxicological 254
gene ontology analysis 259, **260–261**
rDNA 193–194, 198
dauerlarvae 128, 129, 130–131, 137
denaturing gradient gell electrophoresis (DGGE) 191, 197, 199
density
of species 112, 113, 293
see also abundance
detection limits
transgenic indicator strains 237–239
Diplogasterida 159, 162, 173
discriminant analysis 118, 119
dispersal 68–69, 150
disturbance of ecosystems 17, 64–65, 127–128, 130, 134, 153–154
diversity 2, 7–8, 45–75
α, β and γ scale 8, 28, 49, 108
in aquatic sediments 27
controlling factors 67–73
deep-sea 24
described species 57–60
and ecosystem processes 27–28, 107, 108
and heavy metal pollution 148–149, 149–150
and hydrocarbon pollution 147–148
indices 3, 46–49, 108–110, 113–114, 126–127, 132, 293, **306**
influences on 65–67
measurement 45–57
in natural environments 60–65
and organic enrichment 65, 150–151, 152–153
profiles 174
DNA 191, 192, 195–196, 229–230
extraction 198–199, 200–201, 202
see also rDNA
DNA microarrays 103, 253
analysis by 256–257
C. elegans experiments 262–271
cluster analysis 258–259
data mining 262
databases **260–261,** 262, 263
experimental design 255–256
functional annotation 259
normalization 257–258
statistical analysis 258, 264, 265–266, 267, 269
transcript profiling by 255–262
validation of data 259
dominance indices **110,** 114, 293–294, 302–304, **306**
Dorylaimida 48, 70, 129–130, 155, 184
dracunculiasis 15
Dracunculus medinensis 15
dredging 150, 153–154
drugs, nematicidal 15, 20, **21,** 29, 31, 125
resistance to 28
Duddingtonia flagrans 28

ecogenomics 253, 254
ecological impact assessment (EcIA) 276
ecological monitoring surveys, commercial
  background 275–278
  case studies 299–308
  constraints on 286–287
  customers' needs 278–282
  data assessment 293–299
  financial aspects 286–287
  legal aspects 276, 277, 278, 279–280, 281–282
  methods 287–292
  non-standard methods **290,** 291–292
  sampling 279, 292
ecological risk assessment (EcRA) 276
ecosystems
  basal indices 134, **135,** 136–137, **138**
  channel indices 136–137
  disturbed 17, 64–65, 127–128, 130, 134, 153–154
  diversity within 46, 71–72, 73–74, 126–127
  enrichment indices 132, 134, 135–140
  function indices 127–140
  functions in 125–126
  management 28, 29–30, 72, 127–128
  patches within 6, 28, 52, 68, 69
  productivity 71–72
  roles in 1–31, 73–74
  services in 1, 7, 8, 10–12, 28, 31, 125, 140
  stability 129
  structure indices 132, 134, 135–137, **138**
  succession 114–115, 139–140
ecotoxicogenomics 253
  aims 254–255
  using *C. elegans* 262–271
ecotoxicology
  DNA microarray use 253–271
  toxicity tests 208–219
  transgenic biosensors 225–247
EcRA (ecological risk assessment) 276
effluent 301
  monitoring 280, 304–307, 308
  see also sewage pollution; wastewater
EIAs (environmental impact assessments) 275–276
eigenanalysis 119–120
electrical conductivity 158–159, 163
elephantiasis 15, 28
energy balance perturbation *see* metabolic stress induction
Enoploidae 17, 21–22, 148, 152, 155
enrichment
  indices of, in ecosystems 132, 134, 135–140
  opportunists 128, 129, 131–132, 134, 137, 140, 174
  organic 65, 150–151, 152–153, 161–163
entomophilic nematodes 9, 11, 12–13, 15, 60, 70, 74
environmental impact assessments (EIAs) 275–276
environmental indicators, commercial 275–311
environments
  diversity patterns 60–65
  modification by nematodes 25–27
epigrowth communities 283, 289, **290,** 292
ethanol, *C. elegans* response to 263–264
EU Directive 85/337/EEC (Council directive on the assessment of the effects of certain private and public projects) 276
EU soil protection strategy 94–95
eutrophication 152, 161, 163
evenness indices 108–110, 113–114, 132
evolution 24, 48, 60, 69
excretion 11–12, 13
explosives contamination monitoring 301–304
extraction *see* sampling

False Discovery Rate (FDR) analysis 258
Fatty acid methyl ester (FAME) profiles 190
feeding
  indirect effects of 26–27
  regulatory effects on food sources 13, **14,** 15–17, 20, **21**
  strategies 2, 9–10
  trophic groupings 4, 8–10, 49, **50–51,** 94, 129–130, 212, 291, 294, **295, 298,** 308

financial aspects, ecological monitoring surveys 286–287
fish farms 151–153
food resources 8–12
food webs 1, 8, 9, 17
  complexity 129, 134
  connectance 129, 134, 135, 137
  and heavy metal pollution 175, 181
  resource flow 137, 139, 140
forests *see* woodland systems
freshwater nematodes *see* aquatic nematodes
freshwater systems
  bioindication 128, **210,** 216, 217, 218–219, 299–300
  boreholes 301
  diversity in **53, 62,** 64, 71–72
  ecological monitoring surveys 277, 309
  humic substances in 264–265
  pollution 155–164, 175, 180
function, indices of 127–140
functional annotation, DNA microarrays 259
functional approach 7–8
functional guilds 30
  as indicators 134–136, 137, 140
fungi
  nematode trapping 28, 31
  plant pathogens 26–27
fungus feeders 4, 12, 16–17, 18, 125 *see also* trophic groupings

β-galactosidase 226–267
  transgenic strains using 230, 234–235, 237, 239, 240, 242
gene ontology (GO) analysis 259
genes
  C. elegans expression patterns 262–271
  cytochrome P450 (CYP) 235–236
  human homologs in *C. elegans* 225, 246, 264
  lacZ 226–267
  luc 226, 227–278
  marker 103, 225, 309–310
  metabolic stress reporter 226
  metallothionein 232–**234**
  reporter 225–226, 226–229, 230–232
  stress-responsive 225–226, 229, 230–236, 245–246
genetic studies 148–149
  and taxonomy 47, 294, 310
genetically modified organisms
  monitoring 217–218
  see also transgenic indicator strains
genomic response to stress 253–271
geotropism 5
GFP *see* green fluorescent protein
Globodera spp. 8, 17
GO (gene ontology) analysis 259
grassland systems 25–26, 127–128
green fluorescent protein (GFP) 227, 228
  transgenic strains using 227, 230, 232, **233,** 235–236
guilds, functional 30
  as indicators 134–136, 137, 140

H′ diversity index *see* Shannon's Diversity Index
habitats, diversity comparisons 60–65
heat-shock response
  transgenic strains using 225–226, 230–232, 239, 240–241, 243
heavy metals
  C. elegans response to 269–271
  copper field experiment 97–99, 131
  in effluent 304–307
  from explosives 301–304
  freshwater pollution 156, 160, **162**
  marine pollution 148–150
  natural levels 149–150, **306**
  non-toxic 176–177
  quantification in soil 175–176
  soil pollution 172–185, 302–304
  tolerance of 149, 163, 164, 174, 178, 182–183, 212
  toxicity tests 212, 213–214, 214, 215, 217
  transgenic indicator strains **231,** 232–**234,** 234–235, 236, 237, **238,** 239, 240, 242, 246
Heip's Evenness Index **110,** 113, 114
herbivores *see* plant feeders
Heterodera spp. 4, 5, 7, 17
Hill's Diversity Index **110,** 113–114, 126, **127**
Hill's Evenness Index **110**
Hill's Reciprocal of D **110,** 114, 126, **127**

history of nematode studies 2–3, 124–126, 172–173, 275–278
hormones, steroid
  C. elegans response to 262–263
  transgenic indicator strains 246
Hox genes 2
human disease *see* parasites
humic substances
  C. elegans response to 264–265
  effects in ecosystems 264
hydrothermal vents 24, 283

identification of taxa 45, 52–53
  keys 128
  molecular methods 189, 192–202, 310
  by morphology 47, 188–189, 191, 200, 291
  see also taxonomy
Index of Trophic Diversity (ITD) 147, 153
indicator species analysis (ISA) 296
indicators
  biomass **139,** 140, 172
  body size 152
  chemical 158–159, 163
  commercial 275–311
  functional guilds as 30, 134–136, 137, 140
  non-biological 94–95
  see also bioindication
indices
  of abundance 48, 49, 109, 110, 112, 113, 115, 131–132, 293–294, 302
  for assemblage analysis 126–140, 287
  basal 134, **135,** 136–137, **138**
  case studies, freshwater systems 155–164
  case studies, marine systems 146–155
  case studies, soils 172–185
  channel 136–137
  for community analysis 107–122
  customer-specific 299
  of diversity 3, 46–49, 108–110, 113–14, 126–127, 132, 293, **306**
  dominance **110,** 114, 293–294, 302–304, **306**
  of ecological succession 114–115
  of ecosystem enrichment 132, 134, 135–140
  of ecosystem function 127–140
  of ecosystem structure 132, 134, 135–137, **138**
  of evenness 108–10, 113–114, 132
  formulae **110–111**
  identity independent 108–117, 121
  of richness 108–110, 112–114, 293, 294, **306**
  of spatial dependency 297–299
insects
  biocontrol of 15
  entomophilic nematodes 9, 11, 12–13, 15, 60, 70, 74
  as vectors 12–13, 68
intertidal sediments 21–22, 23, 25
invertebrates *see* bioindication
iron 149, 304, **306**
  transgenic indicator strains 242
Ironidae 155
ISA (indicator species analysis) 296
ITD (Index of Trophic Diversity) 147, 153

k-dominance curves 152–153, 294
K-strategists 49, 128, 129

lacZ 226–227
  transgenic strains using 232–**234**
lakes 23, **53, 62,** 64, 71–72, 264–265
lead **306**
  pollution, assemblage analysis 175, 177, 183
  toxicity tests 213–214
  transgenic indicator strains 232–233, 236, 237–239, 240
legal aspects, ecological monitoring surveys 276, 277, 278, **279–280,** 281–282
Leptolaimida 300
lethality tests using *C. elegans* 237–239
life cycle times 284–285
light-limited environments 23–24
luc 226, 227–228
luciferase 226, 227–228, 236–237
luminescence assays 226, 227–228, 236–237, 237–239, 240, 241
lymphatic filariasis 15

macrofaunal bioindicators *see* bioindication
Macropodia 6, 24
management of ecosystems 28, 29–30, 72, 127–128
manganese 149, 304
  transgenic indicator strains 242
Mantel test 114–115
Margalef's Richness Index 6, **110,** 113
marine nematodes 9, 17, 64, 124
  dispersal 68
  diversity **63,** 64
  feeding effects **14**
  populations, controlling factors 21–22, 23–24
marine systems
  ecological monitoring surveys 287–291, 292, 304–307
  heavy metal pollution 148–150
  organic chemical pollution 146–148
  see also sediments
marker genes 103, 225, 309–310
marram grass 20, **22,** 302
Maturity Index (MI) 128, 130–131, **133,** 294
Maturity Index 2-5 (MI2-5) 131, **133**
ΣMaturity Index (ΣMI) 131–132, **133**
ΣMaturity Index 2-5 (ΣMI2-5) 132, **133**
maturity indices 128, 129–132, **133**
  computation 130–132
measurement of diversity 45–57
meiofaunal bioindicators *see* bioindication; indicators
Meloidogyne spp. 4, 17, 27
MELTING software 194
mercury 176
  transgenic indicator strains 232–233
metabolic stress induction
  transgenic strains using 226, 236–237
metabolic stress reporter genes 226
metallothionein induction
  transgenic strains using 232–**234**
metals *see* heavy metals
MI (Maturity Index) 128, 130–131, **133,** 294
MI2-5 (Maturity Index 2-5) 131, **133**
ΣMI (ΣMaturity Index) 131–132, **133**
ΣMI2-5 (ΣMaturity Index 2-5) 132, **133**
microarrays *see* DNA microarrays
microinjection 229, 230
microparticle bombardment 230
midges 216, 307
mineralization 11–12, 16, 18–20
mobility 284
molecular methods 56–57
  assemblage analysis 192–202
  bacterial communities 191
molybdenum 176–177
Monhysterida 152, 155, 163–164, 300
monitoring surveys, ecological *see* ecological monitoring surveys, commercial
Mononchidae 70, 184
mucus 25
multivariate analyses 107–108, 117–122, 163, 294–297, 302, **306,** 307
mycorrhizal fungi **16,** 16–17, 20, **22**

naked gobies 23
natural levels
  of chemical stressors 254, 264–265
  of heavy metals 149–50, **306**
Nebalia plots 294, **296**
nematicides *see* drugs, nematicidal
nematode colonization unit (NCU) 291, 308
nematode to copepod ratio 150
Neodiplogastrida 300
neutral community models 115–117
nickel **306**
  pollution, assemblage analysis 174–175, 177, 178, 179, 181–182
  toxicity tests 213, 215
  transgenic indicator strains 232–233
nitrogen cycling 9, 11–12, 18–20, 24, 74
Nonmetric Multi-dimensional Scaling (NMDS) 119
novel stress promoters
  transgenic strains using 234–236
nutrient cycling 1, 7, 8, 9, 10, 11, 18–20, 25, 74

oil industry, offshore production 307–308
oil pollution *see* chemical stressors
omnivores 4, 9, **14,** 17, 70 *see also* trophic groupings

Onchocerca volvulus 3, 15
Oncholaimidae 152
opportunists 147, 148
  enrichment 128, 129, 131–132, 134, 137, 140, 174
ordination techniques 118, 119–121, 294
organic agriculture 17, 18–20, 28, 29, 65–66, **66**
organic enrichment 65, 150–151, 152–153, 161–163
organic pollution *see* organic enrichment
oxygen
  availability 25, 147, 152
  chemical demand (COD) 158–159, 163
  dissolved, water concentration 158–159, 163

Panagrolaimida 57, 212, 215
paper mills 300–301
parasites
  animal 4–5, 6, 9, 10, **14,** 15, 28, 31, 124
  bird 20, **21**
  dispersal 68–69
  effects on resources 6–8
  human 2, 3, 15, 28, 124
  parasite-host interactions 7
  roles in ecosystems 29
pathogens of crops *see* plant feeders
PCBs (polychlorinated biphenyls) *see* chemical stressors
PCoA (Principal Coordinates Analysis) 119
PCR 191, 229
  identification using 192–202
Pellioditis spp. 148–149
PERMANOVA 151
phoresy *see* vectors
phototaxis 5
physico-chemical analyses 302, **306**
PICT (pollution induced community tolerance) 96
Pielou's Evenness Index 6
Pielou's J' Evenness Index 6, **110,** 126, **127**
plant communities 1, 13, 20, 25–26, 74
plant feeders 4, 10, 12–13, 13–**14,** 18, 25–26, 60, 124, 125
  attractants 5, 264, 265
  indices of abundance 131–132
  indirect effects 26–27
  and plant species diversity 69, 74
  root feeders 20, **22,** 31, 45–46
  taxonomy 47
  see also trophic groupings
Plant-Parasite Index/Maturity Index (PPI/MI) 131, **133**
Plant-Parasite Index (PPI) 131, **133**
Plectidae 194–195, 196
Polar Ordination (PO) 119
pollution
  anthropogenic 146–164, 189, 208, 254
  chemical tests for 208
  commercial bioindicator tests for 275–311
  containment incidents **279,** 281, 287, 307
  by fish farms 151–153
  freshwater systems case studies 155–164, 175, 180
  gradients 157–160
  and human health 225
  marine systems case studies 146–155
  organic *see* organic enrichment
  see also chemical stressors; heavy metals; sewage pollution
pollution induced community tolerance (PICT) 96
polychlorinated biphenyls (PCBs) *see* chemical stressors
population dynamics 49, 69, 96
  red grouse 20, **21**
pore water 285
PPI/MI (Plant-Parasite Index/Maturity Index) 131, **133**
PPI (Plant-Parasite Index) 131, **133**
Pratylenchus spp. 177
PRC (Principal Response Curves) 118, 121
predator-prey interactions 11
predators of nematodes 5–6, 17, 21–22, 71
predatory nematodes 4, 5–6, 10, 12, **14,** 17, 21–22, 30, 70, 140, 173 *see also* trophic groupings
primers 192, 198
  bacterial 191
  design 192–195
  family specific 194–195

Principal Coordinates Analysis (PCoA) 119
Principal Response Curves (PRC) 118, 121
Pristionchus pacificus 212, 213, 215
probes, DNA *see* primers
Procrustean superimposition procedure 115

quantitative PCR (Q-PCR)
  identification using 192–196

r-K scale 128
r-strategists 49, 129, 134
RDA (Redundancy Analysis) 120
rDNA 196–197
  bacteria 191
  identification using 192, 193–194, 198
real-time quantitative PCR *see* quantitative PCR
red fluorescent protein (RFP) 227
red grouse 20, **21**
Redundancy Analysis (RDA) 120
reporter genes 225–226, 226–229, 230–232
reproducibility of assays 241–242
reproductive strategies 285–286
resistance
  to nematicides 28
  to stressors *see* tolerance
response to stimuli 5
RFP (red fluorescent protein) 227
Rhabditida 131, 155, 159–160, 162, 212, 300
ribosomal DNA *see* rDNA
richness
  indices 108–110, 112–114, 293, 294, **306**
  of species numbers 283
river blindness 3, 15, 28
rivers **62,** 64
  pollution 155–160, 175, 180, 217
  transgenic indicator strain assays 242
RNA interference (RNAi) 234, 236, 267
root feeders 20, **22,** 31, 45–6
  indirect effects 27
  see also trophic groupings

ΣMaturity Index (ΣMI) 131–132, **133**
ΣMaturity Index 2-5 (ΣMI2-5) 132, **133**

Sabatieria spp. 154
salt marshes 23, 25
sampling 45, 52–3, **54, 55,** 55–57
  asymmetric design 151
  for ecological monitoring surveys 279, 292
  methods 112–13, 114–15, 117, 189, 198–199, 200–201
  for transgenic indicator assays 242
sand dunes, coastal 20, **22,** 127, 302
Saprobien Index 156
sediment transport analysis (STA) 309
sediments 10
  aquatic 25, 27, 309
  disturbed 153–154
  influence on diversity 65
  dredging 150, 153–154
  freshwater **62,** 64, 72, 158, **210,** 216, 218–219
  intertidal 21–22, 23, 25
  marine **14,** 17, 23–24, 60, **63,** 64, 149–155, **210,** 304–307, 307–308
  monitoring pollution status 102, 103
  oil drilling monitoring 307–308
  influence of physical properties 214–215, 289
  salt marsh 23, 25
  sampling 55–56
  sulphidic 24
  toxicity tests 209, **210,** 214–215, 216, 218–219
  transgenic indicator strain assays 242
selenium 177
  pollution, assemblage analysis 184–185
sensitivity to stressors 283–284
  transgenic indicator strains 237–239
sewage pollution 150–151, 159, 173, 174, 301
Shannon's Diversity Index 48, 108, **110,** 113–114, 126, **127**
Shannon–Wiener Index *see* Shannon's Diversity Index
sheep 15
silver
  transgenic indicator strains 232–233
Similarilty Percentage (SIMPER) 151
Simpson's Diversity Index **67,** 126, **127**